McDONALD INSTITUTE MONOGRAPHS

Modelling the early human mind

Edited by
Paul Mellars and Kathleen Gibson

ISSN 1363-1349
ISBN 0-9519420-1-8

McDonald Institute Monographs
Series Editor: Chris Scarre
Assistant Series Editor: Dora A. Kemp

Publishing Office:
McDonald Institute for Archaeological Research
University of Cambridge
Downing Street
Cambridge
CB2 3ER
(0)(1223) 339336

Printed and bound by Short Run Press, Bittern Rd, Sowton Industrial Estate, Exeter, EX2 7LW.

Distributed by Oxbow Books.
United Kingdom: Oxbow Books, Park End Place, Oxford, OX1 1HN.
Tel: (0)(1865) 241249; Fax: (0)(1865) 794449
USA: The David Brown Book Company, P.O. Box 511, Oakville, CT 06779, USA.
Tel: 860-945-9329; Fax: 860-945-9468

Film produced by Gary Reynolds Typesetting, 13 Sturton Street, Cambridge, CB1 2QG.

Contents

Modelling the Early Human Mind

McDonald Institute for Archaeological Research

22–24 September 1993

List of Participants

Leslie Aiello	*Department of Anthropology, University College London*
Richard Byrne	*School of Psychology, University of St Andrews*
Francesco d'Errico	*McDonald Institute for Archaeological Research, Cambridge University*
David Erdal	*School of Psychology, University of St Andrews*
Robert Foley	*Department of Biological Anthropology, Cambridge University*
Kathleen Gibson	*Department of Anatomical Sciences, University of Texas (Houston)*
Ian Glynn	*Department of Physiology, Cambridge University*
John Gowlett	*Department of Archaeology, University of Liverpool*
Peter Grossenbacher	*Department of Experimental Psychology, Cambridge University*
Robert Hinde	*Sub-Department of Animal Behaviour, Cambridge University*
Nick Humphrey	*Darwin College, Cambridge University*
Tim Ingold	*Department of Social Anthropology, University of Manchester*
Frédéric Joulian	*Collège de France, University of Paris*
Phyllis Lee	*Department of Biological Anthropology, Cambridge University*
Christopher Longuet-Higgins	*Department of Experimental Psychology, University of Sussex*
Peter Matthews	*Department of Linguistics, Cambridge University*
Paul Mellars	*Department of Archaeology, Cambridge University*
John Mollon	*Department of Experimental Psychology, Cambridge University*
Sue Parker	*Department of Anthropology, Sonoma State University*
Paul Pettitt	*Department of Archaeology, Cambridge University*
Camilla Powers	*Department of Anthropology, University College London*
Chris Scarre	*McDonald Institute for Archaeological Research, Cambridge University*
Colin Renfrew	*Department of Archaeology, Cambridge University*
Trevor Robbins	*Department of Experimental Psychology, Cambridge University*
Kate Robson Brown	*Department of Biological Anthropology, Cambridge University*
James Russell	*Department of Experimental Psychology, Cambridge University*
Nathan Schlanger	*Department of Archaeology, Cambridge University*
Christopher Stringer	*Natural History Museum, London*
Elizabeth Whitcombe	*Department of Physiology, University College London*
Mark White	*Department of Archaeology, Cambridge University*

Figures

Tables

Preface

Few topics generate greater anthropological and archaeological debate than the cognitive capacities of biologically 'pre-modern' populations. Few dispute that 'Upper Palaeolithic' cultures required fully modern cognition. Most agree that highly significant developments in intelligence coincided with the emergence of the earliest members of the genus *Homo* and the expansion in hominid brain size approximately 2 million years ago. Critical debates, however, revolve around the nature of intelligence in those hominid forms intermediate between apes and modern humans, the possible presence of fully modern cognition prior to the Upper Palaeolithic, and the selective pressures which may have led to modern intelligence. The crucial challenge is to understand exactly what form these 'pre-modern' patterns of intelligence may have taken, and how they have developed over the course of human evolution.

Our aim in organizing this conference was to adopt the widest possible approach to 'modelling' pre-modern patterns of cognition, by enlisting the contributions of specialists in many different branches of cognitive science — psychologists, neurologists, specialists in child development and artificial intelligence, biological anthropologists and of course archaeologists. The meeting that was held in Corpus Christi College Cambridge from 22–24 September 1993 — under the sponsorship of the McDonald Institute for Archaeological Research — was attended by over 30 participants from these various fields, of which 18 have contributed to the present volume.

It is particularly gratifying that these conference proceedings are being published as the first in the new Research Monograph series of the McDonald Institute for Archaeological Research. We are indebted to the Director of the Institute, Professor Lord Renfrew, the Deputy Director, Dr Chris Scarre, and the members of the Institute's Managing Committee for their support and encouragement in many ways. In particular, we thank the Institute for providing one of us (K.G.) with a visiting fellowship for the planning of the conference in 1993. The development of human cognition was one of the great interests of the founder of the Institute, Dr D.M. McDonald, and remains one of the prime directions of current research in the McDonald Institute.

Further support in organizing and running the conference was provided by research students in the Department of Archaeology at Cambridge, in particular Paul Pettitt, Nathan Schlanger and Mark White. The task of copy-editing the papers and preparing the whole collection for publication was shouldered by Dora Kemp, Assistant Editor of the *Cambridge Archaeological Journal*. Further editorial help was provided by Katie Boyle and Nathan Schlanger. Last, but not least, we are indebted to the conference participants, both for contributing to the meeting and for responding patiently to editorial requests for amendments and clarification.

The present collection resolves, at best, a fraction of the current burning questions in the evolution of human cognition. We hope, however, that it serves to highlight critical issues and to point the way forward to promising lines of future research.

Paul Mellars
Kathleen Gibson
Cambridge, January 1996

Introduction: Approaches to Modelling Early Human Minds

Kathleen Gibson & Paul Mellars

When, in September, 1993, a group of psychologists, neuroscientists, anthropologists, and archaeologists gathered at Corpus Christi College to discuss how to model the early human mind, they confronted an issue which has led to a fascinating multicultural array of origin myths and religious/philosophical explanations of the nature of human mentality. Only, however, within the last 150 years and with the emergence of the Darwinian evolutionary paradigm, has science actively approached this issue. Unfortunately, the result of more than a century of evolutionary modelling is a proliferation of conflicting scientific explanations which seemingly rival the diversity of mythological explanations and which are, in many cases, no more testable. One reason for the diversity of scientific explanations is that throughout the late nineteenth and much of the twentieth century, scientific progress in the area of human cognitive evolution was hampered by a sparse fossil record, the fraudulent Piltdown skull, minimal knowledge of the cognitive capacities of our animal brethren, and poor understanding of brain functions. Today, however, both the fossil record and scientific knowledge in the behavioural and neural sciences are growing by leaps and bounds. Thus, the time may now be ripe for serious approaches to modelling human cognitive and behavioural evolution.

We took it as self-evident in organizing this conference that fully modern intelligence, with its capacity for inventing — and operating — computers, space satellites, electron microscopes and the like — cannot have emerged instantaneously, and that there must inevitably have been a long and essentially gradual progression of mental capabilities over the course of hominid evolution from those of our closest primate ancestors. Even the harshest critic of anthropocentric models would not question that there are indeed very profound contrasts between the mental capacities of apes and those of modern humans. The crucial question is exactly what form these pre-modern patterns of mental development may have taken. Above all perhaps — if we adopt an explicitly scientific stance — where do we look to find plausible working models (or working hypotheses) of exactly what the nature of these earlier patterns of cognition might have been? Our view, as reflected in the papers in the present volume, is that we should adopt the widest possible approach to this question, and look into the potential contributions of the broadest possible range of cognitive disciplines to provide clues to these early stages of human mental development. This is why the present volume contains papers not only from psychologists, but also from neurologists, from specialists in child development and artificial intelligence, from ethologists, from anthropologists and — last but not least — from archaeologists. We must never forget that however ingenious and imaginative we may be in deriving hypothetical models for the course of human cognitive development from a range of more or less 'hard' sciences (i.e. sciences concerned with the observable, present-day world), it is only archaeologists who are capable of providing the direct evidence for the actual patterns of development of different aspects of behaviour over the course of human evolution, and who can therefore engage actively in evaluating the alternative hypothetical models of cognitive evolution against a (more or less) empirical data base. The analysis of human cognitive development hinges, in other words, on two separate procedures: first, scanning the full range of modern cognitive disciplines to discover a range of potential, hypothetical models for the patterns of cognitive evolution; and, second, attempting to evaluate these alternative hypothetical models against the available archaeological evidence.

Effective models of human evolution must include descriptions of the behavioural and cognitive capacities of the last common ancestor of humans and African apes, describe successive stages in the evolutionary development of human mental capacities, and determine the geological time-frames of these evolutionary events. Such models must also delineate the ecological or other selective agents that resulted in the emergence of human capacities and, ultimately, they must explain the genetic basis of human and ape mental differences. The proliferation of models of the origin of such phenomena as language and consciousness suggest that it is very easy to devise evolutionary scenarios which incorporate many of these elements. What is difficult is to determine which methods of evolutionary modelling are most appropriate and which models are the most scientifically credible.

In recent decades, referential models of human behavioural evolution have been quite popular. These models use specific living groups, such as baboons, chimpanzees, or modern hunter-gatherers, to model the behaviour of our hominine ancestors. Tooby & Devore (1987), however, have cogently argued that evolutionary models must be strategic in nature. That is, such models must draw on a diversity of scientific findings to determine predictable behavioural associations with other factors such as ecology, population size, and anatomy. This volume forges paths in the direction of strategic modelling by drawing on modern research findings in the neurobehavioural, archaeological, and evolutionary sciences for insights into methods of modelling prehistoric minds and for discussion of some current scientific knowledge pertinent to our interpretations of prehistoric behaviour.

Part 1: Defining the problem: setting the stage

The first three papers set the stage by delineating critical issues in the emergence of fully modern human minds. Renfrew, in 'The Sapient Behaviour Paradox' reminds us of the diversity of modern material cultures and of the difficulties of determining the emergence of modern cognitive abilities from archaeological remains. For instance, nearly all palaeoanthropologists and archaeologists agree that, by at least 35–40,000 years ago, anatomically modern humans possessed fully modern mental capacities. However, vast differences in technological, scientific, and literary accomplishments distinguish Upper Palaeolithic from later human cultures. Similarly, the material remains of medieval Europeans are quite different from those of their twentieth-century descendants. Yet all evolutionary biologists would agree that both populations were fully modern in their cognitive capacities. How, then, can we judge when our ancestors first attained modern cognitive, linguistic, and other mental capacities?

Mellars, in the second paper, assumes that despite some problems in interpretation, the archaeological record continues to provide the clearest evidence of the emergence of modern capacities. He notes that the onset of the Upper Palaeolithic coincided with an explosion of technology, art, jewellery, long-term planning strategies, and other behaviours which characterize most modern human cultures. This explosion was unprecedented in the earlier archaeological record. It also occurred in synchrony with the dispersion of anatomically modern humans into Europe. Based on these data, Mellars argues that fully modern language and symbolic expression first emerged at or slightly prior to the Upper Palaeolithic (Mellars 1989a,b; 1991). The paper by Gibson derives quite a different conclusion by considering archaeological findings in conjunction with comparative studies of human and non-human primate cognitive skills, fossil brain size and form, and knowledge of the plasticity of human and brains. In Gibson's view, the Upper Palaeolithic represents a technologically-propelled cultural revolution in populations which had long possessed essentially modern neurological and cognitive capacities, but which were motivated to develop new technologies and other socio-cultural pursuits in order to exploit a seasonally migratory life style. Which, if either, of these divergent hypotheses is correct will no doubt be debated for some time. Hopefully, however, the issue will eventually be resolved as our methods of interpreting the fossil and archaeological records improve.

Part 2: Brain size and structure in earlier hominines

Palaeoneurological analyses are potentially among the most important tools for modelling early human minds. Fossil cranial capacities and endocasts yield information about brain structure, and, hence, insights into the cognitive and behavioural capacities of earlier hominines. Certainly, most scholars would agree that the possession of modern brain size and structure also signifies the possession of modern cognitive and behavioural capacities.

As delineated in Byrne's contribution, however, many shaky assumptions underlie brain size analyses and attempts to relate brain size and intelligence. For example, it remains unclear whether the most appropriate measure of brain size is absolute brain size or brain size in relationship to body size. During most of this century, relative brain size has been considered more important, primarily because it is the only measure that invariably results in human superiority over all other animals. A reliance on relative brain size, however, can produce problematical results. Gorillas have a much smaller relative brain size than other great apes, most monkeys, and even some prosimians. However, gorillas share a suite of cognitive skills with other apes which would suggest that gorilla intelligence is far closer to that of orangutans and chimpanzees than to that of prosimians or monkeys. Absolute brain size would predict this result, at least for those scholars who consider that brain size contributes to intelligence — a much debated point.

That brain size is a critical measure of information-processing capacity and intelligence, has been strongly argued by Jerison (1973; 1982) and Gibson (1988; 1990a; 1993). It is also implicit in the arguments of many scholars who have tried to correlate various brain size measures with other attributes such as foraging strategies (Clutton-Brock & Harvey 1980; Eisenberg & Wilson 1978; Gibson 1986; Milton 1988) or social structure (Aiello & Dunbar 1993; Dunbar 1992; 1993). Some scholars have argued, however, that brain size is primarily a by-product of selection for traits other than intelligence, such as bipedalism, heat adaptation, or prolonged maturation (Aiello this volume; Gould 1977), rather than a product of selection for advanced cognitive skills in any behavioural domain. What, then, can we learn from studies of brain size? Foley (this volume) attempts to answer this question by correlating various estimates of brain size with other parameters in the hominine archaeological record, including archaeological complexity, duration of archaeological tool 'cultures', and maturation rate. He concludes that changes in maturation rates and life-history strategies were among the earliest changes in human evolution. Increased technological complexity occurred much later subsequent to major changes in brain size and encephalization quotient.

In addition to other problems in the use of brain size as an estimate of intelligence, Holloway has argued, for some three decades now, that the human brain has been reorganized in comparison to the brains of great apes and that it is this brain reorganization, rather than brain size, which determines human mental capacities (Holloway 1966; 1968). Holloway's neural reorganization hypothesis implies that palaeoanthropologists should analyze not just the size of fossil brains, but also their neural organizations. Unfortunately, however, only external brain structure is amenable to palaeoanthropological analyses. We cannot determine critical issues such as the size and structure of subcortical nuclei and tracts from fossil endocasts. We can, however, determine the position and size of some gyri and sulci, and thus determine some aspects of neural organization. For example, a posterior location of the lunate sulcus implies the possession of a human-like, rather than an ape-like configuration of the parietal lobes (Falk 1985a,b; Holloway 1979; 1981). The identification of critical fissures, sulci, and gyri also allows us to determine the relative size of various cortical lobes and/or to determine the size of neural gyri known to be expanded in humans in comparison to apes. Several papers in this volume focus on these issues.

One area of the brain known to have greatly enlarged during human evolution is the frontal lobe. Indeed, it was long considered axiomatic that one major reorganizational feature of the human brain was a differential enlargement of the frontal lobes in comparison to other neural regions (von Bonin 1963). Although claims for differential frontal lobe enlargement have been disputed (Passingham 1975), Deacon (1988) has resurrected them. In this volume, Roberts, Collins and Robbins focus on the functions of the frontal lobes. These neural regions play critical roles in long-term planning, in the ability to keep several short-term memories in mind simultaneously, and in providing for behavioural flexibility and creativity. As noted by Mellars (this volume), some of these behavioural traits appear to differentiate Upper and Middle Palaeolithic cultures. Relative size of the frontal lobes of Middle and Upper Palaeolithic peoples thus emerges as a critical question in the emergence of modern human cognition (Gibson 1985; this volume). Neanderthal frontal lobes are certainly flatter than those of many modern humans, and earlier palaeontologists generally considered them to be smaller as well (Kochetkova 1978). Holloway (1985), however, has disputed previous claims that Neanderthal frontal lobes were small, albeit without directly studying Neanderthal endocasts. Thus, the issue of Neanderthal frontal lobe size emerges a prime topic of restudy by future investigators.

Other neural regions involved in linguistic and other cognitive functions may also have expanded in human evolution. Tobias (1981; 1987) and Falk (1983) have noted that early *Homo* possessed an expanded Broca's area. Since it has long been considered axiomatic that Broca's area controls speech, they and Gibson (this volume) have argued for the presence of language capacities in these taxa approximately 2 million years ago. Whitcombe (this volume), however, reviews the history of cerebral localization studies, particularly as they relate to the functions of typical language areas such as the inferior parietal lobe and Broca's and Wernicke's areas. As noted previously by Pribram (1971), early evidence in support of the linguistic functions of Broca's and Wernicke's areas was derived from autopsies and was, at best, tenuous. Although modern PET and MRI data continue to support the view that these areas have some language functions, Whitcombe argues that these areas do not function in isolation, but rather as parts of entire anatomical systems which include peripheral nerves and other structures such as the hands and the oral apparatus. This implies that Broca's area would mediate speech only in those species who also possessed the appropriate peripheral anatomy. Hence, the early enlargement of Broca's area in human phylogeny may have functional implications quite different from the emergence of language, such as expansions in overall mental constructional capacities or manual functions (see Greenfield 1991).

Aiello expands upon Whitcombe's argument by presenting detailed analyses of both peripheral and neural structures of various taxa in the hominine line. She suggests two major periods of rapid human brain expansion: the emergence of the genus *Homo* and the emergence of archaic *Homo sapiens*. Her comparative analysis also leads her to conclude that selection for symbolically based cognitive capabilities may not have been important until the transition between *Homo erectus* and archaic *Homo sapiens* taxa. Earlier enlargements in brain size may simply have reflected adaptations of a bipedal hominid to a hot, dry environment.

Part 3: Comparative studies of behaviour

Analysis of the fossil record is just one of a number of potential approaches to evolutionary modelling. Comparative anatomy of living species has long served as a means of modelling the evolution of animal and human form. Comparisons of albumin structure accurately predicted a relatively late origin of hominine taxa (Sarich & Wilson 1967) at a time when palaeoanthropologists still erroneously argued for a long evolutionary separation between apes and humans. Comparative DNA structure and other comparative molecular analyses continue to provide fruitful approaches for charting phylogenies and for delineating times of divergence of various animal and hominine taxa. Some scholars have long considered that comparative animal and human psychology can be similarly fruitful (Hinde & Tinbergen 1958; Parker & Gibson 1977; 1979).

In this volume, Robson Brown describes the comparative approach and notes that, when combined with cladistic analyses, comparative psychology can provide clues to the minds of common ancestors of extant species, even when the fossil remains of those species may not yet have been discovered. Several additional papers use comparative studies of living humans or animals as the primary means of deriving insights into the evolution of mental phenomena. Grossenbacher uses the phylogenetic distribution of the neural basis of colour vision as a means of shedding light on the origin of consciousness. He concludes that consciousness is a phenomenon of newly evolved neurological and mental structures. Papers by Lee and by Erdal & Whiten specifically address a currently popular scientific theme — the importance of social behaviour in human evolution. Lee notes, on the basis of comparative data, that sociality may select for complex cognitive skills such as individual recognition, classification, contingency thinking, and intentionality. Erdal & Whiten note that humans in agricultural and industrial societies share certain behavioural traits such as competitiveness with other primates. In contrast, most modern hunter-gatherer societies are very egalitarian. They discuss this phenomenon in depth from perspectives of genetic and other cost/benefits analyses in relationship to food sharing, leadership roles, and methods of preventing leaders from becoming too dominant. They conclude that humans, prior to the emergence of agriculture, evolved away from the primate strategy of 'scare off others and get what I can' to a strategy of fair play, pleasure in equal relationships, and commitment to sharing with and helping others. The result is a very complex psychology which demands a very high level of cognitive capacity.

It would be tempting to conclude from the papers by Lee and by Erdal & Whiten that human brains and intelligence evolved solely for social strategies. As noted, however, by these authors, as well as by Foley (in this volume), Mithen (this volume) and Gibson (1993), intimate relationships exist between foraging strategies, ecology, and social structure. Thus, however strong the relationship between social intelligence and higher cognitive capacities, other factors such as foraging strategy or technology may, nonetheless, have played major roles in selecting for higher cognitive and linguistic capacities.

Part 4: Ontogeny and phylogeny

Relationships between ontogeny and phylogeny — between patterns of growth and development in individual organisms and those in the course of evolution — have been debated for more than a century.

Although it is now generally agreed that ontogeny does not invariably recapitulate phylogeny, it has nonetheless been recently argued that some aspects of cognitive and neural development parallel cognitive and neural evolution. In particular, the ontogenetic development of information processing capacities and performance on Piagetian tasks appears to partially parallel the phylogeny of those attributes (Gibson 1981; 1990b; Parker & Gibson 1979). Piagetian and neo-Piagetian analyses have also proven of value in analyzing the cognitive capacities of the makers of stone tools (Gibson 1985; Gowlett 1984; Parker 1985; Parker & Gibson 1979; Wynn 1979; 1981).

In this volume, Longuet-Higgins provides theoretical justification for the use of Piagetian and other developmental frameworks in the modelling of early human minds. Russell follows with an in-depth discussion of the development of symbolic functions from a Piagetian perspective. Russell notes that the transition to representational thought, symbolic behaviour and symbolic function which occurs at 18 months of human age occurs in synchrony with an enlargement of working memory and the development of language and other complex cognitive skills. Based on the view that the artistic and other accomplishments of the Upper Palaeolithic represent the first evolutionary occurrences of symbolic expression, he concurs with Mellars that the Upper Palaeolithic provides the first clear reflection of modern cognitive, symbolic and linguistic capacity, as well as the first attainment of a level of cognitive development equivalent to that of the human child at the transition between Piaget's sensorimotor and preoperational stages.

In this context, however, it should be noted that this interpretation contrasts with the interpretations of other Piagetian scholars, that during their development, great apes proceed through all of Piaget's sensorimotor stages and eventually acquire many cognitive skills equivalent to the preoperational stage as manifested by a three- to four-year-old human child (Parker & Gibson 1979). Several scholars have also argued that the makers of Acheulean handaxes exhibited cognitive levels equivalent to Piaget's third stage, that of concrete operations, which emerges at about seven to eight years of age (Gibson 1985; Parker 1985; Parker & Gibson 1979; Wynn 1979). These differences in interpretation reflect Russell's concentration on technologically based art as a primary symbolic medium, versus the views of some other scholars that evidence of advanced Piagetian abilities can be seen in such properties as the construction of symmetry and spatial proportions in stone tools and that linguistic symbolism could have appeared prior to the invention of the technology essential for complex artistic expression. It is also important to note that although an entire suite of cognitive skills does develop in tandem in human children, this does not necessarily mean that each of these skills evolved in tandem. For example, it now seems clear that some monkeys exhibit much higher levels of sensorimotor intelligence in their tool-using behaviours than in their imitative capacities (Gibson 1990b).

Part 5: Interpreting the archaeological record

Comparative studies of living animals and analyses of fossil endocasts provide useful information for modelling early human minds. Only the archaeological record, however, provides direct evidence of early human behaviour and places that behaviour in specific time frames and geographical locations. Improvements in methods of analyzing cognitive development from archaeological remains must, thus, remain a paramount goal of those who would chart the evolution of the human mind. To this end, the final papers focus in-depth on methods of interpreting hominine archaeological remains.

An 'ethoarchaeological' approach is demonstrated by Joulian. This approach compares ethological data derived from chimpanzee tools and tool-using sites with Plio-Pleistocene tools and archaeological sites. Joulian analyzes chimpanzee tool-assisted nut-extraction and the production of Oldowan chopping tools from the perspective of operational chains — 'the continuous system of technical activity from acquisition of raw materials to the discard of tools'. He concludes that chimpanzees and Plio-Pleistocene hominines possessed broadly similar cultures and cognitive skills. His conclusion concords with those of Wynn & McGrew (1989) who suggested that great apes could have been responsible for the Oldowan culture. On the other hand, Toth *et al.* (1993), following detailed studies of the tool-making procedures of the human-taught pygmy chimpanzee, Kanzi, concluded that the Oldowan tool-makers used two-handed percussion techniques which do not come naturally to Kanzi, were more consistently right-handed than the great apes, and were better able to produce sharp cutting edges. Their work suggests that Oldowan hominids did possess some anatomical, neurological, or cognitive advantages which rendered them better tool-makers than modern apes.

Robson Brown (this volume) argues for the use of 'experimental knowledge' to resolve archaeological dilemmas. She suggests that experimental archaeology could help resolve this issue of the cognitive capacities essential for making stone tools. For one, there are many captive apes. Some might prove more adept at acquiring human-like tool-making skills than Kanzi. It would also be of interest to determine the maturational levels of Piagetian and other cognitive skills needed for modern humans to make stone tools by, for example, teaching children of varying ages to make such tools.

Gowlett (this volume) also draws somewhat different conclusions from Joulian. He suggests that by 2.5 million years ago, hominid archaeological sites were more human-like and more complex than chimpanzee tool-using sites, in that Plio-Pleistocene hominids understood the properties of stone, travelled long distances to acquire the best materials, and used learned instruction sets or procedural templates when making tools.

Gowlett advocates a multi-pronged approach to interpreting the archaeological record, including statistical analyses of entire tool-assemblages, analyses of behavioural patterning in space and time, and analyses of tool design. He analyzes Acheulean handaxes from some of these perspectives. Although handaxes vary greatly in size, they nearly always exhibit the same geometrically accurate proportions. This strongly suggests that the makers of Acheulean handaxes possessed fairly advanced mathematical skills. The regularity of handaxe shape and proportion also effectively refutes Davidson & Noble's (1993) claims that handaxes were an accidental by-product of flake production, rather than purposely shaped tools. Since human children acquire a sense of mathematical proportion only during Piaget's late pre-operational and early concrete operational stages, these findings could be taken to suggest that, contrary to Russell's conclusions (this volume), those levels of Piagetian development essential for symbolism may have been reached long prior to the Upper Palaeolithic.

In the final paper, Mithen again turns to the Neanderthal problem. His paper echoes portions of Gibson's paper in its emphasis on a variety of cognitive skills and in its caution that changes in cognitive capacities during human evolution may have involved skills other than language. Mithen also addresses issues of domain-specific intelligences versus domain-general intelligence. He argues that Neanderthals exhibited high levels of social intelligence, natural history intelligence, and technical intelligence, but that they showed little ability to integrate the varied intellectual domains. In contrast, Upper Palaeolithic peoples did integrate these domains. The implicit conclusion of Mithen's paper is that this difference reflected biological differences in domain-general capacities. If, in fact, Neanderthal frontal and parietal lobes were smaller than those of Upper Palaeolithic peoples, then it is entirely possible that such cognitive differences actually occurred. The alternative would be that Middle Palaeolithic peoples simply lacked the degrees of technological advancement necessary to integrate the various domains via such behaviours as art, depiction, and record keeping.

Summary and conclusion

All of this brings us back full circle to the issues defined in the opening papers in the volume: how do we set about investigating the evolution of mental *capacities* for behaviour, as opposed to the changing *patterns* of behaviour themselves? Renfrew has reminded us of the ultimate intransigence of this paradox in behavioural and archaeological terms. Yet we cannot escape the fact that the genetic capacities for intelligent behaviour must have expanded fairly dramatically over the course of the past two million years, and that this expansion almost certainly occurred at many different stages during the course of human evolution. Mellars has emphasized that even in populations as chronologically close to ourselves as the European Neanderthals, we could be dealing with people who had pursued a genetically separate trajectory of evolutionary development from that which led to behaviourally and biologically modern populations over a time span of at least 300,000 years. The issue of potential, innate biological factors in the development of even relatively recent, archaeologically defined behavioural patterns therefore cannot be avoided.

In this volume we have argued that a productive approach to this question demands the cooperation of the widest possible range of cognitive disciplines — of psychologists, neurologists, ethologists, primatologists, biological and social anthropologists, and linguists. Ultimately, it may also require the co-operation of other fields such as artificial intelligence, molecular biology and philosophy. We have also argued that the eventual testing and evaluation of alternative, hypothetical models of cognitive evolution must rely heavily on the 'hard' behavioural evidence which only the archaeological record can provide. The current volume incorporates views from most of the pertinent disciplines. It demonstrates a strong, but surprising affinity, between the fields of neuroscience, psychology, and biological anthropology and indicates that the merging of these perspectives is one important 'way-forward' in modelling early human minds and in charting the evolution of human mentality.

References

Aiello, L. & R.I.M. Dunbar, 1993. Neocortex size, group size and the evolution of language. *Current Anthropology* 34, 184–93.

Bonin, G. von, 1963. *The Evolution of the Human Brain.* Chicago (IL): University of Chicago Press.

Clutton-Brock, T.H. & P. Harvey, 1980. Primates, brains, and ecology. *Journal of Zoology* (London) 190, 309–32.

Davidson, I. & W. Noble, 1993. Tools and language in human evolution, in Gibson & Ingold (eds.), 363–88.

Deacon, T.W., 1988. Human brain evolution II. Embryology and brain allometry, in *Intelligence and Evolutionary Biology*, eds. H.J. Jerison & I. Jerison. Berlin: Springer-Verlag, 383–416.

Dunbar, R.I.M., 1992. Neocortex size as a constraint on group size in primates. *Journal of Human Evolution* 22, 469–93.

Dunbar, R.I.M., 1993. Co-evolution of neocortex size, group size, and language in primates. *Behavioral and Brain Sciences* 16, 681–735.

Eisenberg, J.F. & D.E. Wilson, 1978. Relative brain size and feeding strategies in Chiroptera. *Evolution* 32, 740–51.

Falk, D., 1983. Cerebral cortices of East African early hominids. *Science* 221, 1072–4.

Falk, D., 1985a. Apples, oranges, and the lunate sulcus. *American Journal of Physical Anthropology* 67, 313–15.

Falk, D., 1985b. Hadar AL 162-28 endocast as evidence that brain enlargement preceded cortical reorganization in hominid evolution. *Nature* 313, 45–7.

Gibson, K.R., 1981. Comparative neuro-ontogeny; its implications for the development of intelligence, in *Infancy and Epistemology*, ed. G. Butterworth. Brighton: Harvester Press, 52–82.

Gibson, K.R., 1983. Comparative neurobehavioral ontogeny: the constructionist perspective in the evolution of language, object manipulation and the brain, in *Glossogenetics*, ed. E. de Grolier. New York (NY): Harwood Academic Publishers, 52–82.

Gibson, K.R., 1985. Has the evolution of intelligence stagnated since Neanderthal man?, in *Evolution and Developmental Psychology*, eds. G. Butterworth, J. Rutkowska & M. Scaife. Brighton: Harvester Press, 102–14.

Gibson, K.R., 1986. Cognition, brain size, and the extraction of embedded food resources, in *Primate Ontogeny, Cognition and Social Behavior*, eds. J. Else & P.C. Lee. Cambridge: Cambridge University Press, 93–105.

Gibson, K.R., 1988. Brain size and the evolution of language, in *The Genesis of Language: a Different Judgement of Evidence*, ed. M. Landsberg. Berlin: Mouton de Gruyter, 149–72.

Gibson, K.R., 1990a. New perspectives on instincts and intelligence: brain size and the emergence of hierarchical mental constructional skills, in Parker & Gibson (eds.), 97–128.

Gibson, K.R., 1990b. Tool use, imitation, and deception in a captive cebus monkey, in Parker & Gibson (eds.), 205–18.

Gibson, K.R., 1993. Tool use, language and social behavior in relationship to information processing capacities, in Gibson & Ingold (eds.), 251–69.

Gibson, K.R. & T. Ingold (eds.), 1993. *Tools, Language and Cognition in Human Evolution.* Cambridge: Cambridge University Press.

Gould, S.J., 1977. *Ontogeny and Phylogeny.* Cambridge (MA): Harvard University Press.

Gowlett, J.A.J., 1984. Mental abilities of early man: a look at some hard evidence, in *Hominid Evolution and Community Ecology*, ed. R. Foley. London: Academic Press, 167–92.

Greenfield, P.M., 1991. Language, tools and the brain: the development and evolution of hierarchically organized sequential behavior. *Behavioral and Brain Sciences* 14, 531–95.

Hinde, R. & N. Tinbergen, 1958. The comparative study of species-specific behavior, in *Behavior and Evolution*, eds. A. Roe & G.G. Simpson. New Haven (CT): Yale University Press, 251–68.

Holloway, R.L., 1966. Cranial capacity, neural reorganization, and hominid evolution: a search for more suitable parameters. *American Anthropologist* 68, 103–21.

Holloway, R.L., 1968. The evolution of the primate brain: some aspects of quantitative relationships. *Brain Research* 7, 121–72.

Holloway, R.L., 1979. Australopithecine endocasts, brain evolution in the Hominoidea, and a model of hominid evolution, in *The Functional and Evolutionary Biology of Primates*, ed. R. Tuttle. Chicago (IL): Aldine Press, 185–204.

Holloway, R.L., 1981. Revisiting the South African Taung Australopithecine endocast: the position of the lunate sulcus as determined by the stereoplotting technique. *American Journal of Physical Anthropology* 56, 43–58.

Holloway, R.L., 1985. The poor brain of *Homo sapiens neanderthalensis*: see what you please, in *Ancestors: the Hard Evidence*, ed. E. Delson. New York (NY): Alan R. Liss, 319–24.

Jerison, H.J., 1973. *Evolution of the Brain and Intelligence.* New York (NY): Academic Press.

Jerison, H.J., 1982. Allometry, brain size, cortical surface, and convolutedness, in *Primate Brain Evolution*, eds. E. Armstrong & D. Falk. New York (NY): Plenum, 77–84.

Kochetkova, V.I., 1978. *Paleoneurology.* Washington (DC): V.H. Winston.

Mellars, P.A., 1989a. Major issues in the emergence of modern humans. *Current Anthropology* 30, 349–85.

Mellars, P.A., 1989b. Technological changes across the Middle–Upper Paleolithic transition: economic, social, and cognitive perspectives, in *The Human Revolution: Behavioural and Biological Perspectives on the Origins of Modern Humans*, eds. P. Mellars & C. Stringer. Princeton (NJ): Princeton University Press, 338–65.

Mellars, P.A., 1991. Cognitive changes and the emergence of modern humans in Europe. *Cambridge Archaeological Journal* 1(1), 63–76.

Milton, K.T., 1988. Foraging behaviour and the evolution of primate intelligence, in *Machiavellian Intelligence*, eds. R. Byrne & A. Whiten. Oxford: Clarendon Press, 284–305.

Parker, S.T., 1985. Higher intelligence as an adaptation for social and technological strategies in early *Homo sapiens*, in *Evolution and Developmental Psychology*, eds. G. Butterworth, J. Rutkowska & M. Scaife. Brighton: Harvester Press, 83–100.

Parker, S.T. & K.R. Gibson, 1977. Object manipulation, tool use and sensorimotor intelligence as feeding adaptations in cebus monkeys and great apes. *Journal of Human Evolution* 6, 623–41.

Parker, S.T. & K.R. Gibson, 1979. A model of the evolution of language and intelligence in early hominids. *Behavioral and Brain Sciences* 2, 367–407.

Parker, S.T. & K.R. Gibson (eds.), 1990. *'Language' and Intelligence in Monkeys and Apes: Comparative and Developmental Perspectives*. Cambridge: Cambridge University Press.

Passingham, R.E., 1975. Changes in the size and organization of the brain in man and his ancestors. *Brain, Behavior, and Evolution* 11, 73–90.

Pribram, K.H., 1971. *Languages of the Brain*. Englewood Cliffs (NJ): Prentice Hall.

Sarich, V. & A. Wilson, 1967. Immunological time scale for hominid evolution. *Science* 158, 1200–203.

Tobias, P.V., 1981. The emergence of man in Africa and beyond. *Philosophical Transactions of the Royal Society of London* Series B 292, 43–56.

Tobias, P.V., 1987. The brain of *Homo habilis*: a new level of organization in cerebral evolution. *Journal of Human Evolution* 16, 741–62.

Tooby, J. & I. DeVore, 1987. The reconstruction of hominid behavioral evolution through strategic modeling, in *The Evolution of Human Behavior: Primate Models*, ed. W.G. Kinsey. Albany (NY): State University of New York Press, 183–237.

Toth, N., K.D. Schick, E.S. Savage-Rumbaugh & D. Rumbaugh, 1993. Pan the tool-maker: investigations into the stone tool-making and tool-making capabilities of a Bonobo. *Journal of Archaeological Science* 20, 81–91.

Wynn, T.G., 1979. The intelligence of later Acheulean hominids. *Man* 14, 371–91.

Wynn, T.G., 1981. The intelligence of Oldowan hominids. *Journal of Human Evolution* 10, 529–41.

Wynn, T. & W.C. McGrew, 1989. An ape's view of the Oldowan. *Man* 24, 383–98.

Part I

Defining the Problem: Setting the Stage

1

The Sapient Behaviour Paradox: How to Test for Potential?

Colin Renfrew

The central paradox is examined that while the most significant step in human evolution, the emergence of Homo sapiens sapiens, *occurred more than 40,000 years ago, the immediate consequence of that step seems to have been little more than the emergence of a* potential *for change, not any very significant initial behavioural innovation. Indeed recent hunter-gatherers have followed a life style differing in its archaeology only marginally from that of their Palaeolithic predecessors.*

Are we framing our questions correctly? Is there not something disturbingly teleological about explanations that lay such weight upon the potential *for innovation? Why is there such a separation between genetic cause and behavioural effect? That is the sapient behaviour paradox.*

My aim in this paper is to comment on the apparent paradox involving the difficulty, perhaps the impossibility, of observing potential or capacity or ability until it is revealed in performance/actuality/achievement. Until a way can be found around this difficulty there are grave difficulties in understanding adequately the process of the emergence of sapient cognitive abilities and capacities.

The central problem is that we are interested in examining the threshold between pre-sapient and fully sapient existence. There is the underlying assumption that the pre-sapiens hominids, such as *Homo erectus*, would not have been capable, even after millennia of cultural evolution, of the technical and cultural advances which are familiar to us as part of the human story, such as the Neolithic revolution, the development of pyrotechnology, the urban revolution, the development of writing, the industrial revolution and so forth. It is also widely assumed that the fundamental genetic changes which produced modern humankind were already largely accomplished with the appearance of *Homo sapiens sapiens* in different parts of the world some 40,000 years ago. From then on, it is held, we are all one species. But it is manifestly evident that many of the elements in the human story since 40,000 years ago of developing technology and social organization have only been documented in restricted areas and periods since. In one sense that is obvious enough. But, if we are speaking in causal terms, how satisfactory an explanation would it be (if we had one) that succeeded in explaining the changes which took place over the preceding millennia that led to the emergence and widespread dispersal of our species? If it is the case that genetically we today differ rather little from our *sapiens* ancestors of forty millennia ago, how does that genetic composition which emerged then explain the cultural differences between then and now? The usual answer is that from that time the human animal had the skill, the intelligence, the *potential* to achieve its later accomplishments. But what kind of explanation is it that lays such weight upon so apparently teleological a concept as potential?

In genetics the distinction between genotype (as determined by the molecular genetic structures, by the genes) and phenotype (as actually apparent in the organism) is a very familiar one. But the distinction being drawn here is a rather different one: not so much in appearance of form (i.e. the phenotype) as in behaviour, action (which we should perhaps designate 'praktotype'). For it is generally held that the main genetic contributions to our species were present in *Homo sapiens sapiens* from early times. But many aspects of behaviour emerge only very much later. They are based upon abilities which must be genetically determined, but they are called into effect, or triggered, by culturally determined antecedent circumstances.

First it may be useful to make explicit two commonly held assumptions, which are widely accepted.

Hypothesis 1. Limited contemporary diversity

It is generally accepted that, despite some degree of diversity among humans today, all human groups share broadly the same innate cognitive capacities and abilities. Geneticists make the point that there is relatively low genetic diversity among all living humans: we are one species. Documented cases of identical twins, separated at birth and raised in very different cultural contexts, show how readily genetically identically individuals can adapt to very different cultural environments. The same point is supported, of course, by the competent and effective part undertaken by members of all ethnic groups not only in traditional undertakings, but also in many aspects of modern urban and literate society.

This 'Contemporary Unity Hypothesis' is, in addition, a political tenet for most people today, and a tenet of faith for many. That, however, does not necessarily call into question its empirical validity. The converse view is often asserted by groups which one may, in some cases, characterize as racist. I accept that the weight of evidence is in favour of the hypothesis, and that the increasing amount of genetic evidence now becoming available tends to support it.

Hypothesis 2. The 40 ky ability hypothesis

It is widely accepted that with the appearance of fully modern *Homo sapiens* over much of the globe between 60,000 and 20,000 years ago, the genetic make-up, the hardware, of contemporary *Homo sapiens sapiens* was effectively established, and that there have been few subsequent changes of major significance. The Great Transition from *Homo erectus* to *Homo sapiens* had taken place.

This hypothesis can certainly be questioned, and among sociobiologists to have done so are E.O. Wilson and C.H. Darlington, who claim that the differences between the cultures and civilizations in documented different places over the past few millennia are in large measure due to significant differences in genetic composition between them. But such has not yet been shown to be the case, and the time intervals involved seem rather short for very radical genetic changes to have occurred.

The sapient behaviour paradox. Part I

To a casual observer, the archaeological record for the Upper Palaeolithic (and thus to be associated with *Homo sapiens sapiens*) is not, despite a few restricted exceptions, amazingly different in character from that from the Lower and Middle Palaeolithic (associated with *Homo erectus*, and in part for Europe and the Near East with *Homo sapiens neanderthalensis*). Certainly this is so if we contrast the artefacts preserved from these periods — mainly of stone — with those from the Neolithic period or from the early civilizations.

Mellars (1991, 63) has conveniently listed the behavioural changes (in the above terminology 'praktotypic') which characterize the archaeological transition from Middle to Upper Palaeolithic:

1. a shift in stone production from 'flake' technologies to more regular and standardized forms of 'blade' manufacture;
2. increase in the variety and complexity of stone tools, with more obvious standardization etc.;
3. appearance of extensively shaped bone, antler and ivory artefacts;
4. increased tempo of technological change, and increased degree of regional diversification;
5. appearance of a wide range of beads, pendants and other 'personal ornaments';
6. appearance of representational or 'naturalistic' art;
7. changes in both the economic and social organization of human groups.

These do indeed constitute a significant range of features, but they are, with the exception of no. 6 (art) more impressive to the specialist than to the casual observer, especially since the practice of deliberate burial is well documented for the Middle Palaeolithic.

It would not be difficult to develop a comparable series of attributes which could serve in a very similar and general way to distinguish between the Upper Palaeolithic period in Europe (the main focus of Mellars' study) and the succeeding Neolithic period. An initial listing might run as follows:

1. development of food production in the use of a well-defined range of domestic plants;
2. use of querns, grindstones and other tools suitable for the processing of such plants;
3. intensive exploitation of a narrow range of domesticated animals;
4. intensive use of animal bones for the production of a new range of artefact types;
5. settled village life, documented by the construction of permanent dwellings, often using pisé

(mud) or mud brick, or in regions of higher rainfall, complex timber structures;
6. developed pyrotechnology, including the use of ovens for parching cereals and for producing bread;
7. widespread production of pottery vessels, and baked clay representations of humans and animals;
8. production and use of polished stone tools — axes, adzes and other artefacts, some of them utilized for specialized wood-working;
9. ritual practices involving human representations, often in baked clay;
10. systematic disposal of the dead in cemeteries, sometimes with monumental constructions;
11. development of long-distance procurement systems, for obsidian and other raw materials, which in coastal areas involved the systematic practice of seafaring;
12. development of local stylistic zones, seen, for instance, in the decoration of pottery. Such style zones may sometimes be associated with social distinctions, such as ethnic groupings.

I appreciate that this is in some senses a rather superficial observation, falling rather short in terms of analytical coherence, of the clear statements which have been made dealing with the differences seen with the introduction of the Upper Palaeolithic (e.g. by Mellars), which are not in question here. But at a superficial level the observation has some validity.

The paradox here then is that the big changes in behaviour seem to have taken place many millennia *after* the alleged genetic changes which are said to have 'caused' them. The genetic changes apparently had little immediate effect. What then is their explanatory power?

The sapient behaviour paradox. Part II

The above statements considered variations in human behaviour in time. It is worth making a similar observation in spatial terms, using the ethnographic present (which in reality really implies the recent past).

If one considers only the surviving material evidence, as it would appear to archaeologists some centuries later, the materials remains of many recent hunter-gatherer communities would not differ dramatically in kind from those available from the Upper Palaeolithic.

This circumstance constitutes, of course, the insight which has informed much of the recent development of ethnoarchaeology. Ethnographers have indeed emphasized the richness of various aspects of hunter-gatherer life. But it is the case, nonetheless, that what has, in the main, motivated prehistoric archaeologists to undertake ethnoarchaeology among the Eskimos or the Bushmen, for instance, is the perception of a basic structural similarity among hunter-gatherer life styles and procurement practices. There is no assumption that these modern hunter-gatherers are the same as their prehistoric predecessors: simply that by virtue of being hunter-gatherers they are conditioned to have many things in common. Formulated so baldly the assumptions may appear questionable, but they have nonetheless been there. To say this does not assume that modern hunter-gatherers are 'surviving' Upper Palaeolithic hunters, frozen in aspic. Obviously they are separated by as many millennia as literate urban communities from the Upper Palaeolithic. But the differences are not immediately obvious from the durable material record.

The paradox here is that recent hunter-gatherers (genetically modern) have often manifested behaviour not greatly more complex than that of their genetically pre-modern predecessors. So just what is the genetic change conceived of as explaining?

Conclusion 1

There would seem to be a need to define very much more precisely what, in the Upper Palaeolithic in general, is so impressive that it can be seen to reflect what is generally regarded as the most decisive step in human evolution. (Localized phenomena, such as French cave art, cannot be allowed to obscure the broad picture.) If the emergence of our species was of such decisive significance, in a causal sense, for the subsequent unfolding of human history, why are the consequences of that emergence evident only in such a patchy way in space and time?

Conclusion 2

If we are considering human evolution and the emergence of *Homo sapiens sapiens*, it is necessary to call into evidence also the trajectory of cultural evolution after the crossing of the 40 ky threshold. If the hypotheses above are correct (and they are very widely accepted) the achievements of the subsequent evolutionary trajectory were already there in potential soon after the threshold was crossed. But only with the development of the 'software' — that is to say the cultural component as contrasted with the genetic 'hardware' component — was that potential actualized.

All of this amounts to a striking paradox. We are studying events which led up to a very significant threshold some 40,000 years ago. But the material record of that time or of the immediate aftermath does not bring out at all clearly that significance. The momentous nature of the transition can only be comprehended in the light of 40 millennia of subsequent cultural evolution. As implied earlier, it is a feature of teleological or goal-direct explanations that they explain changes at one time in terms of their consequences at a later time. But they are generally felt to be poor explanations, opening the way to circular arguments.

One is left here with the impression that there is something inappropriate with the way we are formulating these problems. Of course, in the future, developments in molecular genetics may well allow us to focus much more closely upon the relatively few genetic changes which were central to the great transition from *Homo erectus* to *Homo sapiens sapiens*. But even here, if we could define more closely the genotype, in terms of molecular composition, it is difficult to predict from that the nature of the phenotype. And it is just as difficult, having established the physical form of the phenotype, to proceed to predict its behaviour, the praktotype, and then further again, to predict group behaviour involving a number of individuals of the species.

The conclusion to which one may be led is that more effective methods have to be developed for the study of human behaviour, in the past as well as the present. For it is doubtful whether molecular sequences will ever give us any clear insights into such behaviour, even if they ultimately determine the physical framework within which it develops and operates.

Colin Renfrew
McDonald Institute for Archaeological Research
University of Cambridge
Downing St
Cambridge
CB2 3ER

Reference

Mellars P.A., 1991. Cognitive changes and the emergence of modern humans in Europe. *Cambridge Archaeological Journal* 1(1), 63–76.

2

Symbolism, Language, and the Neanderthal Mind

Paul Mellars

The essence of the contrast between the behavioural and cognitive patterns of Neanderthal and anatomically modern populations is often perceived as a 'symbolic explosion', involving the sudden appearance of many new forms of explicitly symbolic expression, and most probably a closely related change in the structure and complexity of language. This paper re-examines the basis for this generalization with special reference to the archaeological records from western Europe. Particular attention is focused on the much greater element of morphological standardization and visually imposed form apparent in Upper Palaeolithic as opposed to Middle Palaeolithic tool forms, as a potential reflection of more highly structured language and related social communication patterns. How far these changes reflect significant contrasts between the cognitive capacities of Neanderthal and anatomically modern populations remains extremely difficult to assess. The possibility of significant divergences in neurological capacities during the long period of evolutionary separation of the Neanderthal and modern human lineages, however, cannot be ruled out.

The cultural and mental capacities of the Neanderthal populations of Europe and western Asia have been a topic of debate throughout this century (Trinkaus & Shipman 1993; Stringer & Gamble 1993). Poised between the earlier populations of *Homo erectus*/*Homo heidelbergensis* and the succeeding populations of anatomically modern (or 'Cro-Magnon') humans, the Neanderthals were large bodied and in many ways rather primitive looking individuals, lacking many of the impressive cultural achievements (art, personal ornamentation, elaborate bone technology etc.) of their modern human successors. Recent research in both molecular genetics and hominid fossil anatomy suggests that the Neanderthals became largely if not entirely extinct by around 35,000 BP, and contributed few if any genes to later populations (Stoneking & Cann 1989; Wilson & Cann 1992; Cann *et al.* 1994; Stringer 1990; Stringer & Gamble 1993). And yet they had brains which, in terms of volume, were just as large as ours.

The question of the cognitive capacities of Neanderthals is addressed in two other papers in this volume, those of Gibson and Mithen. My own approach to this question rests on two basic premises. The first is that as an initial working hypothesis we should assume that there were no significant contrasts between the mental abilities of Neanderthals and modern populations, and that such contrasts should only be inferred after a careful and essentially sceptical scrutiny of the behavioural evidence appears to make this conclusion at least highly plausible if not inescapable. As both Gibson and Renfrew (this volume) have stressed, we should beware of equating the notions of simplicity of technology with simplicity of mind — a correlation which would be as absurd in studies of human cultural development as in anthropological studies of modern societies (see Mellars 1989a, 387–8; 1991, 70–72). The second premise however is that we can in no way rule out on *a priori* grounds the possibility that there *were* significant contrasts between the neurological capacities of Neanderthal and modern populations. One of the implications of the recent molecular genetic research is that the evolutionary trajectories of Neanderthal and anatomically modern populations in Europe are likely to have been separated over a period of at least 300,000 years, if not much longer (Stoneking & Cann 1989; Stoneking *et al.* 1992; Wilson & Cann 1992; Cann *et al.* 1994; Stringer & Gamble 1993; Harpending *et al.* 1993). Clearly, it would be wrong to assume that there can have been no significant change in the structure or complexity of the human brain over this period — at a time when other aspects of biological evolution were proceeding very rapidly. In this context it should be recalled that Humphrey (1976) and others have argued that some of the main stimuli to increasing intelligence throughout human evolution (*cf.* Aiello this volume, Fig. 8.1) are likely to have come not from the pressures of purely biological survival in the face of competing species and changing environmental conditions, but from the pressures of competing in social and personal terms with members of one's own species (see also Byrne & Whiten 1988; 1992; Alexander 1989). Seen in these terms it would be remarkable if the process of mental evolution had effectively come to an end at precisely the point when we know that most forms of human technological, economic and social development — and therefore the overall scale of intra-species social competition — were becoming most intense.

Neanderthal brains

How far we can draw reliable conclusions about intelligence or other aspects of mentality from the size and morphology of endocranial casts of human brains has been debated frequently in the literature over the past century (e.g. Holloway 1985; Passingham 1989; Byrne this volume). It is now agreed that in terms of absolute volume Neanderthal brains are certainly no smaller than those of anatomically modern populations, and perhaps on average slightly larger — ranging from *c.* 1200 to 1700 cc, with an average of *c.* 1400 cc, as compared to a modern average of around 1350 cc (Holloway 1985; Stringer & Gamble 1993). Of course, absolute brain volume alone is a notoriously poor indicator of cognitive abilities in different species (as the variation in modern animal brains clearly indicates: Passingham 1989; Byrne this volume) and it is still difficult to produce precise ratios of the 'allometric' relationship of Neanderthal brain sizes to overall body size, mainly owing to the difficulties of estimating Neanderthal body mass with any precision from the surviving skeletal evidence (Foley this volume). Dunbar and others have recently suggested that a more sensitive reflection of the relative intelligence of different primate species may be provided by the so-called 'neocortical ratio' — i.e. the ratio of the outer, neocortical region of the brain to total brain volume (Dunbar 1992; Aiello & Dunbar 1993). When calculated in these terms Neanderthal brains may perhaps be slightly less complex than those of anatomically modern populations (see Foley this volume). When allowance is made for potential biases in the samples available and the difficulties of making accurate estimates of the parameters involved, however, it is debatable whether any real significance can be attached to these variations in terms of the evolution of mental complexity.

Similar ambiguities surround the nature of any variations in the precise forms of Neanderthal brain casts, and the relative morphology and development of different areas of the brain. Holloway, who has probably written more than anyone else on this topic, has summarized his most recent conclusions as follows:

> I have no confidence in unambiguously identifying any convolutional patterns that are suggestive of a 'primitive condition' . . . I believe Neanderthal brains were fully *Homo*, with no significant differences in their organization compared to our own. (Holloway 1985, 320–21)

Other workers however have been prepared to see more significant variations. Several authors (e.g. Kochetcova 1978) have noted that the frontal lobe region of Neanderthal brains does seem to be significantly smaller than that of modern brains, and have interpreted this as potentially indicating a significantly lesser ability for both long-term planning and goal-oriented behaviour, and perhaps more limited creative capacities (Gibson this volume). Others have occasionally identified variations in the parietal areas, in the position of the lunate sulcus, and in the patterns of the sylvian and parieto-occipital fissures (Boule & Anthony 1911; Connolly 1950).

Other speculations have focused not on the overall size and morphology of the brain but on the possibility of increasingly complex linkages in the internal circuity between separate processing areas (Gibson 1988; 1990; 1993). Tooby & Cosmides (1992) in particular have argued that over the course of primate and human evolution a number of specialized, genetically determined information processing networks are likely to have developed to cope with specific problems of survival — such as mapping the distribution of food resources in space and time, similar mapping of interpersonal social structures and relationships and (possibly most significant) the emergence of specialized patterns of communication involving various forms of symbolic and verbal communication. The classic example of the latter is embodied in Chomsky's (1986) notion of a genetically programmed 'universal grammar' in human language (see also Pinker 1994). In the present volume Mithen has argued that the eventual cross-linking of these different areas of so-called 'domain-specific intelligence' could have led to the dramatic behavioural and symbolic revolution associated with the replacement of Neanderthal by anatomically modern populations in Europe.

Behavioural evidence for the intelligence of Neanderthals

While the various lines of evidence discussed above underscore that we can in no way rule out the possibility of significant contrasts between the mental capacities of Neanderthal and biologically modern populations, attempts to argue for such contrasts from the available archaeological evidence lead in most cases to equally ambiguous conclusions. As Renfrew and Gibson have argued in their papers, it remains effectively impossible to make direct correlations between actual patterns of behaviour and the underlying neurological *capacities* for behaviour (see also Mellars 1989a, 387–8; 1991, 70–72). There is no difficulty in drawing up a long and impressive list of the various ways in which the technology and behaviour of Upper Palaeolithic communities in Europe was demonstrably far more complex and highly structured than that of Middle Palaeolithic groups — embracing contrasts in lithic and bone technology, hunting equipment, personal ornamentation, art, notation systems, burial practices, musical instruments, complex living structures and, apparently, more specialized and 'strategically' organized subsistence practices (Mellars 1989a,b; 1991; Kozlowski 1990; Knecht *et al.* 1993). But whether any of these changes can be taken to reflect significant changes in the intellectual capacities for behaviour, as opposed to a simple, cumulative and mutually reinforcing increase in the overall complexity of various behavioural patterns — perhaps analogous to that of the subsequent Neolithic revolution or industrial revolution — is, as Gibson and Renfrew point out, extremely difficult to evaluate in objective terms.

The most explicit argument for a major cognitive component in these behavioural changes has been put forward in several papers by Lewis Binford (1985; 1987; 1989; 1992). His suggestion, in brief, is that many of the documented behavioural shifts between the Middle and Upper Palaeolithic seem to reflect a significant increase in the degree of time depth or long-range planning in the organization of activities. He argues that in the behavioural records of Upper Palaeolithic groups, as well as in those of all recent hunter-gatherers, we can document an impressive capacity to plan and structure activities over a relatively long time span, and with a clear perception of ultimate goals. Thus tools and equipment tend to be produced long before they are actually needed for use, and with a clear anticipation of extended periods of use. Occupation sites are similarly planned and structured with a view to relatively long periods of occupation, and with an anticipation of the needs that will arise during these periods of extended use. Systematic movements of camp sites are made with a view to intercepting or harvesting food resources at long distances from earlier camps, and so on. These capacities for long-term planning, forethought, anticipation etc. are seen by Binford as central to the behavioural adaptations of all present-day hunter-gatherers; he believes they are equally explicit in the behavioural records of most Upper Palaeolithic groups (Binford 1989; see also Gibson 1985; 1988; Parker & Milbraith 1993).

The main thrust of Binford's argument is that this dimension of long-term planning, forethought, prediction etc., appears to be largely lacking from the archaeological records of the Lower and Middle Palaeolithic periods (Binford 1987; 1989; 1992). He argues, for example, that Lower and Middle Palaeolithic tools were normally manufactured, used and discarded within a short space of time and, even if occasionally transported over considerable distances, rarely remained in use for more than at most a few days. In other words, Lower and Middle Palaeolithic tools, in contrast to those of the Upper Palaeolithic, show little evidence for long-term 'curation' (Binford 1973). He similarly argues that most food resources exploited by Middle Palaeolithic groups were processed and consumed close to, if not actually on, the point of procurement, with little attempt to anticipate regular shifts in the distribution of resources at different times of the year. In the same vein he argues that all of the claimed occupation sites of the Lower and Middle Palaeolithic appear to reflect very short-term activities (mostly by small human groups), with little if any organization of the occupation surfaces with a view to long periods of occupation.

Many of Binford's conclusions could no doubt be contested from several aspects of the archaeological data. As Geneste and others have pointed out, there is ample evidence that certain varieties of high-quality raw materials were systematically carried over long distances by Middle Palaeolithic groups, in some cases apparently from specialized quarry or 'procurement' sites (Geneste 1985; 1989; 1990; Turq 1989; Féblot-Augustins 1993). There is equally clear evidence that some specialized quarry sites were devoted to the production of one or more specific artefact forms (such as Levallois flakes or cordiform handaxes) which were subsequently removed from the sites for later use or shaping elsewhere (Geneste 1985). And as Tavoso (1984) has argued, there are many cases where fully-shaped tools were carried for long distances into areas where lithic raw materials were entirely lacking, with an anticipation of their use far removed from the original point of production. All of this evidence could be held up as proof that Neanderthal groups did have a clear perception of time-structured technologies and of potential future needs, in at least some of their activities.

Nevertheless the central point of Binford's argument is that whatever degree of forethought, prediction, long-term planning etc. can be identified in the behaviour of Middle Palaeolithic groups would seem to be greatly exceeded in the ensuing Upper Palaeolithic. In this respect, my own perception of the archaeological evidence coincides closely with his. In the Upper Palaeolithic, for example, we have evidence not only that high-quality lithic raw materials were transported over the landscape in much more substantial quantities than in the Middle

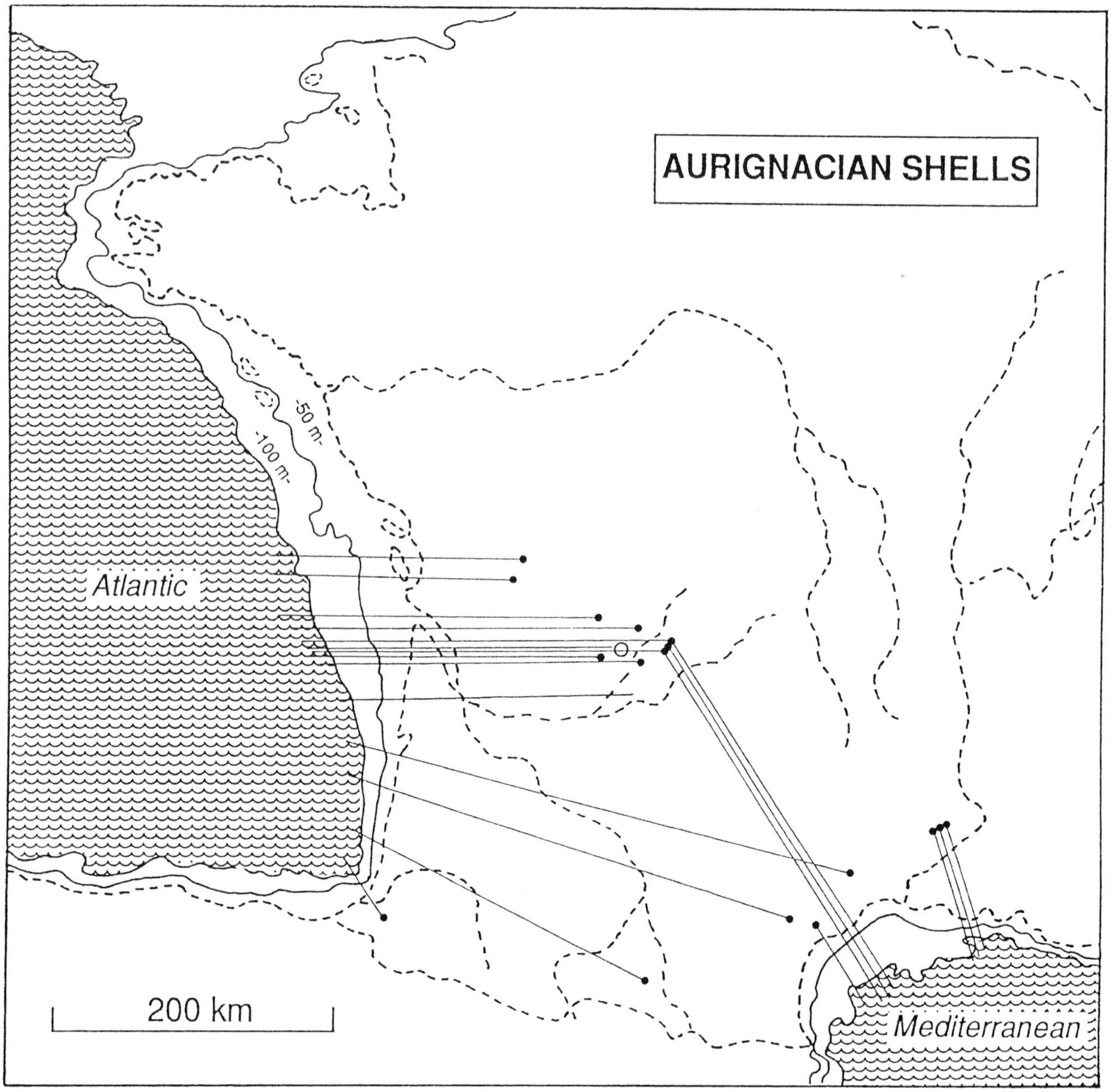

Figure 2.1. *Long-distance transport of marine shells documented in early Upper Palaeolithic Aurignacian sites (c. 35–30,000 BP) in southern France. (After Taborin 1993.)*

Palaeolithic (Geneste 1989; 1990), but also that certain specific commodities (most notably marine shells) were in some cases transported over much greater distances (see Fig. 2.1) of 500 km or more (Roebroeks *et al.* 1988; Gamble 1986; Taborin 1993). In the same vein there is no doubt that certain Upper Palaeolithic occupation sites (particularly those of the Upper Perigordian and later periods) show much clearer evidence for highly structured internal organization than any of the documented Lower and Middle Palaeolithic sites, in a way which almost certainly reflects the anticipation of longer periods of occupation (Mellars 1973; Soffer 1985a,b; Gamble 1986; Combier 1988). Most significant of all, perhaps, is the apparent evidence for the deliberate storage of food resources on some Upper Palaeolithic sites — if some of the large pit features recorded at open-air sites in Central and Eastern Europe can be reliably interpreted in these terms (Soffer 1985a,b). Finally, I have argued that some of the strategies of hunting documented in the Upper Palaeolithic sites of southwestern France appear to reflect a much more sharply focused pattern of exploitation of reindeer resources than anything so far documented from the preceding Mousterian sites of the same region (see Mellars 1989a, 357–8). This seems to be reflected not only in the highly specialized composition of the faunal assemblages from many sites (often with up to 99 per cent of reindeer remains — even in sites dating from the earliest stages of the Aurignacian) but also by the conspicuous clustering of these specialized reindeer-hunting sites along certain stretches of the Vézère and Dordogne valleys, which are likely to have formed the major and most predictable migration routes for reindeer herds. I would argue that it is this aspect of the evidence which provides the most explicit demonstration of the capacity of Upper Palaeolithic groups to plan and organize their subsistence activities in a more efficient and strategic way than the Middle Palaeolithic groups in the same region.

The critical question is what, if anything, does this apparently clear evidence for increased time perspectives, long-range planning etc. imply for the relative intelligence of Neanderthals and modern humans? The answer could be argued in two different ways. Following the criteria outlined by Gibson (1990), one could certainly argue that the appearance of more long-term planning strategies — as well as the integration within these strategies of larger numbers of cooperating individuals and, no doubt, more complex technological operations — is fully consistent with a significant increase in human cognitive abilities across the Middle–Upper Palaeolithic transition — and in this context the apparently increased neocortical ratio and expanded volume of the prefrontal area of the brain in anatomically modern populations might well be significant (*cf.* Gibson 1985; 1988; this volume; Parker & Milbraith 1993; Roberts *et al.* this volume). The alternative, which Binford himself apparently favours (1987, 692; 1989), would be to attribute this increased capacity for long-range planning of activities not to any increase in innate cognitive *capacities*, but simply to the emergence of new cognitive *structures*, in the form of more complex and highly developed language and associated social communication patterns. It is this particular scenario which will be explored in the remaining sections of the paper.

The symbolic explosion

Few would dispute that the most conspicuous single feature of the classic Middle–Upper Palaeolithic transition in Europe was the remarkable 'explosion' of explicitly symbolic behaviour (e.g. Pfeiffer 1982; White 1989; 1993; Mellars 1989a; 1991). This explosion is reflected most dramatically in the sudden appearance of representational art, personal ornamentation, musical instruments, decorated bonework and ceremonial burials, all apparently associated with the archaeological dispersal of 'Aurignacian' technologies across Europe, and all apparently associated with the earliest populations of anatomically modern humans (Mellars 1992). The questions to be addressed here are: first, to what extent can we identify convincing evidence for any comparable forms of symbolic expression among the preceding Middle Palaeolithic communities in Europe? And second, how far can this 'symbolic revolution' be assumed to correlate with a corresponding 'linguistic revolution' over the period of the Neanderthal to modern human transition?

Middle Palaeolithic symbolism

The topic of Middle Palaeolithic symbolism has generated a large and lively literature over the past decade (e.g. Chase & Dibble 1987; 1992; Chase 1991; Dibble & Chase 1990; Dibble 1989; Lindly & Clark 1990, with comments; Marshack 1972; 1981; 1988; 1990; Hayden 1993; Byers 1994). The first and most basic question is what exactly we mean by symbolism? The most widely accepted definition of a symbol is anything, be it an object, sign, gesture, vocal expression etc., which in some way refers to or represents something beyond itself (Chase 1991; Hodder 1982). This leads to the fundamental distinction between the symbol itself (the 'signifier') and the object or idea signified (the 'referrent' or 'signified'). Most authors go on to make further distinctions between different forms of symbols, at increasing levels of abstraction between the symbol and the referrent. Thus Chase (1991, 195) differentiates between what he refers to as 'iconic' symbols, which relate in a very direct and obvious way between the symbol and the referrent (as a road sign of a jumping deer might refer to the presence of wild game in the vicinity), 'index' symbols (which might use, say, the symbol of smoke to indicate fire), and totally abstract or abitrary symbols, which have no obvious relationship to the idea being symbolized beyond that attached in a totally arbitrary way by the group of individuals sharing this particular pattern of symbolic expression. The classic illustrations of the latter are most words, which can only be related to the objects or ideas being symbolized in essentially abstract, arbitrary ways.

Most discussions of symbolism go on to make some direct connections between the existence of clear, explicit symbolic expression and the existence of at least rudimentary forms of language (e.g. Holloway 1969; 1983; Dibble 1989; Davidson & Noble 1989; 1993). Whether these correlations are as direct and obvious as many authors assume (in the sense of necessarily implying the existence of language) is clearly debatable. Even so, no one would question that the whole of present-day culture and behaviour is highly symbolic in character, and that the heavily symbol-laden nature of culture is ultimately fundamental not only to modern language patterns, but to all other aspects of society, technology, communication etc. (Gellner 1989; Alexander 1989; Binford 1987; 1989; Donald 1991; Pinker 1994). There is equal agreement that the same patterns of complex symbolic — and probably linguistic — expression can be traced back to the beginning of the Upper Palaeolithic, and arguably provide the most dramatic contrast with the behaviour and organization of Middle Palaeolithic communities (Pfeiffer 1982; Chase & Dibble 1987; 1992; White 1989; Binford 1987; 1989; Mellars 1989a,b; 1991; Davidson & Noble 1989; Knight 1991).

Viewed in these terms, therefore, the existence and character of any clearly symbolic behaviour among Middle Palaeolithic communities must be seen as central to an understanding of the mentality and cognition of these groups and to the nature and significance of the documented behavioural contrasts between the Middle and Upper Palaeolithic. In the following sections I will review the principal areas where the existence of symbolic expression in Middle Palaeolithic/Neanderthal contexts has often been claimed, and then assess what significance can be attached to this evidence.

Use of pigments

The occurrence of what are almost certainly colouring materials in Middle Palaeolithic contexts is now effectively beyond dispute (Bordes 1952; 1972; Wreschner 1980; Marshack 1981; Demars 1992). The evidence comes in two forms: first, fragments of iron oxide or red ochre which, depending on the source, can provide a range of colours from yellow through to deep maroon or red-brown; and black manganese dioxide. Fragments of ochre have now been recorded from at least a dozen Mousterian sites in southwest France, while occurrences of manganese dioxide are even more frequent (Bordes 1952; Demars 1992). Evidence that the materials were used as pigments seems hard to dispute. Many of the individual fragments show clear signs of either scraping (presumably to yield a powder) or well-developed facets, which suggest that they were applied directly to a hard or soft surface (Bordes 1952; 1972). Significantly, both materials were used extensively as pigments for the production of Upper Palaeolithic cave art. While the possibility of other uses has occasionally been suggested (such as the use of ochre in tanning hides), the status of both ochre and manganese dioxide as colouring pigments in many Middle (as well as some Lower) Palaeolithic contexts seems hard to deny.

The critical issue here is whether the simple use of colouring materials can be regarded as an explicitly symbolic act. What is conspicuously lacking from the Middle Palaeolithic is any indication of exactly how these colours were employed. There are several reports of fragments of stone apparently stained with ochre, and traces of ochre smeared on stone tools or fragments of bone (e.g. Marshack 1988; 1990). But in none of these cases is there any evidence that the pigments were used to produce explicit designs or other identifiable markings. In other words, the possibility remains that colour was employed in the Middle Palaeolithic purely to change the surface appearance of things, as one might paint a piece of furniture or dye clothing. Captive chimpanzees have been taught to use paint, and in many cases obviously enjoy (or at least are intrigued by) the visual effects of the paint on the canvas or other objects provided. It therefore remains highly debatable whether the mere existence and use of pigments in

Neanderthal contexts need reflect anything more than natural curiosity about transforming the appearance of things, or at best perhaps a rudimentary form of aesthetic appreciation (Chase & Dibble 1987). Whether the pigments were applied to human skin or to some other objects (such as skin clothing, wooden artefacts etc.) remains equally unclear from the archaeological evidence.

Symbolic artefacts

Most of the arguments about symbolism in Middle Palaeolithic contexts have centred on a small number of (mostly) bone fragments which are held to show either deliberate patterns of incisions or, in a few cases, perforated holes (Chase & Dibble 1987; 1992; Marshack 1972; 1988; 1990; Lindly & Clark 1990; Bednarik 1992). The most controversial pieces are the fragments of bone showing scratches or linear markings on one or more surfaces, such as those recovered from one of the Rissian levels at Pech de l'Azé II (Bordes 1969; 1972), the later Mousterian levels at Cueva Morín (Gonzalez Echegaray 1988), and the earlier, Middle Pleistocene levels at Bilzingsleben (Mania & Mania 1988) (see Fig. 2:2). Among the small group of perforated objects are the swan's vertebra and wolf metapodial from the Micoquian levels at Bocksteinschmiede in western Germany, the reindeer phalanges from La Quina, the flattened bone fragment with a small perforation from Repolusthöhle, and the two highly typical but remarkably isolated specimens of perforated animal teeth recovered from supposedly Mousterian levels at La Quina and Repolusthöhle. Finally, and most significantly, there is the intriguing nummulite fossil from Tata in Hungary, on which a natural crack through the centre of the fossil has been supplemented by a further, finely incised line engraved almost exactly at right angles to form a neat symetrical cross (see Marshack 1988; 1990; and Bednarik 1992 for details of these pieces, and full references to the relevant literature).

The various problems with accepting these objects as convincing examples of explicitly symbolic or even decorative objects have been spelled out fully by Chase, Dibble and others (Chase & Dibble 1987; 1992; Dibble & Chase 1990; Davidson & Noble 1989; 1993). The main ambiguities concern the grouping of supposedly engraved bones. As Chase & Dibble (1992) point out, deep and in some cases quite regularly-spaced incisions on bone fragments can be produced easily in the course of skinning and butchering animal carcasses, and are frequent occurrences on most Palaeolithic and Mesolithic sites with well preserved faunal material. In my own opinion, and that of Chase and Dibble, none of the markings so far described from Lower and Middle Palaeolithic sites in Europe shows anything approaching the degree of clarity, regularity, organization, or obvious intentionality that one would reasonably expect from consciously symbolic or 'decorative' engravings. In this respect they contrast strikingly with the majority of engravings from Upper Palaeolithic sites, where the clear patterning and intentionality of the engravings cannot be questioned (Marshack 1972). The status of these and similar pieces has been summed up aptly by Chase as follows:

> [Most] archaeological bone assemblages contain very large numbers of specimens with tool marks, many of them with many such marks. This means a large number of patterns of marks. It is all too easy to pick out a few specimens from the eye-catching end of the natural variation among these patterns and to attribute them to intentional symbol making rather than to the laws of probability. (Chase 1991, 210)

The occasional specimens of perforated bones and teeth reported from Middle and Lower Palaeolithic contexts pose similar problems. It is now generally accepted that the claimed perforation on the bone fragment from Pech de l'Azé II, and the perforated reindeer phalanges from La Quina are almost certainly the results of natural damage to the bones, caused by chemical erosion or by carnivore gnawing (Chase 1990). The two, isolated specimens of perforated animal teeth from the Mousterian levels at La Quina and Repolusthöhle (a fox canine and a wolf incisor respectively) on the other hand clearly cannot be explained in these terms. As White (1989, 386) has pointed out, the critical question is whether these pieces were truly associated with the Mousterian levels on the sites or whether (as seems more likely in these particular contexts) they were misplaced during the original excavations from later, Upper Palaeolithic levels. We are left with the intriguing perforated bones (a swan's vertebra and a wolf metapodial) from the Bocksteinschmiede cave which, if truly associated with the Micoquian levels, do seem to represent deliberately perforated objects (Marshack 1990). Whether these two, isolated specimens can be taken to represent explicitly symbolic or decorative items is still an open question. Chase & Dibble (1992) have recently suggested the possibility of a purely utilitarian function for these pieces, for example as parts of toggles or other components of clothing or thongs.

The problem with accepting any of these objects as convincing evidence for explicitly symbolic

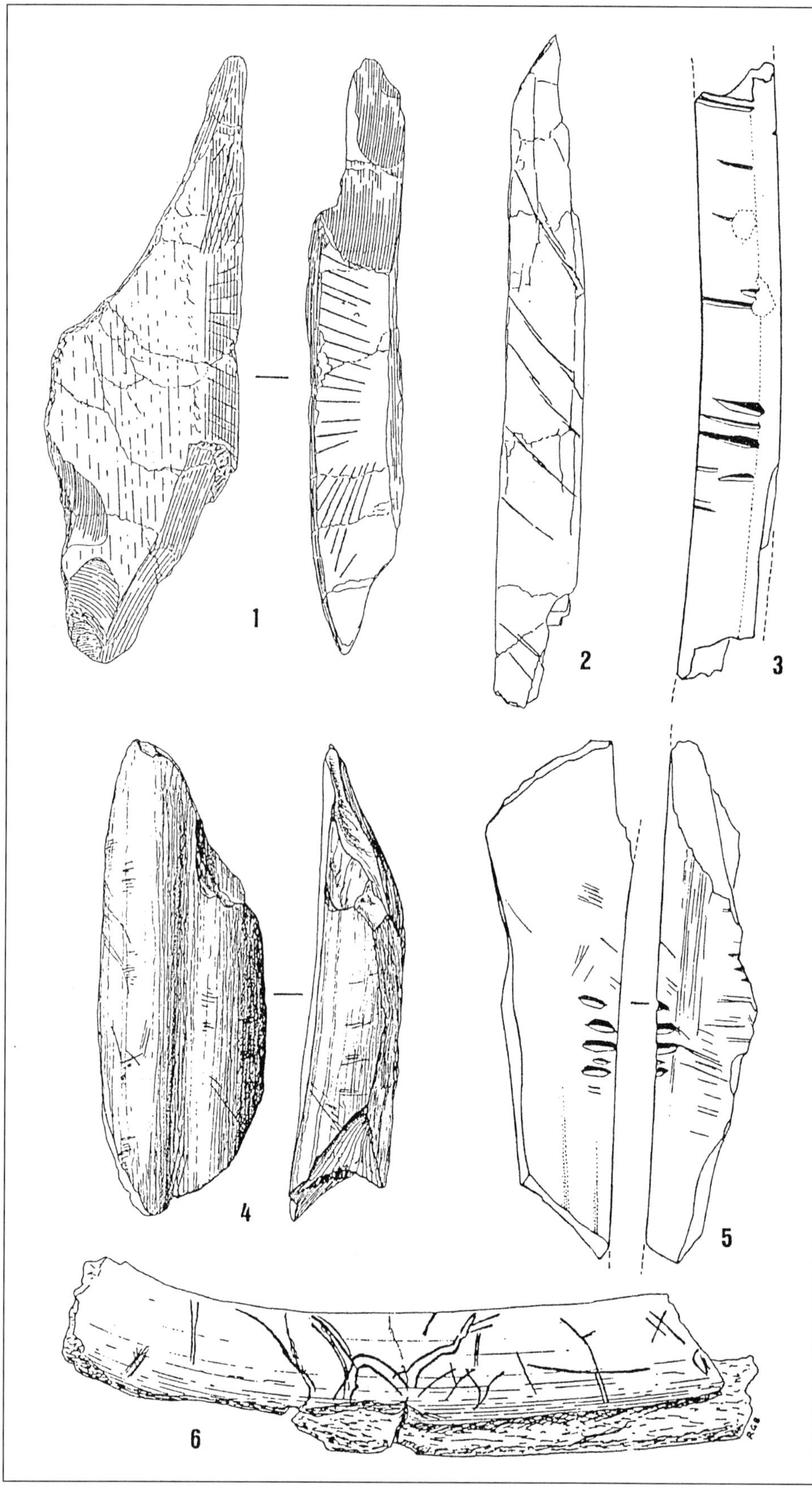

expression in Middle Palaeolithic contexts lies in their extreme rarity — especially seen in the context of the long time-span of the Middle Palaeolithic (over 100,000 years) and the large numbers of sites and occupation levels investigated throughout Europe. As Chase & Dibble (1987; 1992) and others have pointed out, in order to qualify in any meaningful sense as 'symbols' there must presumably be evidence that these particular forms of symbolic expression were shared amongst the members of the societies in question, and served in some way as a medium for either explicit or implicit *communication* between the group members. How this kind of significance can be attached to objects which are effectively unique within the total space and time framework of the European Middle Palaeolithic (such as the perforated bones from Bocksteinschmiede, or the two highly questionable perforated teeth from La Quina and Repolusthöhle) is by no means clear. Again,

Figure 2.2. *Incised bones from various Lower and Middle Palaeolithic contexts in Europe: nos. 1, 2. Bilzingsleben (East Germany, Middle Pleistocene); 3, 5. Abri Suard (southwest France, Mousterian); 4. Cueva Morín (northern Spain, Mousterian); 6. Pech de l'Azé II (southwest France, Acheulean). (From Bednarik 1992 and other sources.)*

the contrast with the Upper Palaeolithic could hardly be more dramatic. From the earliest stages of the Aurignacian, for example, we not only have large numbers of perforated animal teeth recovered from many different occupation levels (such as the 53 teeth from the Aurignacian I levels at La Souquette, or the 29 specimens from the similar levels at the Grotte de Spy: White 1989; Otte 1979) but also a clear *repetition* of this pattern from many early Aurignacian sites distributed across the length and breadth of Europe, from the Balkans to Belgium, southern France and northern Spain (Hahn 1972; 1977; Otte 1979; White 1989; 1993). Regardless of the significance attached to any individual objects (such as the enigmatic cross on the Tata pebble), the sheer scarcity and isolation of these objects within the Middle Palaeolithic universe as a whole makes it difficult to see this kind of symbolic expression as a real and significant component of Neanderthal behaviour.

Burial practices

The long-debated topic of Middle Palaeolithic burials and associated burial rituals raise issues similar to those discussed above. Is the archaeological evidence reliable? If the evidence is reliable, what can we legitimately infer from it about the symbolic or other nature of the behaviour represented?

The whole issue of Middle Palaeolithic burial practices has been raised and debated at length in a *Current Anthropology* review article by Robert Gargett (1989). Here, Gargett sets out the case systematically against the acceptance of most of the reported incidences of deliberate burials in Neanderthal contexts, followed by a variety of positive and negative responses from other commentators. What emerged from these exchanges is that while the case for the existence of intentional burial practices in Neanderthal contexts remains strong, the evidence for potential rituals or symbolic offerings associated with these burials (at least in Europe) remains very weak.

The arguments for accepting the majority of the claimed burials were set out persuasively in the comments by Bricker, Trinkaus, Frayer, Montet-White and others on Gargett's article. In the case of the southwest French sites, for example, it seems impossible to dismiss the evidence for deliberately excavated graves in at least three sites — La Chapelle-aux-Saints, La Ferrassie and Le Moustier. At La Chapelle-aux-Saints, for example, there are clear accounts of a rectangular burial pit excavated at the base of the Mousterian deposits, which is difficult if not impossible to attribute to any natural geological processes on the site (Boule 1909; Bouyssonie *et al.* 1908; 1913). The same is suggested by the original accounts of at least one of the burials at La Ferrassie (see Fig. 2.3; Peyrony 1921; 1934; 1939; see especially the comments by Bricker on these burials in Gargett's paper), and by the similar accounts of an apparently clear and deep grave pit containing the remains of a young child in the later Mousterian levels at Le Moustier (Peyrony 1930; Bordes 1959).

Perhaps the most persuasive arguments in favour of the intentionality of these Neanderthal burials, however, is provided by purely taphonomic considerations. As Trinkaus and others pointed out, it seems inconceivable that so many Neanderthal remains could have survived as largely intact, articulated skeletons unless the bodies had been deliberately protected in some way (from both predators and natural decay processes) shortly after death. This is especially true in the view of the large numbers of individual corpses recorded from certain sites (such as the remains of at least seven individuals at La Ferrassie and nine at Shanidar) and the fact that many of these are the remains of very young children, whose bones are especially delicate. Combined with the stratigraphic and other observations discussed above, the case for deliberate interment of at

Figure 2.3. *Photograph of burial of Neanderthal skeleton no. 2 at La Ferrassie (southwest France), taken at the time of discovery. (From Peyrony 1934.)*

least the majority of these skeletons appears virtually beyond dispute.

The arguments for deliberate grave offerings or other forms of ritual associated with these burials are much weaker. As Chase & Dibble (1987) and others have pointed out, most if not all of the supposed 'grave-goods' reported in association with Middle Palaeolithic burials in Europe can be interpreted more convincingly as objects which were incorporated accidentally into the graves at the time of interment. This applies most obviously to stone tools and animal bones, which are of course some of the commonest objects in Mousterian occupation levels, and which would almost inevitably become incorporated into the infill of graves from material simply lying on the surface at the time of burial. At best, all the reported occurrences of supposed grave offerings from European sites must be regarded as unproven. Whether the claimed ring of Ibex horns in association with the Neanderthal burial at Teshik Tash in Uzbekistan (Movius 1953), or the curious concentration of flower pollen in one of the Neanderthal graves at Shanidar in Iraq (Leroi-Gourhan 1975) can be interpreted in similar terms is no doubt more controversial (see for example the comments by A. Leroi-Gourhan on Gargett's paper). Equally controversial is the supposedly ritualistic setting of the isolated Neanderthal skull from the Grotta Guattari Cave in central Italy, which has recently been interpreted as a product of hyena activity (Stiner 1991). If these conclusions are accepted, we are left with only two fully convincing examples of intentional, and presumably symbolic, grave offerings recorded from well documented Middle Palaeolithic contexts — the large fallow-deer antler and the complete boar's jaw associated respectively with the burials at Qafzeh and Skhul in Israel (Vandermeersch 1970). The critical interest of these finds, of course, is that they were both associated with skeletons of essentially anatomically modern, as opposed to Neanderthal, physical type (Vandermeersch 1981; 1989).

The final and no doubt most controversial question is how far the existence of clearly intentional burials can be seen as a reliable reflection of symbolic behaviour among Neanderthals. In other words, does the mere act of human burial need to be seen as explicitly symbolic in nature, or could it be seen as a reflection of much simpler and more basic motivations? The argument that Neanderthal burials were a purely functional response to the need to dispose of dead bodies seems to me rather slim — especially in view of the large numbers of burials recorded in sites such as La Ferrassie and Shanidar and the fact that there are much simpler and less labour-intensive ways of disposing of corpses, as the ethnographic record clearly shows. At the very least I think we must assume that the act of deliberate burial implies some strong social or emotional bonds in Neanderthal societies, which dictated that the remains of relatives or other close kin should be carefully protected, and perhaps preserved, in some way after death. But to go beyond this and suggest that the act of burial must be seen as inherently and explicitly symbolic seems to me difficult to sustain. In the absence of either clear ritual or unambiguous grave offerings associated with Neanderthal burials in Europe it must be concluded that the case for a clear symbolic component in burial practices remains at best unproven.

Symbolism and style in tool manufacture

The possibility that evidence for symbolism may be reflected in the morphology and patterns of manufacture of stone tools has been debated in one form or another for the past 30 years. Holloway (1969; 1983), for example, argued that the mere existence of clearly differentiated forms of stone tools at Olduvai Gorge and elsewhere could be taken as reasonable *a priori* evidence for the existence of essentially linguistic mental concepts as far back as two million years. The same idea has been repeated in the publications of many other workers (e.g. Isaac 1969; Parker & Gibson 1979; Foster 1990; see Hewes 1993 for a review).

Recently, these notions of an explicitly symbolic element in early Palaeolithic stone tools have come under attack from several directions. Dibble (1987; 1989) has argued that almost all the documented variation in Lower and Middle Palaeolithic tools can be attributed either to the different tasks for which the tools were intended, or to successive modifications in the shape and apparent design of the tools through repeated resharpening and reworking in the course of use. His idea, in other words, is that there were no discrete 'concepts' or 'mental templates' for the different tool forms which went beyond at most a simple mental categorization of the tools in terms of their intended functions (see also Chase & Dibble 1987; Chase 1991).

Broadly similar arguments have been expressed in a different form in my own publications. In discussing the range of side-scraper forms from the Greek Mousterian site of Kokkinopilos I commented: 'the edge alone is the important part of the tool; the overall shape is of little importance' (Mellars 1964,

231). I went on to add: 'despite this variety of form, it is difficult to identify any clearly intentional subtypes amongst the racloirs, since so much of the secondary working applied to a piece depends on the original form of the flake chosen and the natural edges on it'. In a later paper I argued that the forms of the majority of Mousterian tools, in contrast to those of the Upper Palaeolithic, showed a high degree of fluidity and lack of morphological standardization, with the result that Middle Palaeolithic tool assemblages were far more difficult to classify into discrete, clearly separated forms, and that any attempt to categorize Middle Palaeolithic industries would require at best a much more limited range of discrete forms than would taxonomies of Upper Palaeolithic industries (Mellars 1989b; 1991).

As discussed elsewhere, the essence of these contrasts between the morphology of Lower and Middle versus Upper Palaeolithic tools lies in the notion of 'imposed form' (Mellars 1989b; 1991; see also Chase 1991). The central proposition is that in the majority of Upper Palaeolithic tools the artisan seems to have invested a good deal of effort to control and modify the overall shapes of the original flake blanks in order to achieve a distinctive, and in most cases relatively standardized, appearance in the forms of the finished tools (see Fig. 2.4). In the majority of Lower and Middle Palaeolithic tools, by contrast, this element of clearly imposed form seems to be lacking; most attention during tool manufacture seems to have been placed on the strictly functional properties of the working edges, with little if any attention to the overall shape or appearance of the resulting tool (Fig. 2.4). I have summed up this contrast by suggesting that Upper Palaeolithic artisans adhered to the maxim 'this is an endscraper: I use it as an endscraper, I call it an endscraper, and it must therefore *look* like an endscraper', whereas Middle Palaeolithic groups seem to have been content with the notion 'who cares what this tool looks like, so long as it performs adequately for the job in hand' (Mellars 1991, 66).

This same element of tightly imposed form and standardization in visual morphology is apparent in many other aspects of Upper Palaeolithic technology. It is arguably even more explicit in the form and technology of Upper Palaeolithic bone and antler tools, and may also be reflected in the highly patterned organization of many Upper Palaeolithic living sites (e.g. Combier 1988; Soffer 1985a), as well as in the similar regularity and standardization of art and decorative motifs (Mellars 1989b; 1991).

The central assumption is that this morphological standardization and imposed form in tool manufacture does indeed have some clear symbolic significance (see for example discussions in Chase 1991 and Byers 1994). The argument is that this element of standardization and imposed form goes far beyond the purely functional or utilitarian requirements of the tools, and must necessarily imply some symbolic *concept* of the individual tool forms, which in turn was reflected, either consciously or unconsciously, in their specific shapes and visual appearance (see also Wynn 1991, 203–4). Exactly what symbolic significance or associations this patterning might have had inevitably remains more speculative. Two obvious possibilities are that it could be related either to the social and ethnic identity of the groups who produced the tools (reflecting what Wiessner (1983; 1984) and others have referred to as 'emblemic style' in tool manufacture) or alternatively that it could be related to the particular economic, environmental or social contexts in which the different tools were employed. In an earlier paper I commented:

> Eskimo groups, for example, make sharp conceptual distinctions between tools used for processing different forms of plant versus animal products, between tools used on different social occasions or at different seasons of the year, or between tools used by males and females (e.g. Birket-Smith 1945;

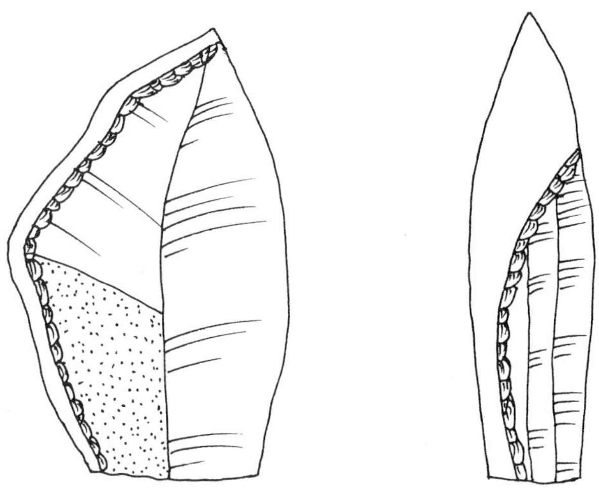

Figure 2.4. *Diagram to illustrate the notion of 'imposed form' in tool manufacture, illustrated by the comparison between a Mousterian backed knife (left) and an early Upper Palaeolithic Châtelperron point (right). Whereas in the Mousterian backed knife the retouch is relatively light and adheres closely to the original outlines of the flake, in the Upper Palaeolithic Châtelperron point much of the original blank is chipped away to impose a visually distinctive, repetitive and standardized form.*

> McGhee 1977). Potentially, therefore, formalized, perceptually defined differences in the forms of stone and bone artefacts could have been tied into a much wider framework of symbolism and symbolically defined behaviour embracing many different aspects of the social and economic organization of Upper Palaeolithic groups. (Mellars 1989b, 359; see Byers 1994 for a discussion of similar ideas)

Whether this element of imposed form is entirely lacking from Middle Palaeolithic tools is more debatable. A strong case for the existence of clear mental templates could be argued for some of the distinctive forms of bifacial tools recorded in certain Middle Palaeolithic industries — such as some of the cordiform, triangular or subtriangular handaxe types which characterize the Mousterian of Acheulean tradition industries of Western Europe, or the symmetrical, bifacial leaf-points in the later Mousterian industries of Central and Eastern Europe (Bordes 1961; 1984). Despite some recent protests (e.g. Dibble 1989; Davidson & Noble 1993), it is difficult to escape the conclusion that these forms were produced with a clear mental template in the minds of the artisans, which dictated that the tools should not only perform adequately for some specific tasks, but also conform to specific, predetermined, and presumably preconceived forms (see Gowlett this volume).

In this context there would seem to be two possibilities. Either that the occurrence of these pieces reflects at least an incipient measure of symbolic patterning in Middle Palaeolithic stone tools, directly heralding that which emerged on a greatly increased scale in the ensuing Upper Palaeolithic. Or alternatively that the apparent regularity, symmetry and standardization of these particular tool forms can be attributed to some more basic motivation, reflecting either simple aesthetic appreciation (i.e. simple curiosity or satisfaction in the production of regular and symmetrical shapes), or perhaps even (as Binford has suggested: 1989, 28) to a more long-term adaptation to the production of these symmetrical, bifacial tool forms extending (in the form of the Acheulean) over at least a million years (see also Wynn 1979; 1985; Gowlett this volume). At present I can see no way of choosing between these alternatives. What must be accepted is that whatever degree of standardization, symmetry, imposed form etc. can be recognized in the morphology of Middle Palaeolithic tools emerges on a much greater, more complex and more rapidly changing scale in the ensuing tool inventories of Upper Palaeolithic groups. It is this dramatic shift in tool production patterns that suggests that there was indeed a major change in the symbolic and cognitive aspects of tool manufacture between the Middle and Upper Palaeolithic periods, which may have equally significant implications for the general mental and cognitive dimensions of the populations involved.

Neanderthal language

Debates over the origins and development of language have filled several books over the past decade (e.g. Chomsky 1986; Landsberg 1988; Lieberman 1990; Bickerton 1990; Parker & Gibson 1990; Gibson & Ingold 1993) and spawned a society devoted exclusively to the study of language origins (Wind *et al.* 1989). It sometimes seems that there are almost as many different views on the origins of language as there are linguists, psychologists and palaeoanthropologists who have written on the issue. No one doubts that even the earliest hominids must have possessed some rudimentary form of language, since several of the great apes and even monkeys are known to employ relatively complex forms of vocal communication between individuals in the groups (e.g. Smuts *et al.* 1987; Cheney & Seyfarth 1990; Parker & Gibson 1990). The question is how these primitive forms of communication were transformed into fully developed language as observed in all present-day human populations.

The essence of most of these debates lies in the dichotomy between the 'evolutionary' views of language origins (which envisage a fairly gradual, step-by-step increase in the overall complexity and structure of language throughout human evolution), and the alternative of a relatively abrupt or 'catastrophic' transition at some point in language development, reflecting either a major genetic shift in the neurological organization of the brain, or at least a radical cognitive invention which revolutionized the whole structure of conceptual thought and associated communication patterns (e.g. Ragir 1985; Bichakjian 1989; Foster 1990; Lieberman 1990; Bickerton 1981; 1990; Pinker 1994). The most explicit account of the 'catastrophist' hypothesis has been set out in the publications of Derek Bickerton (Bickerton 1981; 1990). Basically, Bickerton argues for a distinction between what he refers to as 'proto-language' and fully developed, 'true' language. The theoretical basis for this distinction derives partly from studies of the patterns of language development in young children, and partly from similar studies of documented transitions between various forms of 'pidgin' languages and more complex 'creole' languages observed in many recent contact situations between European and non-European populations — all of which appear to reflect a relatively abrupt transition from very simple, rudimentary forms of language to much more complex and highly structured language within a short space of time. To Bickerton, these studies suggest that there are probably no potentially intermediate patterns between these two forms (Bickerton 1990, 140, 164–74, 190 etc.). The crucial distinction between these two language patterns lies in the structure of complex syntax or grammar which is the diagnostic hallmark of all present-day languages. Proto-languages by contrast (as in the language of very young children and in various forms of pidgin) are characterized by a virtual lack of clear grammatical structure or syntax (such as tense constructions, embedded clauses, use of null elements, auxiliary verbs etc.) and consist merely of short sequences of words, usually referring either to immediate here-and-now situations and relationships or, at best, to events in the immediate past or future. While the structure of proto-languages does not necessarily rule out the possibility of occasional reference to objects or events at some distance in time or space (i.e. some notion of 'displacement'), Bickerton argues that the lack of clear syntactical structure would dramatically reduce the potential complexity of such discussions — as well as introducing potential ambiguities — and would therefore have a drastically limiting effect on the efficiency of the language. He argues, more controversially, that this would influence not only the actions of the human groups, but also their capacity for structured, coherent thought:

> Thinking of the kind that humans do is at best extremely difficult in the absence syntax, since it depends crucially on the existence of structures like: x happened because y happened; whenever x happens, y happens; unless x happens, y will not happen . . . Complex sentences of any kind are impossible in protolanguage because there is no way in which one clause can be inserted into another . . . Without any reliable way of determining 'who did what to whom', ambiguities will quickly accumulate until they are too numerous to process. This would pose almost as much of a problem for coherent thought as it would for speech. (Bickerton 1990, 162–3)

Whether the emergence of language during the course of human evolution did take this dramatic, punctuated form remains the major issue in language origins (e.g. Ragir 1985; Chomsky 1986; Bichakjian 1989; Foster 1990; Bickerton 1981; 1990; Pinker 1994). The argument which has been advanced by Binford (1987; 1989), Whallon (1989), Clark (1981; 1992), myself (1989b; 1991) and several others is that if there *were* a relatively abrupt shift in the overall complexity and structure of language patterns associated with the transition from archaic to anatomically modern populations (even if on a less dramatic scale than that envisaged by Bickerton), this might well help to explain some of the radical transformations in human behavioural patterns over this period, documented in the archaeological records. The arguments are as follows:

1. First, Lieberman, Laitman and others have advanced a series of purely anatomical arguments for questioning the linguistic capacities of Neanderthals (e.g. Lieberman 1989; Lieberman & Crelin 1971; Laitman *et al.* 1979). Their argument, in essence, is that the detailed form of the mandible and basicranial region of several Neanderthal skulls (notably La Chapelle-aux-Saints) seem to rule out the possibility that these hominids possessed a fully developed vocal tract and the associated ability to form a full range of vowel sounds. These conclusions have been contested (for example by Arensburg (1989) based on the recently discovered Neanderthal from Kebara in Israel; see also Falk 1975; DuBrul 1977), and Lieberman (1989) admits that more highly developed capacities for speech appear to be reflected in some earlier, pre-Neanderthal remains (e.g. Petralona and Steinheim). Even if the anatomical arguments are accepted, it could be argued that the mere inability to form a full range of vowel sounds need not rule out the possibility of relatively advanced language among the Neanderthals — although as Leiberman points out it would inevitably restrict the complexity of vocal communications and might introduce serious ambiguities into language patterns.

To take a slightly different perspective, it would be possible to reverse the basic line of Lieberman's argument and see the lack of the full anatomical ability for highly developed speech as a reflection of the limited need for this ability among the Neanderthals — in other words that the lack of the full anatomical capacity for language *reflects* rather than *causes* the absence of highly developed language. Either way, the evidence cited by Lieberman and Laitman could provide an initial argument against the presence of fully developed language among the European Neanderthals.

2. In terms of the *behavioural* evidence, by far the most significant observation is the virtual lack of convincing evidence for explicit symbolic behaviour or expression in Neanderthal contexts. The relevant evidence has already been discussed above, and is

reflected not only in the conspicuous lack of all kinds of clearly 'decorative' or 'artistic' items in Mousterian contexts, but in the similar lack of convincing evidence for ceremonial burials, and in the lack of any obvious symbolic component in the forms of at least the majority of Middle Palaeolithic tools. The potential links between language and various forms of visual and other symbolism can be argued in different ways (e.g. Chase & Dibble 1987; Davidson & Noble 1989; Donald 1991; Knight 1991; Gibson & Ingold 1993). No one would question, however, that complex symbolic thought and expression is one of the defining hallmarks of all fully developed languages. The virtual lack of convincing evidence for symbolism in Mousterian contexts is at least consistent with the lack of highly developed language, even if this observation alone cannot be taken as concrete proof.

3. Potentially the most significant aspect of symbolic expression was raised earlier in the discussion of the morphology and taxonomy of Middle and Upper Palaeolithic tools. The point is not only that Middle Palaeolithic tools exhibit far less variety and diversity in shapes than those of the Upper Palaeolithic, but also, apparently, a much weaker element of morphological standardization and deliberately imposed form (see Fig. 2.4). As I have argued elsewhere (1989b; 1991), these patterns would suggest that the inferred 'mental templates' which lay behind the production of the majority of Middle Palaeolithic tools were far less sharply defined than those of the Upper Palaeolithic (see also Dibble 1989). This would fall naturally into place if the pattern of mental and associated *linguistic* categorization applied to different tool forms in the Middle Palaeolithic was much less sharply structured than that in the Upper Palaeolithic, and could be taken to suggest that even if Neanderthal groups did have discrete mental and linguistic categories for certain groups of tools, the total number of these categories was much less than in the Upper Palaeolithic — arguably reflecting a general increase in the scale and complexity of linguistic vocabularies over this transition.

4. Finally, we can return to the arguments advanced by Binford and others relating to the varying degrees of long-range planning and foresight implied by the archaeological records of the Middle and Upper Palaeolithic. As discussed in a preceding section, a significant shift in the time depth and overall complexity of these planning strategies over the Middle–Upper Palaeolithic transition can be argued from several aspects of the archaeological record — the patterns of procurement and distribution of raw materials, evidence for increased storage of food resources, more highly structured patterns of seasonal mobility and settlement strategies, and a more specialized and sharply focused pattern of animal exploitation, apparently based in the Upper Palaeolithic on an increased ability to predict and anticipate the movements of animal herds. As Binford points out (1987; 1989), such complex, logistical planning and organization would virtually *demand* the existence of highly structured language, in which the ability to discuss events far removed in both space and time would be essential to the effective organization and coordination of economic and social strategies. Essentially the same point has been made by Whallon (1989), Soffer (1994) and others, who argue that the increased ability to share information about the distribution of essential economic resources, and the increased ability to coordinate and integrate the social and economic activities of both individuals and widely dispersed local groups — which only fully developed language could provide — could well have been the critical factor which allowed the colonization of some of the more extreme and unpredictable periglacial environments in Central and Eastern Europe which seem to have been effectively and permanently colonized for the first time during the Upper Palaeolithic:

> The capacity to plan ahead certainly exists on an individual basis even among primates. The capacity to do this on a group basis, however, requires an ability to communicate among the members of the group, and in so doing refer to and discuss anticipated events in the future . . . Groups lacking such language capabilities would have been incapable of regularly and reliably organizing themselves for logistically exploiting their environments, and therefore would have operated with less and less efficiency in environments of increasing differential resource distribution and decreasing resource variety. (Whallon 1989, 442–3)

As Whallon (1989, 443) and Bickerton (1990, 158–60) point out, complex language could be equally indispensable to the long-term storage of information about past events and experiences (both economic and social), which could be equally critical to the survival prospects of human groups in many environments.

On present evidence the arguments about the character of Neanderthal language can hardly be taken any further. Few would seriously question that Neanderthals, as well as much earlier hominids, must have possessed some reasonably effective form of communication and, as Bickerton has argued, this must have been far more complex and highly structured than that of any of the living primates. The question is whether Neanderthal language was of essentially modern form, or whether (as Bickerton and others have argued) it was radically simpler in its grammatical and syntactical structure. In common with many other prehistorians I take the view that a radical restructuring in language patterns would be not only consistent with the available archaeological records of behavioural changes over the Middle–Upper Palaeolithic transition, but would provide the most economical explanation for at least the most striking of these changes. Whether or not these changes in language were associated with equally fundamental changes in the neurological organization of the human brain is of course an entirely separate question.

As a final point we should perhaps return to the proposition advanced by Chomsky (1986), Bickerton (1981; 1990) and others that the emergence of fully modern language must, by its nature, have been a relatively sudden event, rather than the product of gradual, step-by-step mental and linguistic evolution. If this conclusion is valid then we must presumably expect to see some fairly dramatic reflection of this transition in the behavioural records of human development, and arguably in the whole spectrum of behaviour ranging from technology, through subsistence and social patterns, to the more overtly symbolic domains of the human groups (Pinker 1994, 367–9). The question is where, in the available archaeological records of Europe, we might identify such a watershed, if not over the period of the Middle–Upper Palaeolithic transition?

Paul Mellars
Department of Archaeology
University of Cambridge
Downing Street
Cambridge
CB2 3DZ

References

Aiello, L.C. & R.I.M. Dunbar, 1993. Neocortex size, group size and the evolution of language. *Current Anthropology* 34, 184–93.

Aitken, M., C.B. Stringer & P.A. Mellars (eds.), 1992. The origins of modern humans and the impact of chronometric dating. *Philosophical Transactions of the Royal Society* series B, 337(1280).

Alexander, R.D., 1989. The evolution of the human psyche in Mellars & Stringer (eds.), 455–513.

Arensburg, B., 1989. New skeletal evidence concerning the anatomy of Middle Palaeolithic populations in the Middle East: the Kebara skeleton, in Mellars & Stringer (eds.), 165–71.

Bednarik, R.G., 1992. Palaeoart and archaeological myths. *Cambridge Archaeological Journal* 2(1), 27–57.

Bichakjian, B., 1989. Origine du langage et évolution linguistique. *Société d'Etudes et de Recherches Préhistoriques des Eyzies* 39, 93–111.

Bickerton, D., 1981. *Roots of Language.* Ann Arbor (MI): Karoma.

Bickerton, D., 1990. *Language & Species.* Chicago (IL): University of Chicago Press.

Binford, L.R., 1973. Interassemblage variability: the Mousterian and the 'functional' argument, in Renfrew (ed.), 227–54.

Binford, L.R., 1985. Human ancestors: changing views of their behavior. *Journal of Anthropological Archaeology* 4, 292–327.

Binford, L.R., 1987. An interview with Lewis Binford (edited by A.C. Renfrew). *Current Anthropology* 28, 683–94.

Binford, L.R., 1989. Isolating the transition to cultural adaptations: an organizational approach, in *The Emergence of Modern Humans: Biocultural Adaptations in the Later Pleistocene,* ed. E. Trinkaus. Cambridge: Cambridge University Press, 18–41.

Binford, L.R., 1992. Hard evidence. *Discover* February 1992, 44–51.

Birket-Smith, K., 1945. *Ethnographical Collections from the Northwest Passage.* (Report of the Fifth Thule Expedition 1921–1924 6, no. 2.) Copenhagen.

Bordes, F., 1952. Sur l'usage probable des peintures corporelles dans certaines tribus moustériennes. *Bulletin de la Société Préhistorique Française* 49, 169.

Bordes, F., 1959. Le contexte archéologique des Hommes du Moustier et de Spy. *L'Anthropologie* 63, 154–7.

Bordes, F., 1961. *Typologie du Paléolithique Ancien et Moyen.* (Publications de l'Institut de Préhistoire de l'Université de Bordeaux, Mémoire 1.) Bordeaux: Delmas.

Bordes, F., 1969. Os percé moustérien et os gravé acheuléen du Pech de l'Azé II. *Quaternaria* XI, 1–6.

Bordes, F., 1972. *A Tale of Two Caves.* New York (NY): Harper & Row.

Bordes, F., 1984. *Leçons sur le Paléolithique,* vol. 2: *Le Paléolithique en Europe.* (Institut du Quaternaire, Université de Bordeaux I, Cahiers du Quaternaire 7.) Paris: Centre National de la Recherche Scientifique.

Boule, M., 1909. L'homme fossile de la Chapelle-aux-Saints (Corrèze). *L'Anthropologie* 20, 257–71.

Boule, M. & R. Anthony, 1911. L'encephale de l'homme fossile de la Chapelle-aux-Saints. *L'Anthropologie* 22, 129–96.

Bouyssonie, A., J. Bouyssonie & L. Bardon, 1908. Découverte d'un squelette humain moustérien à la Bouffia de la Chapelle-aux-Saints (Corrèze). *L'Anthropologie* 19, 513–18.

Bouyssonie, A., J. Bouyssonie & L. Bardon, 1913. La station moustérienne de la 'Bouffia' Bonneval, à la Chapelle-aux-Saints. *L'Anthropologie* 24, 609–34.

Byers, A.M., 1994. Symboling and the Middle–Upper Palaeolithic transition: a theoretical and methodological critique. *Current Anthropology* 35, 369–99.
Byrne, R.W. & A. Whiten, 1988. *Machiavellian Intelligence: Social Expertise and the Evolution of Intellect in Monkeys, Apes and Humans.* Oxford: Oxford University Press.
Byrne, R.W. & A. Whiten, 1992. Cognitive evolution in primates: evidence from tactical deception. *Man* 27, 609–27.
Chase, P.G., 1990. Sifflets du Paléolithique moyen(?). Les implications d'un coprolite de coyote actuel. *Bulletin de la Société Préhistorique Française* 87, 165–7.
Chase, P.G., 1991. Symbols and Paleolithic artifacts: style, standardization, and the imposition of arbitrary form. *Journal of Anthropological Archeology* 10, 193–214.
Chase, P.G. & H.L. Dibble, 1987. Middle Paleolithic symbolism: a review of current evidence and interpretations. *Journal of Anthropological Archeology* 6, 263–96.
Chase, P.G. & H.L. Dibble, 1992. Scientific archaeology and the origins of symbolism: a reply to Bednarik. *Cambridge Archaeological Journal* 2(1), 43–51.
Cheney, D.L. & R.M. Seyfarth, 1990. *How Monkeys see the World: Inside the Mind of Another Species.* Chicago (IL): University of Chicago Press.
Chomsky, N., 1986. *Knowledge of Language: its Nature, Origin, and Use.* New York (NY): Praeger.
Clark, J.D., 1981. 'New men, strange faces, other minds'. An archaeologist's perspective on recent discoveries relating to the origins and spread of modern man. *Proceedings of the British Academy* 67, 163–92.
Clark, J.D., 1992. African and Asian perspectives on the origins of modern humans, in Aitken *et al.* (eds.), 201–16.
Combier, J., 1988. L'organisation de l'espace habité des hommes du Paléolithique Supérieur en France. *Espacio, Tiempo y Forma, Series 1, Prehistoria* 1, 111–24.
Connolly, C.J., 1950. *External Morphology of the Primate Brain.* Springfield (IL): C.C. Thomas.
Davidson, I. & W. Noble, 1989. The archaeology of perception: traces of depiction and language. *Current Anthropology* 30, 125–55.
Davidson, I. & W. Noble, 1993. Tools and language in human evolution, in Gibson & Ingold (eds.), 363–87.
Demars, P.-Y., 1992. Les colorants dans le Moustérien du Périgord: l'apport des fouilles de F. Bordes. *Préhistoire Ariégeoise* 47, 185–94.
Dibble, H.L., 1987. The interpretation of Middle Paleolithic scraper morphology. *American Antiquity* 52, 109–17.
Dibble, H.L., 1989. The implications of stone tool types for the presence of language during the Lower and Middle Palaeolithic, in Mellars & Stringer (eds.), 415–32.
Dibble, H.L. & P.G. Chase, 1990. Comment on J.M. Lindly and G.A. Clark 'Symbolism and modern human origins'. *Current Anthropology* 31, 241–3.
Donald, M., 1991. *Origins of the Modern Mind: Three Stages in the Evolution of Culture and Cognition.* Cambridge (MA): Harvard University Press.
DuBrul, E.L., 1977. Origins of the speech apparatus and its reconstruction in fossils. *Brain and Language* 4, 365–81.
Dunbar, R.I.M., 1992. Co-evolution of neocortex size, group size, and language in humans. *Behavior and Brain Sciences* 16, 681–735.
Falk, D., 1975. Comparative analysis of the larynx in man and the chimpanzee: implications for language in Neanderthals. *American Journal of Physical Anthropology* 43, 123–32.
Féblot-Augustins, J., 1993. Mobility strategies in the late Middle Palaeolithic of Central Europe and Western Europe: elements of stability and variability. *Journal of Anthropological Archaeology* 12, 211–65.
Foster, M.L., 1990. Symbolic origins and transitions in the Palaeolithic, in Mellars (ed.), 517–39.
Gamble, C., 1982. Interaction and alliance in palaeolithic society. *Man* 17, 92–107.
Gamble, C., 1986. *The Palaeolithic Settlement of Europe.* Cambridge: Cambridge University Press.
Gargett, R.H. 1989. Grave shortcomings: the evidence for neandertal burial. *Current Anthropology* 30, 157–90.
Gellner, E. 1989. Culture, constraint and community: semantic and coercive compensations for the genetic underdetermination of *Homo sapiens sapiens*, in Mellars & Stringer (eds.), 514–25.
Geneste, J-M., 1985. Analyse Lithique d'Industries Moustériennes du Périgord: une approche technologique du comportement des groupes humains au Paléolithique moyen. Unpublished Doctoral dissertation, University of Bordeaux I.
Geneste, J-M., 1989. Economie des ressources lithiques dans le Moustérien du sud-ouest de la France, in *L'Homme de Néandertal*, vol. 6, ed. M. Otte. (Etudes et Recherches Archéologiques de l'Université de Liège.) Liège: Université de Liège, 75–97.
Geneste, J-M., 1990. Développement des systèmes de production lithique au cours du paléolithique moyen en Aquitaine septentrionale, in *Paléolithique Moyen Récent et Paléolithique Supérieur Ancien en Europe*, ed. C. Farizy. (Mémoires du Musée de Préhistoire d'Ile de France 3.) Nemours: Musée de Préhistoire d'Ile de France, 205–7.
Gibson, K.R., 1985. Has the evolution of intelligence stagnated since Neanderthal Man? (Commentary on Parker), in *Evolution and Developmental Psychology*, eds. G. Butterworth, J. Rutkowska & M. Scaife. Brighton: Harvester Press, 102–14.
Gibson, K.R., 1988. Brain size and the evolution of language, in *The Genesis of Language: a Different Judgement*, ed. M.E. Landsberg. Berlin: Mouton de Gruyter, 149–72.
Gibson, K.R., 1990. New perspectives on instincts and intelligence: brain size and the emergence of heirarchical mental constructional skills, in Parker & Gibson (eds.), 97–128.
Gibson, K.R., 1993. Tool use, language and social behavior in relation to information processing capacities, in Gibson & Ingold (eds.), 251–69.
Gibson, K.R. & T. Ingold (eds.), 1993. *Tools, Language and Cognition in Human Evolution.* Cambridge: Cambridge University Press.
Gonzalez Echegaray, J., 1988. Decorative patterns in the Mousterian of Cueva Morín, in Otte (ed.), 37–42.
Gowlett, J.A.J., 1984. Mental abilities of early man: a look at

some hard evidence, in *Hominid Evolution and Community Ecology*, ed. R.A. Foley. London: Academic Press, 167–92.

Hahn, J., 1972. Aurignacian signs, pendants, and art objects in Central and Eastern Europe. *World Archaeology* 3, 252–66.

Hahn, J., 1977. *Aurignacien: Das Altere Jungpalâolithikum in Mittel- und Osteuropa.* (Fundamenta Series A9.) Köln.

Harpending, H., S. Sherry, A. Rogers & M. Stoneking, 1993. The genetic structure of ancient human populations. *Current Anthropology* 34, 483–96.

Hayden, B., 1993. The cultural capacities of Neandertals: a review and re-evaluation. *Journal of Human Evolution* 24, 113–46.

Hewes, G.W., 1993. A history of speculation on the relation between tools and language, in Gibson & Ingold (eds.), 20–31.

Hodder, I., 1982. *Symbols in Action: Ethnoarchaeological Studies of Material Culture.* Cambridge: Cambridge University Press.

Holloway, R.L., 1969. Culture: a human domain. *Current Anthropology* 10, 395–412.

Holloway, R.L., 1983. Human brain evolution: a search for units, models and synthesis. *Canadian Journal of Anthropology* 3, 215–30.

Holloway, R.L., 1985. The poor brain of *Homo sapiens neanderthalensis*: see what you please, in *Ancestors: the Hard Evidence,* ed. E. Delson. New York (NY): Alan D. Liss, 319–24.

Humphrey, N., 1976. The social functions of intellect, in *Growing Points in Ethology*, eds. P.P.G. Bateson & R.A. Hinde. Cambridge: Cambridge University Press, 303–17.

Isaac, G.Ll., 1969. Studies of early culture in East Africa. *World Archaeology* 1, 1–28.

Knecht, H., A. Pike-Tay & R. White (eds.), 1993. *Before Lascaux: the Complex Record or the Early Upper Paleolithic.* Boca Raton (FL): CRC Press.

Knight, C.D., 1991. *Blood Relations: Menstruation and the Origins of Culture.* New Haven (CT): Yale University Press.

Kochetcova, V.I., 1978. *Paleoneurology.* Washington (DC): V.H. Winston.

Kozlowski, J.K., 1990. A multi-aspectual approach to the origins of the Upper Palaeolithic in Europe, in Mellars (ed.), 419–37.

Laitman, J.T., R.C. Heimbuch & E.S. Crelin, 1979. The basicranium of fossil hominids as an indicator of their upper respiratory systems. *American Journal of Physical Anthropology* 51, 15–34.

Landsberg, M.E. (ed.), 1988. *The Genesis of Language: a Different Judgement of Evidence.* Berlin: Mouton de Gruyter.

Leroi-Gourhan, A., 1975. The flowers found with Shanidar IV, a Neanderthal burial in Iraq. *Science* 190, 562–4.

Lieberman, P., 1989. The origins of some aspects of human language and cognition, in Mellars & Stringer (eds.), 391–414.

Lieberman, P., 1990. *Uniquely Human: the Evolution of Speech, Thought, and Selfless Behavior.* Cambridge (MA): Harvard University Press.

Lieberman, P. & E.S. Crelin, 1971. On the speech of Neanderthal man. *Linguistic Inquiry* 2, 203–22.

Lindly, J.M. & G.A. Clark, 1990. Symbolism and modern human origins. *Current Anthropology* 31, 233–61.

McGhee, R., 1977. Ivory for the Sea Woman: the symbolic attributes of a prehistoric technology. *Canadian Journal of Archaeology* 1, 141–9.

Mania, D. & U. Mania, 1988. Deliberate engravings on bone artefacts of *Homo erectus*. *Rock Art Research* 5, 91–107.

Marshack, A., 1972. *The Roots of Civilization*. New York (NY): McGraw-Hill.

Marshack, A., 1981. On Paleolithic ochre and the early uses of color and symbol. *Current Anthropology* 22, 188–91.

Marshack, A., 1988. The Neanderthals and the human capacity for symbolic thought: cognitive and problem-solving aspects of Mousterian symbol, in Otte (ed.), 57–92.

Marshack, A., 1990. Early hominid symbol and evolution of the human capacity, in Mellars (ed.), 457–98.

Mellars, P.A., 1964. The Middle Palaeolithic surface artifacts at Kokkinopilos. *Proceedings of the Prehistoric Society* 30, 229–44.

Mellars, P.A., 1973. The character of the Middle–Upper Palaeolithic transition in southwest France, in Renfrew (ed.), 255–76.

Mellars, P.A., 1989a. Major issues in the emergence of modern humans. *Current Anthropology* 30, 349–85.

Mellars, P.A., 1989b. Technological changes across the Middle–Upper Paleolithic transition: technological, social, and cognitive perspectives, in Mellars & Stringer (eds.), 338–65.

Mellars, P.A. (ed.), 1990. *The Emergence of Modern Humans: an Archaeological Perspective.* Edinburgh: Edinburgh University Press.

Mellars, P.A., 1991. Cognitive changes and the emergence of modern humans in Europe. *Cambridge Archaeological Journal* 1(1), 63–76.

Mellars, P.A., 1992. Archaeology and the population-dispersal hypothesis of modern human origins in Europe, in Aitken *et al.* (eds.), 225–34.

Mellars, P.A. & C. Stringer (eds.), 1989. *The Human Revolution: Behavioural and Biological Perspectives on the Origins of Modern Humans.* Princeton (NJ): Princeton University Press.

Movius, H.L., 1953. The Mousterian cave of Teshik-Tash, southeastern Uzbekistan. *Bulletin of the American School of Prehistoric Research* (Peabody Museum, Harvard University) 17, 11–71.

Otte, M., 1979. *Le Paléolithique Supérieur Ancien en Belgique.* (Monographies d'Archéologie Nationale 5.) Brussels: Musées Royaux d'Art et d'Histoire.

Otte, M. (ed.), 1988. *L'Homme de Néandertal*, vol. 5. (Etudes et Recherches Archéologiques de l'Université de Liège.) Liège: Université de Liège.

Parker, S.T. & K.R. Gibson, 1979. A developmental model for the evolution of language and intelligence in early hominids. *The Behavioural and Brain Sciences* (1979) 2, 367–408.

Parker, S.T. & K.R. Gibson (eds.), 1990. *'Language' and Intelligence in Monkeys and Apes: Comparative Developmental Perspectives.* Cambridge: Cambridge University Press.

Parker, S.T. & C. Milbraith, 1993. Higher intelligence, propositional language, and culture as adaptations for planning, in *Tools, Language and Cognition in Human Evolution*, eds K.R. Gibson & T. Ingold. Cambridge: Cambridge University Press, 314–33.

Passingham, R.E., 1989. The origins of human intelligence, in *Human Origins*, ed. R. Durant. London: Clarendon Press, 123–36.

Peyrony, D., 1921. Les moustériens inhumaient-ils leurs morts? *Bulletin de la Société Historique et Archéologique du Périgord* 1921, 132–9.

Peyrony, D., 1930. Le Moustier: ses gisements, ses industries, ses couches géologiques. *Revue Anthropologique* 40, 48–76, 155–76.

Peyrony, D., 1934. La Ferrassie: Moustérien, Périgordien, Aurignacien. *Préhistoire* 3, 1–92.

Peyrony, D., 1939. Le Comte Begouën en Périgord, in *Mélanges de Préhistoire et d'Anthropologie offerts par ses Collègues, Amis et Disciples au Professeur Comte H. Begouën*. Toulouse, Editions du Museum, 235–41.

Pfeiffer, J.E., 1982. *The Creative Explosion: an Enquiry into the Origins of Art and Religion*. New York (NY): Harper & Row.

Pinker, S., 1994. *The Language Instinct: the New Science of Language and Mind*. London: Penguin.

Ragir, S., 1985. Retarded development: the evolutionary mechanism underlying the emergence of language. *The Journal of Mind and Behavior* 6, 451–68.

Renfrew, C. (ed.), 1973. *The Explanation of Culture Change: Models in Prehistory*. London: Duckworth.

Roebroeks, W., J. Kolen & E. Rensink, 1988. Planning depth, anticipation and the organization of Middle Palaeolithic technology: the 'archaic natives' meet Eve's descendants. *Helinium* 28, 17–34.

Smuts, B.B., D.L. Cheney, R.M. Seyfarth, R.W. Wrangham & T.T. Struhsaker (eds.), 1987. *Primate Societies*. Chicago (IL): University of Chicago Press.

Soffer, O., 1985a. *The Upper Paleolithic of the Central Russian Plain*. Orlando (FL): Academic Press.

Soffer, O., 1985b. Patterns of intensification as seen from the Upper Paleolithic of the Central Russian Plain, in *Prehistoric Hunter-Gatherers: the Emergence of Cultural Complexity*, eds. T.D. Price & J.A. Brown. Orlando (FL): Academic Press, 235–70.

Soffer, O., 1994. Ancestral lifeways in Eurasia — the Middle and Upper Paleolithic records, in *Origins of Anatomically Modern Humans*, eds. M.H. Nitecki & D.V. Nitecki. New York (NY): Plenum Press, 101–9.

Stiner, M.C., 1991. The faunal remains at Grotta Guattari: a taphonomic perspective. *Current Anthropology* 32, 103–17.

Stoneking, M. & R.L. Cann, 1989. African origin of human mitochondrial DNA, in Mellars & Stringer (eds.), 17–30.

Stoneking, M., S.T. Sherry, A.J. Redd, & L. Vigilant, 1992. New approaches to dating suggest a recent age for the human DNA ancestor, in Aitken *et al.* (eds.), 167–76.

Stringer, C.B., 1990. The emergence of modern humans. *Scientific American* 263, 68–74.

Stringer, C. & C. Gamble, 1993. *In Search of the Neanderthals: Solving the Puzzle of Human Origins*. London: Thames & Hudson.

Taborin, Y., 1993. Shells of the French Aurignacian and Perigordian, in Knecht *et al.* (eds.), 211–29.

Tavoso, A., 1984. Réflexion sur l'économie des matières premières au Moustérien. *Bulletin de la Société Préhistorique Française* 81, 79–82.

Tooby, J. & L. Cosmides, 1992. The psychological foundations of culture, in *The Adapted Mind*, eds. J.H. Barkow, L. Cosmides & J. Tooby. Oxford: Oxford University Press, 19–136.

Trinkaus, E. & P. Shipman, 1993. *The Neandertals: Changing the Image of Mankind*. London: Jonathan Cape.

Turq, A., 1989. Exploitation des matières premières lithiques et occupation du sol: l'exemple du Moustérien entre Dordogne et Lot, in *Variations des Paléomilieux et Peuplement Préhistorique*, ed. H. Laville. Paris: Centre National de la Recherche Scientifique, 179–204.

Vandermeersch, B., 1970. Une sépulture moustérien avec offrandes découverte dans la grotte de Qafzeh. *Comptes Rendus de l'Académie des Sciences de Paris* 270, 298–301.

Vandermeersch, B., 1976. Les sépultures néanderthaliennes, in *La Préhistoire Française*, ed. H. de Lumley. Paris: Centre National de la Recherche Scientifique, 725–7.

Vandermeersch, B., 1981. *Les Hommes Fossiles de Qafzeh (Israël)*. Paris: Centre National de la Recherche Scientifique.

Vandermeersch, B. 1989. The evolution of modern humans: recent evidence from southwest Asia, in Mellars & Stringer (eds.), 155–64.

Whallon, R., 1989. Elements of cultural change in the later Palaeolithic, in Mellars & Stringer (eds.), 433–54.

White, R., 1989. Production complexity and standardization in early Aurignacian bead and pendant manufacture: evolutionary implications, in Mellars & Stringer (eds.), 366–90.

White, R., 1993. Technological and social dimensions of 'Aurignacian-age' body ornaments across Europe, in Knecht *et al.* (eds.), 277–300.

Wiessner, P., 1983. Style and social information in Kalahari San projectile points. *American Antiquity* 48, 253–76.

Wiessner, P., 1984. Reconsidering the behavioral basis for style: a case study among the Kalahari San. *Journal of Anthropological Archaeology* 3, 190–234.

Wilson, A.C. & R.L. Cann, 1992. The recent African genesis of humans. *Scientific American* 266(4), 22–7.

Wind, J., E.G. Pulleybank, E. de Grolier & B. Bichakjian (eds.), 1989. *Studies in Language Origins*, vol. 1. Amsterdam: John Benjamins.

Wreschner, E., 1980. Red ochre and human evolution: a case for discussion. *Current Anthropology* 21, 631–44.

Wynn, T., 1979. The intelligence of later Acheulean hominids. *Man* 14, 379–91.

Wynn, T., 1985. Piaget, stone tools and the evolution of human intelligence. *World Archaeology* 17, 32–43.

Wynn, T., 1991. Tools, grammar and the archaeology of cognition. *Cambridge Archaeological Journal* 1(2), 191–206.

3

The Biocultural Human Brain, Seasonal Migrations, and the Emergence of the Upper Palaeolithic

Kathleen R. Gibson

The fundamental foundations of human cultural abilities rest on a suite of interacting cognitive and learning skills each of which is present in rudimentary form in great apes but greatly expanded in our species. Human abilities also reflect our highly plastic 'biocultural brains' which are functionally responsive to environmental input. These findings have important implications for interpreting the cognitive and linguistic skills of fossil hominids. They suggest that human cognition and language may have evolved in a gradual step-wise fashion over a lengthy period of time and that no single aspect of human behaviour such as language or tool-making could have reached totally modern form until all other human cognitive and learning skills had also achieved modern capacities. Specific applications of these principles to questions of Neanderthal versus modern human intelligence suggest that most aspects of the modern cognitive-learning suite were present in both Neanderthals and early anatomically modern humans. Thus, the Upper Palaeolithic is better interpreted as an explosion in technology and in technologically-assisted symbolism rather than the emergence of linguistic or symbolic capacities, per se. *It is suggested that cultural adaptations to a seasonally migratory life style initiated this technological explosion.*

The identity of the Neanderthals, their fate, and their phylogenetic relationships to modern humans have perplexed generations of anthropologists. To some scholars, Neanderthals were dim-witted, speechless, side-branches of the human evolutionary tree, more ape-like than human-like in their cognitive capacities, and, hence, readily driven to extinction by more intelligent, linguistically-proficient, modern human invaders of Middle-Eastern or African origins. To others, Neanderthals were the highly intelligent, perhaps even linguistically competent, ancestors of modern Europeans. This controversy began in a scientific age far different from our own — long prior to the discovery of *Australopithecus*, tool-making apes, or DNA — at a time when Neanderthal finds were confined to Europe and Darwinian evolution was still a novel idea. It has continued unabated in much the same form through a century of scientific revolutions.

Throughout this debate, most scholars, irrespective of their opinion of Neanderthal abilities and phylogenetic position, have assumed a relatively late origin of modern humans, invariant associations between Neanderthal fossils and Middle Palaeolithic toolkits, and similarly invariant relationships between Anatomically Modern Human (AMH) fossils and Upper Palaeolithic toolkits. As long as these assumptions held, strong arguments could be made for both sides of the Neanderthal controversy. It now appears, however, that AMHs arose more than 100,000 years ago and that early modern humans in the Levant produced a Middle Palaeolithic toolkit for approximately 60,000 to 70,000 years (Allsworth-Jones 1993; Bar-Yosef 1989; 1992; Bräuer 1989; Klein 1992; Mellars 1989b; Shea 1989; 1993). It is also now clear that Neanderthals sometimes manufactured Upper Palaeolithic tools and sporadically engaged in artistic endeavours (Harrold 1989; Marshack 1989; 1994; Mellars 1989b; 1992; Vandermeersch 1989). These findings challenge all models which assume ancestral–

descendant relationships between the two groups or invariant relationships between fossil taxa and technological, cognitive, or cultural capacities. Clearly, the time is ripe for new approaches to the Neanderthal–AMH paradox.

Although this controversy has repeatedly focused on questions of Neanderthal versus modern human cognitive and linguistic abilities, the findings of modern neural and behavioural scientists have rarely been incorporated into the debates. This paper attempts to remedy this gap by delineating neural and cognitive capacities which differentiate modern humans from modern apes and by applying our knowledge of modern cognition and brain function to an analysis of the fossil and archaeological records. It suggests that the fundamental foundations of modern human cognition rest, not upon one specific ability, such as language or symbolism, but rather upon a highly-plastic, environmentally-responsive 'biocultural brain' and a suite of interacting, mutually-reinforcing, neurological capacities each of which is present in rudimentary form in other animals, but greatly expanded in our species.

Both Neanderthals and early AMH would have possessed an essentially modern cognitive suite. To the extent, however, that the evolutionary divergence of the two taxa was accompanied by a divergence in life styles, slightly different cultural and cognitive capacities could eventually have developed in the two groups. In particular, recent findings suggest that early AMHs exhibited a more seasonally migratory life style than did Neanderthals (Lieberman 1993; Lieberman & Shea 1994; Soffer 1989). Seasonal migrations would have encouraged cultural inventions which enhance long-term planning and prediction abilities, the ability to communicate and to maintain kin and social ties across the boundaries of space and time, and effective means of long distant transport. The migratory life style may also have selected for advanced frontal lobe size. The eventual result of mobility-enhancing cultural innovations, with or without cognitive advance, would have been a quantum leap in the ability of AMHs to exploit seasonally abundant food resources and to displace less culturally advanced, but otherwise highly 'intelligent', populations.

Human culture and the human cognitive-learning suite

Classic theories of Neanderthal cognition derive from Western views of the nature of animal versus human minds (Gibson 1993a). Traditionally, Western thought has defined cognitively based behaviours as all-or-none phenomena which are either entirely present or entirely absent in a given species. According to this view, language, culture, and self-awareness are uniquely human traits present in their entirety in our species and totally absent in other animals. Imitation, in contrast, is considered 'monkey-like', hence, unintelligent. For a time, tool-making was also considered a uniquely human trait, hence, an intelligent behaviour. With the discovery of tool-making chimpanzees, however, tool-making fell from its cognitive pedestal and became, in some minds, a mere animal behaviour. Theories of Neanderthal versus modern behavioural capacities arose in this climate of all-or-none interpretations of behaviours.

Modern behavioural research has overturned these classic views (Parker & Gibson 1990; Gibson & Ingold 1993). We now know that most cognitively based behaviours exhibit varying levels or degrees of development. Elements of these behaviours are present in non-human primates, but human abilities in each domain greatly exceed those of our primate brethren (Gibson 1990; 1991; 1993c; 1994). This is as true for traditional 'monkey' behaviours, such as imitation, as it is for traditional human behaviours such as learned vocalizations, language, tool-making and self-awareness. For example, despite popular conceptions, non-human primates have minimal imitative skills and do not begin to approach the human ability to mime or to learn new motor behaviours via imitative means (Visalberghi & Fragaszy 1990; Tomasello 1990). On the other hand, despite common views of the uniqueness of language and self-awareness, some captive apes have clearly demonstrated the ability to comprehend and use simple syntactical and symbolic constructions as well as to recognize themselves in mirrors (Greenfield & Savage-Rumbaugh 1990; Savage-Rumbaugh & Rumbaugh 1993; Parker *et al.* 1994).

Consequently, the presence or absence of a specific behaviour, such as language, is not the distinguishing feature of humanity. Rather, human culture depends upon an entire suite of interacting cognitive and learning capacities, all of which are present in rudimentary form in the great apes, but highly expanded in our own species. These include, but are not necessarily limited to, procedural and declarative learning processes, symbolic and syntactic abilities, imitation, tool-making, self-awareness, social intelligence and advanced sensorimotor capacities. Most of these capacities are discussed at length in the primatological and anthropological literature (Byrne & Whiten 1988; Gibson & Ingold 1993; Parker

& Gibson 1990; Parker *et al.* 1994). Two critical abilities, however, have largely been ignored by evolutionary biologists: procedural and declarative learning. Procedural learning refers to the acquisition of habits or routines, which, once mastered, are performed automatically, while declarative learning represents the mastery of factual knowledge or knowledge of events. Procedural knowledge tends to be subconscious and relatively context-specific, whereas declarative knowledge is conscious and can be divorced from the contexts in which it was first learned in order to be used for other purposes (Mishkin *et al.* 1984; Squire & Zola-Morgan 1988).

Both forms of learning were first identified in rats and rhesus monkeys, and, hence, are not unique to humans. Indeed, some circus animals, such as sea-lions, are extremely adept at mastering new motor routines, and many mammals acquire significant factual knowledge of their environments. However, in part, because human learning skills are enhanced by our expanded linguistic, imitative, and tool-using abilities, human achievements in both procedural and declarative domains greatly exceed those of most animals and play fundamental roles in generating modern cultural capacities. Linguistically-mediated, declarative learning, for example, underlies the cultural accumulation and transmission of large bodies of naturalistic, historical, mythological, religious, scientific, and other knowledge. Procedural learning in conjunction with expanded human imitative and technical skills serves as the foundation for the mastery of cultural rituals, such as saluting the flag or making the sign of the cross and for the learning of new, culturally-based motor routines including dancing, typing, riding a bicycle, driving a car, playing a musical instrument and using everyday tools such as knives, forks, chopsticks, screwdrivers, and toothbrushes.

Considerable interaction occurs between human procedural and declarative learning domains. Human procedural skills such as dance, gesture, speech, art, and writing serve as means of acquiring and transmitting declarative knowledge. Conversely, culturally-accumulated declarative knowledge permits the manufacture of complex tools, such as pens, computers, and automobiles, which, in turn, require the learning of new procedures and permit the acquisition of additional factual knowledge. These interactive processes permit human technical innovations and accumulated factual knowledge to reverberate throughout society producing additional cultural innovations and social change. As a result, the cultural differences between human and ape are far greater than can be explained simply on the basis of differences in individual components of the cognitive suite. The essence of human, as opposed to ape culture, thus, reflects both the expansion of varied intellectual capacities and the greater interaction between the various components of the human cognitive suite which occurs in our species (Gibson 1993c). The result is the generative interplay between tools, language, and society which characterizes our species far more than any other (Gibson 1993b).

Increased neural information processing capacity and the biocultural human brain

Despite a century of search, no structure has been found within the human brain that does not also exist in the brains of other primates. Rather, the differences between human and non-human primate brains appear to be primarily quantitative in nature (Gibson 1988; 1990; 1993c; Jerison 1982). With the exception of the olfactory areas, almost all regions of the human brain are absolutely larger than similar regions in the brains of apes and monkeys. Certain parts of the brain, however, have enlarged more than others. These include the basal ganglia, cerebellum, and premotor areas which play critical roles in procedural learning, the hippocampus which is important for declarative learning and spatial memory, and Broca's area which is critical for speech. Two neocortical association areas are also greatly enlarged in our species: the prefrontal lobes which mediate long-term planning and creativity and the parietal association areas which are important for behaviours which require the synthesis of auditory, visual, and/or somatosensory information such as object naming, map reading, drawing, block constructions and dressing. Thus, the expansion of the human brain has provided increased information processing capacity in precisely those neural areas essential for human culture.

The information processing capacity and synaptic connectivity of enlarged brains provides increased capacities to break holistic sensory perceptions, motor actions, and concepts into discrete component parts and to flexibly combine and recombine these discrete elements into large, information-rich, novel constructs (Gibson 1981; 1983; 1990; 1993c; Greenfield 1991; Reynolds 1983). All large-brained primates possess some degree of this mental combinatorial skill, but the greatly enlarged human brain endows our species with much more mental constructional ability than that of even our closest phylogenetic kin, the great apes. For example, great apes can make simple tools and use rudimentary symbolic and syntactic constructs. Humans, however, flexibly combine and recombine individual technical and symbolic elements to construct highly variable tools, architectural edifices, drawings, sentences, stories, and musical compositions each of which may be comprised of large numbers of previously constructed subcomponents. Similarly, humans construct more information-rich theories of mind and concepts of social behaviour than do great apes (Byrne & Whiten 1988; Whiten 1991). Humans are also much more able to combine discrete motor units into new motor routines such as dance or gymnastics.

This mental combinatorial ability accounts, in part, for the phenomenon that both non-human primates and humans may exhibit quite different cognitive styles in differing environmental circumstances. For example, apes reared in captivity often manifest greater linguistic, tool-using, imitative, and mirror-recognition capacities than those reared in the wild (Savage-Rumbaugh & Rumbaugh 1993; Parker *et al.* 1994; Tomasello 1993). Similarly, many modern humans drive cars, ride bicycles, fly in airplanes, use mechanical can openers, electricity and computers, listen to radios, and watch television. A century or so ago, no humans did any of these things. Today, most residents of economically-developed countries read and write, but as recently as a few hundred years ago, only a small percentage of humans were truly literate. All of these behaviours affect our thought processes and our cognitive styles. None were present during our evolutionary history or can be considered components of the 'natural' human environment. All depend upon our ability to make flexible and novel combinations of perceptual, motor, and cognitive events.

Mental combinatorial ability is, thus, one primary factor leading to the great flexibility of human behaviour. Another is the developmental plasticity which human brains share with other mammals. Indeed, neuroanatomical plasticity in response to environmental input is so wide-spread and integral to the functioning of many mammalian brains that they could well be considered bio-environmental or biosocial organs (Gibson & Peterson 1991). For example, the size of a rat's brain and its degree of synaptic complexity reflect the richness of its environmental input (Diamond 1988). Similarly, in some mammals the extent to which fingers, vibrissae and other organs of manipulation are used during development influences the ultimate size of their neocortical representation (Merzenich & Kaas 1982). In humans, developmental plasticity is quite obvious with respect to clinical disorders (Finger 1991). For example, childhood damage to the left hemisphere results in transfer to the right hemisphere of linguistic functions normally mediated by the left hemisphere. Similarly, temporal lobe areas which subserve hearing in most humans, acquire visual functions in congenitally deaf children trained in sign language (Neville 1991).

The combination of developmental plasticity, flexible recombination of behavioural elements, and the generative interplay between tools, language, and society renders the human brain a truly biocultural organ. Thus, all developing human brains are profoundly shaped by environmental inputs, and all are capable of creating novel artistic, linguistic, technical, motor, and other behaviours. Most importantly, many human creations enter the cultural environment and help shape future developing human brains. Hence, although all modern populations have comparable intelligence, cultures have traditionally differed profoundly in their cognitive approaches to the world, often as a result of the reverberating effects of a few key technological innovations such as writing, the printing press, the typewriter, and the computer. Indeed, it has been argued that literacy (a form of technologically assisted language), in particular, may have profoundly changed the human cognitive approach to the world, rendering possible for the first time such intellectual skills as abstract mathematics, symbolic logic, and scientific discourse (Goody 1977).

Implications for modelling early human minds

These findings have clear implications for modelling early human minds.

1. The recognition of strong continuities between human and ape cognition implies strong continuities in behavioural capacities among hominid fossil populations.
2. The recognition that human cognitively based behaviours differ from those of the apes primarily in the degree of human information-processing capacity suggests that language, social intelligence, tool-making, and other cognitive skills may have evolved in a gradual, step-wise fashion in synchrony with the enlargement of pertinent neural areas, rather than appearing all at once at some point in the phylogenetic record (Gibson 1988). Thus, hominids who had diverged from the apes, but who had not yet reached fully modern neurological capacities, would have possessed neurologically based linguistic, technical

and other cognitive capacities intermediate between those of modern humans and great apes.

3. The multifaceted and interactive nature of the human cognitive suite suggests that the emergence of modern cultural capacities awaited the evolution of fully modern abilities across all domains — learning, sensorimotor, linguistic, social, imitative, and technical. A time-lag in the evolution of any component on the human cognitive-learning suite would have retarded the development of modern cultural systems. Hence, a narrow focus on language as the key human evolutionary phenomena or as the critical difference between modern humans and Neanderthals may be short-sighted. For example, increased technological accomplishments could reflect increased procedural learning and imitative skills, rather than the emergence of language.
4. The biocultural nature of the human brain and the generative interplay between varied learning, cognitive, and technical domains indicates that modern cognitive and neurological capacities may well have been present long before such capacities manifested themselves in the archaeological record in the form of advanced artistic and technological accomplishments. Indeed, it implies that, at each stage in the evolutionary process, hominids may have possessed the capacity for more advanced cultural systems than those actually exhibited. Finally, it suggests that profound differences in technology could have distinguished fossil populations of very similar intellectual abilities, just as they distinguish modern human populations.

Neanderthal versus modern human minds

The appearance of the genus *Homo*, in Africa, about 2,000,000 years ago represents the first clear indication of evolution towards human-like, as opposed to ape-like, neural form and cognition. The brains of early *Homo* were significantly enlarged in comparison to the brains of great apes. They also possessed human-like expansions of Broca's area and of the parietal lobes — neural areas which in modern humans mediate language, spatial concepts, and body image (Falk 1980; 1983; 1992; Tobias 1987). Moreover, unlike apes, early *Homo* made simple stone flake and core tools. These possessed sharp cutting edges useful for butchering game and for the preparation of plant materials such as digging sticks (Harris & Capaldo 1993). The toolkit of early *Homo* was very simple, but, if accompanied by an extensive factual knowledge of the environment, it would have permitted a highly varied and relatively complex foraging strategy including not only foraging for readily obtained foods such as fresh fruits, vegetables, and small game, but also scavenging for meat, brains, and bone marrow, digging for underground plants, termites, and burrowing vertebrates, and opening hard-shelled nuts, fruits, and shellfish. Previous works suggested that higher intelligence and tool-use would have first arisen in omnivorous extractive foragers: that is, in hominids who foraged on highly varied foods which required extraction from other sources, such as shells, bones, tough skin, or burrows (Gibson 1986; Parker & Gibson 1979). Early *Homo* was technologically well-equipped for such a life style.

Despite these advances, early members of the genus *Homo* still possessed brains far smaller than those of modern humans, and they made no tools which would suggest the presence of complex imitative, constructive, or procedural learning skills. By 1.5 million years ago, however, the descendants of early *Homo* had radiated into new, larger-brained *Homo erectus* forms, who, in Africa and Europe, used multi-stepped operations to produce bilaterally symmetrical Acheulean handaxes. Many of the later handaxes were finely made and beautifully shaped. The ability of *Homo erectus* to make this tool reflects imitative, procedural learning, and planning skills far above those exhibited by apes. It may also indicate that *Homo erectus* possessed relatively advanced intelligence at the level of Piaget's concrete operations (Gibson 1985; Gowlett 1984; Parker 1985; Wynn 1979).

The handaxe evidently served its makers quite well, because they continued to produce it for approximately one million years. Although the production of a stereotyped tool over a period of a million years causes some to question the creativity of the tool-makers, this persistence suggests that the tools were functionally versatile (Kuhn 1993). Handaxes were admirably designed to serve as chopping tools. They also provided sharp penetrating tips not present in earlier stone tools, and their shape rendered them excellent throwing tools which could readily have disabled prey (Calvin 1982; 1993). Thus, the handaxe, in combination with stone flakes, cores, pounding stones, and digging sticks, would have permitted *Homo erectus* to continue the highly varied diet very likely practiced by earlier hominids, and to expand that diet by disabling large game or by driving them over cliffs. Not only would *Homo erectus* technology have provided the capacity for a highly flexible foraging strategy, it would have done so in one relatively light-weight package easily carried by highly mobile gatherers, scavengers, and hunters. Given the versatility of this technology, perhaps the critical question is not why our ancestors produced handaxes for a million years, but why, approximately 200,000 years ago, they stopped doing so.

The demise of the European and African handaxe cultures coincided with the emergence of Middle Palaeolithic traditions about 200,000 kya (Klein 1992). Middle Palaeolithic tools reflected several technological advances. For one, Middle Palaeolithic peoples prepared stone cores from which multipurpose flakes could be readily produced. This technique permitted them to produce larger numbers of tools with greater amounts of cutting surface per unit stone than had been previously possible, a practice that may have been useful to peoples foraging in areas which lacked readily available stone or to peoples who needed to carry large numbers of relatively light-weight, multipurpose tools (Kuhn 1993; 1995). For another, one typical Middle Palaeolithic tool, the Levallois or Mousterian point, was sometimes hafted to wooden shafts to form spears (Churchill 1993; Shea 1989; 1993).

Hafting represents the first clear evidence of composite tools in the hominid archaeological record. It also represents a definite advance over ape tool-manufacturing techniques as no ape constructs composite tools of two or more sub-components joined together to form a single, manipulatable unit (Reynolds 1993). Neanderthals apparently manufactured these points throughout much of their existence and AMHs in the Levant did so from approximately 100,000 kya to approximately 40,000 kya (Allsworth-Jones 1993), a clear indication that both Neanderthals and early Levantine modern humans possessed human, rather than ape-like, mental constructional skills in the technological domain. Since hafting requires the human ability to plan a complex sequence of tool-making steps involving the extraction of resin, the production of lashing materials, and the carving of stone points and of accompanying receptacles on wooden shafts (Lieberman & Shea 1994), its presence also suggests that Middle Palaeolithic peoples possessed extensive factual knowledge of environmental resources and advanced procedural and imitative skills.

No evidence exists for spear-throwers in the Middle Palaeolithic. Hence, Middle Palaeolithic spears were probably used in thrusting or throwing actions of the arms and hands alone. Surveys of modern cultures suggest that such spears function most effectively, when used to disable prey that have already been placed in a compromised position, via some other mechanism, such as driving over cliffs or into corrals (Churchill 1993). Thus, Middle Palaeolithic technologies may have provided superior abilities for some types of hunting. Otherwise, from the functional standpoint, the Middle Palaeolithic toolkit, like those that preceded it, is probably best viewed as one that provided maximum foraging versatility from a limited range of readily-carried, multipurpose tools. This interpretation is consistent with the archaeological record that at some European Middle Palaeolithic sites, Neanderthals focused primarily on the hunting of large bovids (Chase 1989; Mellars 1989a), while, at others a more generalized scavenging, gathering, and extractive strategy was clearly practiced (Kuhn 1993; 1995; Stiner 1994).

The onset of the Upper Palaeolithic about 40,000 to 50,000 years ago in the Near East, Europe, and Siberia and of similar technologies somewhat earlier in Africa (Brooks *et al.* 1995; Deacon 1989; Klein 1992; Mellars 1989b; 1992; Morell 1995; Soffer 1989; Yellen *et al.* 1995) continued and expanded trends begun by some Middle Palaeolithic populations in the direction of specialized foraging strategies (Chase 1989). In particular, for the first time, many human populations exhibited intensive seasonal specialization on such dietary items, as migratory caribou, fish, water fowl, and beached fur seals. These specialized foraging activities were accompanied by greater long-range planning of foraging activities and by more precisely-timed visits to seasonal foraging sites than had previously been the case (Binford 1992; Klein 1989; Mellars 1973; 1989a).

Major technological advances accompanied these shifts to more seasonally, specialized foraging strategies: an explosion in the variety of tools including the production of tools for specialized tasks such as fish hooks and projective weapons, an improvement in the finished quality of tools, an increased reliance on blade technologies, the first widespread utilization of worked bone, ivory and antler tools, and major improvements in hafting techniques (Mellars 1989a,b). The manufacture of these tools required enormous time, creativity, procedural skill, and factual knowledge of the environment. Clearly Upper Palaeolithic peoples and, perhaps, some of their African predecessors possessed fully modern intelligence and regarded high quality, specialized tools as essential components of their life style.

The onset of the Upper Palaeolithic coincided with the dispersal of modern humans into Europe about 40,000 kya, with the disappearance of the Neanderthals about 35,000 kya, and with an explosion in the production of symbolic art, sculpture. and jewellery. The versatility and quality of Upper Palaeolithic art and technology has led to much speculation that Upper Palaeolithic peoples were much more intelligent and creative than their predecessors and that they were the first to possess language and symbolic thought (e.g. Byers 1994; Davidson & Noble 1993; Mellars 1991). The correspondence between the onset of the Upper Palaeolithic and the dispersal of AMHs into Europe has also led to the view that Upper Palaeolithic cultures were products of AMHs and beyond the intellectual capacity of the Neanderthals. A more recent and even more radical suggestion, is that the brains of early AMHs may have experienced a sudden functional reorganization about 40 to 50,000 kya that ushered in Upper Palaeolithic cognitive skills in a species which had already existed for tens of thousands of years (Klein 1992).

That the Upper Palaeolithic represents sudden reorganizational changes in the brains of AMHs is implausible in the absence of changes in external neural morphology. Recent evidence for the production of Upper Palaeolithic-like tools, such as bone harpoons and fish hooks at least 80 to 90,000 years ago also renders it quite clear that the neurological capacity to produce such tools existed long prior to the Upper Palaeolithic (Brooks in press; Brooks & Yellen 1995; Brooks *et al.* 1995; Yellen *et al.* 1995).

Many considerations also argue against the views that Neanderthals and other archaic *Homo sapiens* strongly differed from early modern humans in levels of intelligence or linguistic capacity. Both Neanderthals and AMHs in the Levant produced very similar Mousterian technologies, including hafted tools, from approximately 100,000 to 40,000 years kya (Allsworth-Jones 1993; Mellars 1989a; Shea 1989; 1993). The archaeological record suggests that during this period, neither AMHs nor Neanderthals regularly produced durable art or jewellery, although some evidence indicates that European Neanderthals sometimes did so (Marshack 1989; 1994). Hence, no evidence from this 60,000 to 70,000 year time period suggests that early Levantine modern humans were technologically or artistically more sophisticated than the Neanderthals (Allsworth-Jones 1993). It is also clear that Neanderthals and other archaic *Homo sapiens* possessed the procedural, declarative, and intellectual skills to produce Upper Palaeolithic technologies. Blade tools sometimes occur at Middle Palaeolithic sites, and Upper Palaeolithic Châtelperronian tools have now been found in association with Neanderthal fossils (Harrold 1989). As a result, the Châtelperronian has been reclassified as an Upper Palaeolithic Neanderthal culture (Mellars 1992).

Other lines of evidence also suggest that the cognitive and linguistic abilities of Neanderthals and AMHs would have differed only minimally. The two groups shared a common phylogeny and, hence, a common cognitive evolutionary history for several million years subsequent to the divergence of the hominid and great ape lineages. Neanderthals possessed brain sizes within or exceeding the modern human range, and, like those of predecessor early *Homo* and *Homo erectus* populations, Neanderthal brains possessed human-like, rather than ape-like, gyral and fissural patterns, including a human-like external configuration of the classic language areas such as Broca's, Wernicke's, and the inferior parietal association areas (Holloway 1985). Moreover, at a time when AMHs were still confined to moderate climates, Neanderthals flourished in cold European climates and did so for tens of thousands of years. In depth, archaeological analyses indicate that European Neanderthals often intensely exploited relatively impoverished areas, that they readily adapted their foraging styles to changing environmental conditions, and that they opportunistically moved short distances in search of changing foraging opportunities (Kuhn 1993; Lieberman & Shea 1994; Roebrocks *et al.* 1992; Soffer 1989; Stiner 1994). The entire Neanderthal life style, thus, suggests intensive factual knowledge of the environment and hence, highly developed declarative learning skills. Modern hunter-gatherers in areas as diverse as northern Canada and South Africa, readily use language to inform each other of changing hunting and gathering opportunities at distant sites, a habit which saves energy that might otherwise be expended in futile walks to non-productive areas (Marshall 1976; Smith 1981). Whether or not Neanderthal declarative learning and foraging strategies were also linguistically mediated is likely to be debated for decades to come. Considering, however, that the external markings of the language areas of Neanderthal brains were modern in form and that language-like communications would have greatly benefited them, it is almost inconceivable that they would have possessed no language-like behaviours.

Taken as whole, these findings cast serious doubt on assumptions that Neanderthals lacked language or other essential components of the human cognitive suite, such as language. Rather, Neanderthals appear to have been a highly intelligent group, who inherited most elements of modern human cognition from the last common Neanderthal–Modern

Human ancestor. Nonetheless, Neanderthals and AMHs could have differed somewhat in their cognitive skills. They were, after-all, distinct populations for at least 60,000 to 70,000 years, possibly much longer, if AMHs emerged in Africa long prior to reaching the Levant. No matter how similar the initial cognitive capacities of Neanderthals and modern humans, some evolutionary divergence could have occurred during this lengthy time frame, especially if the two groups were subject to differing selective pressures. Thus, it is not surprising that subtle differences in external shape distinguished most Neanderthal brains from most modern human brains. Neanderthals possessed a larger, more protuberant occipital region and flatter frontal and superior parietal lobes. Neanderthal frontal lobes may also have been somewhat smaller than those of modern humans (Kochetcova 1978; but see Holloway 1985 for a contrary view). These considerations imply that Neanderthals may have differed somewhat from most modern humans in their complex visualization skills, in their ability to map body positions in space, and in their ability for long-term planning and anticipation of future events. Such differences, however, would have been minimal, probably no greater than the cognitive and neural differences which distinguish modern, living individuals. Any significant neurological differences between the two forms are also likely to have evolved after a long period of independent evolution and would not account for initial phylogenetic and behavioural divergences.

Seasonal migrations, technological change, and the onset of the Upper Palaeolithic

Thus, the time has come to confront the possibility that the Upper Palaeolithic may represent a technologically-assisted revolution in populations whose predecessors had possessed essentially modern neurological capacities for tens of thousands of years and whose linguistic abilities and intelligence differed only minimally from those of Neanderthals and other archaic *Homo sapiens*. The biocultural nature of the human brain and the generative interplay between technology, language, and society render such revolutions both possible and probable.

That technologically-assisted revolutions can occur in the absence of biological evolution is evident from recorded history. Key inventions have repeatedly reverberated throughout modern human societies propelling technological, cultural, and social revolutions. Many such inventions come readily to mind: agriculture, the domestication of animals, the invention of wheels, chariots, ships, bronze, iron, automobiles, airplanes, steam engines, electricity, atomic energy, the transistor, writing, the printing press, television, the typewriter, and the personal computer. Some technological breakthroughs have impacted all aspects of modern cultures including art, music, dress, family structure, the spatial structure of our cities and towns, medicine, warfare, and communication. Some have facilitated major population explosions and wide-spread population dispersals. Others have produced major cultural revolutions within the course of individual life-times and allowed entire cultures to move from stone age to computer age life styles within a few decades. The cumulative effect has been an ever-increasing rate of change in technology and an ever-increasing range and breadth of technical and artistic expression.

The events of the Upper Palaeolithic may represent the first of this long line of technologically-assisted cultural revolutions. Although precursors of the Upper Palaeolithic technologies are found at some Eurasian Middle Palaeolithic or African Middle Stone Age sites, Upper Palaeolithic peoples greatly expanded these technologies and put them to new uses. Thus, hafted Middle Palaeolithic spears required joining techniques and materials such as rope, cordage, string and adhesives. During the Aurignacian, however, Upper Palaeolithic peoples produced a wide variety of very finely carved junctional surfaces on stone and bone spear tips (Knecht 1983), and they extended the use of cordage or string to the making of jewellery (White 1989). With the perfection of techniques for creating firm junctions between objects and quality cordage came many other possibilities such as tying bundles or fitting together logs to make ladders and rafts, the construction of animal harnesses and tethers, and the manufacture of fish nets and bows (Barber 1994). Most of these highly useful inventions probably appeared some time during the Upper Palaeolithic.

Similarly, although Middle Palaeolithic Eurasians as well as Middle Stone Age Africans (Clark 1989; Mellars 1989b) produced blade tools, an expanded use of blades almost defines the Upper Palaeolithic. Much of the functional utility of blades related to the ease with which they could be carved into spear points. Once blade production became routine, however, blades served as raw materials for many other tools, such as burins, borers and chisels. These tools rendered possible the carving of bone, antler, and ivory tools, the boring of holes in jewellery, and the production of sculpture.

These considerations suggest that the emergence of Upper Palaeolithic art forms may reflect the use of new technology to create new symbolic media and forms, rather than the first emergence of the biological capacity for symbolism, language, or culture. That the great symbolic achievement of the Upper Palaeolithic may have been an expansion of technologically assisted symbolism rather than the emergence of symbolic capacity *per se*, is supported by an analysis of the technological demands of Upper Palaeolithic artistic expression. The production of cave art demanded lamps or torches (de Beaune & White 1993), pigments, palettes, means of applying pigments, and scaffolding or ladders. Sculpture and parietal art required burins and chisels. Bead production demanded borers and polishing techniques as well as cordage or string. Language, mime, the use of ochre and other pigments, the wearing of animal skins, drawing in dirt or other perishable media, and other forms of symbolism could well have existed long prior to the complex technologically assisted art of the Upper Palaeolithic. Indeed, clear evidence exists for the presence of some of these symbolic activities at earlier times (Bednarik 1995; Marshack 1989; Wreschner 1980).

The interpretation that the essential genius of the Upper Palaeolithic was expanded technologically assisted symbolism *per se*, rather than the emergence of the biological capacity for symbolism shifts the focus of potential archaeological query. Rather than seeking cognitive or neurological causes of the Upper Palaeolithic, the critical issue becomes a search for the ecological, cultural, or climatic factors that led to sporadic early inventions of Upper Palaeolithic tools and art forms in Africa and Europe, but resulted in only one widespread and long-lasting Eurasian, Upper Palaeolithic revolution coincident in time with the expansion of the modern humans into Europe and the disappearance of the Neanderthals.

Keys to resolving these issues may reside in new archaeological evidence that Neanderthals and AMHs practiced different patterns of environmental exploitation. In the Levant and in Russia, during Middle Palaeolithic times, Neanderthals were highly mobile. However, they remained year round in core geographic areas and radiated out from a central home base in search of seasonal food resources (Lieberman 1993; Lieberman & Shea 1994; Soffer 1989). The Neanderthal pattern of radiating mobility, when practiced in relatively impoverished areas, would have rapidly depleted environmental resources and necessitated intensive exploitation of any available food resources, including foods generally considered less desirable or more difficult to find and capture (Leiberman & Shea 1994). Thus, this life style would have encouraged foraging flexibility, ingenuity in the finding of scarce foods, and a flexible technology. It would have strongly discouraged dietary or technological specialization and could have supported only relatively low population densities.

In contrast, AMHs in the Levant, Europe, Russia, and parts of Africa practiced a form of seasonal circulating mobility which involved relatively long distance residential migration on a seasonal basis in search of choice foods (Klein 1989; Lieberman 1993; Lieberman & Shea 1994; Soffer 1989). Once at a preferred seasonal location, however, AMHs apparently remained at the same residential site for relatively long periods.

One can only speculate on why early modern humans, but not Neanderthals, initiated the seasonally migratory life style. Climatic conditions may have played a major role. Most Neanderthals inhabited areas in which they would have encountered hazardous travel conditions during the cold winter months. Even in modern times, inhabitants of cold climates, such as many North American Indians, were able to migrate seasonally only with the aid of advanced transportation systems such as canoes or dog sleds (Wissler 1957). For many Indians of the Canadian North, long-distance seasonal travel via non-motorized canoe remains hazardous even today and is only possible when accompanied by abilities to predict seasonal weather conditions sufficiently far in advance to begin and end a journey of a hundred or more miles prior to the onset of ice, blizzards and other hazardous conditions. If European Neanderthals lacked transportation devices and/or the ability to predict weather and foraging conditions in advance, a seasonally migratory life style would not have been open to them.

In contrast, many early AMHs occupied areas of Africa and the Levant which presented less hazardous seasonal climatic conditions and, thus, would have been conducive to an initial adoption of a migratory life style by populations which possessed no tool-assisted means of transportation and only poorly developed abilities to predict seasonal changes. Once adopted, however, this life style could have selected for advanced frontal lobe mediated planning capacities and would have encouraged all technological innovations and cultural traditions which facilitated it, including methods of predicting and recording seasonal changes in weather and food resources, methods of maintaining social ties and communicating information across boundaries of space and time, the ability to plan long-distance travel in advance, and improved methods of transportation. Once in place such adaptations would have rendered the seasonally migratory life style possible even in inhospitable climates and permitted, possibly for the first time, the occupation of Europe by a seasonally migratory people.

When practiced in regions of unusually abundant seasonal foods, the migratory life style would also have permitted intense seasonal specialization on specific foods and encouraged the development of specialized foraging technologies. The utility of specialized technology in environments presenting unusually abundant seasonal resources is such that many migratory populations may have developed such technologies. However, only those technologies designed for the exploitation of widely distributed food resources could have propagated beyond the local areas in which they were first invented. These considerations may explain the sporadic appearance of specialized technologies long before their widespread distribution during Aurignacian times.

The Aurignacian, which occurred throughout much of Europe, Western Russia, and the Levant, represents the first widely distributed Upper Palaeolithic culture and technology (Mellars 1992). Aurignacian technological advances included improved hafting methods and improved projectile weapons (Knecht 1993; Straus 1993). These appear to have been adaptations for the seasonal hunting of migratory reindeer or other herd animals (Binford 1968; Mellars 1973; 1989b; 1992). The Aurignacian culture may, thus, represent the culmination of a seasonally migratory life style based on the hunting of seasonally migratory prey and assisted by a suite of interacting and complementary cultural (and possibly biological) skills developed over tens of thousands of years.

One cultural behaviour of great benefit to a migratory life style, and well-developed by Aurignacian times, involved the practice of inscribing records on stone, bone, and other materials (Marshack 1991). Inscriptions paved the way for the communication of symbolic, social, or factual knowledge to individuals not present at the immediate space and time. They may also have facilitated the cultural accumulation of environmental and social information in external storage devices whose memory capacity exceeded that of any individual human brain (Donald 1991). Perhaps, most importantly, Aurignacian peoples used inscriptions to record cycles of the moon and other proto-calendrical events (Marshack 1991). The recording of absolute time would have facilitated the advanced prediction of seasonal events such as caribou and avian migrations, salmon spawns, the beaching of seals, the birth of mammals, the ripening of plant foods, and the advent of hazardous weather. In addition, a proto-calendrical system would have enabled dispersed sub-groups to meet at pre-arranged times and locations for festivals, marriage and initiation ceremonies, group hunting expeditions, and trade, thereby facilitating behaviours which sustain kinship bonds and tribal affiliations.

Once equipped with efficient means of inscribing records, predicting events, communicating symbolic and social information across the bounds of space and time, and capturing migratory prey, Upper Palaeolithic peoples would have found themselves far more prepared to exploit the seasonally harsh European environments than any predecessor population. The only additional technological factor needed to render such peoples the most formidable seasonal predators ever to inhabit Europe would have been the possession of efficient and rapid means of transporting food, objects, and people, such as snowshoes, hand-pulled sledges or packs, baskets, sledges or packs harnessed to hand-reared ungulates or wolves, and rafts or other means of water travel. No direct evidence of such transportation devices exists. Clearly, however, Aurignacian peoples were producing cordage or other lashing materials and had the ability to carve or otherwise create interlocking junctional surfaces. Thus, the technology for the production of sledges, harnesses, and water craft clearly existed. It is also clear that humans reached Australia, during or before the European Upper Palaeolithic, and that American Indians with Stone Age cultures no more sophisticated than those of the Upper Palaeolithic manufactured canoes capable of both highly efficient river travel and moderate distance ocean travel. Given that Aurignacian peoples possessed both the technology and the motivation to invent means of technologically assisted transportation, it seems quite probable that they did. If not, surely their successors invented such devices by relatively early Upper Palaeolithic times.

Whallon (1989) and White (1989) have suggested that the Upper Palaeolithic was, at heart, a social revolution. Although the current model posits that the essence of the early Upper Palaeolithic was the culmination of a technologically assisted, seasonally migratory life style, such a life style would both depend upon and propel social change. For example, seasonal migrations would have paved the way for long-distance trade, often considered a central facet of the Upper Palaeolithic. When practiced in seasonally abundant areas, seasonal migrations would also have led to seasonal foraging bonanzas and, thus, to the population explosions known to have characterized the Upper Palaeolithic (Mellars 1989b). Population explosions, in turn, would have provided opportunities for greater social divisions of labour in the production of technology and art. Foraging bonanzas and population explosions would also have permitted, for the first time, the accumulation of individual 'wealth' and economic status, and, thus, a motivation for the production of jewelry. Finally, Upper Palaeolithic migratory peoples would, no doubt, have practiced patterns of seasonal population aggregations and dispersals as do modern populations exhibiting similar life-styles (Conkey 1980). In addition, they would have depended upon each other more heavily than ever before for information about weather and foraging opportunities at distant locations, and they would often have found themselves in unfamiliar territories inhabited by strange or hostile groups. These circumstances would have propelled the invention of kin and social systems which transcended spatial and temporal boundaries (Rodseth *et al.* 1991) and provided the motivation for seasonal festivals, perhaps accompanied by artistic and other symbolism, for the renewal of group ties (Conkey 1981).

Zubrow (1989) has calculated that a mere 1 or 2 per cent difference in mortality rates per generation will allow one population to entirely replace another within the course of a thousand years. The seasonally migratory Aurignacian life style may well have provided much more than a 1 or 2 per cent survival advantage, because it permitted mass exploitation of seasonally abundant resources, not as readily utilized by less migratory peoples, such as early Neanderthals or other archaic *Homo sapiens*. Thus, Aurignacian peoples may have driven Neanderthals to extinction simply by out-competing them. Alternately, Neanderthals may have been absorbed into much larger modern human populations, but contributed so few genes to these populations that their contributions are difficult to distinguish on skeletal grounds. Future genetic research is needed to resolve these issues. What is clear, however, is that many technological and artistic inventions of the Upper Palaeolithic can be interpreted as adaptations to a seasonally migratory way of life, rather than as the products of populations who far exceeded Neanderthals or other Middle Palaeolithic peoples in linguistic or intellectual acumen. It is possible, however, somewhat expanded frontal lobe capacities eventually evolved in AMHs as an adaptation for the advanced planning functions so integral to a migratory life style. Further research is needed on the frontal lobe size of Neanderthals and early modern humans to resolve this issue.

Summary

The human brain is fundamentally a biocultural organ whose functioning reflects both genetic and environmental input. It differs from the brains of great apes primarily in the expansion of information processing capacities which are applied to each element of a large cognitive suite composed of imitation, language, social intelligence, tool-use, and declarative and procedural learning skills. Interactions between the biocultural brain and the human cognitive suite permit a generative interplay between technology, language, and society, such that changes in any one behavioural domain propel comparable changes in others. As a result, the behavioural differences between ape and human are far greater than can be accounted for by differences in brain size or structure alone.

Similarly, cultural differences between fossil populations may have been far greater than differences in biologically based cognitive skills. In particular, phylogenetic and anatomical evidence indicate that Neanderthals and early AMHs both possessed an essentially modern cognitive suite. This suggests that the Upper Palaeolithic represents a flowering of technologically assisted symbolism rather than the first appearance of biologically based symbolic capacity and that a search for the causes of the Upper Palaeolithic must focus primarily on ecological rather than on neurological factors.

Recent evidence suggests that AMHs were more seasonally migratory than Neanderthals. An understanding of migratory life styles may, thus, provide the key to unlocking the mysteries of the Upper Palaeolithic. A migratory life style would have provided incentives for the invention of efficient modes of transportation, of cultural means of anticipating future events, and of methods of communicating information and maintaining social ties across the bounds of space and time. It may also have selected for slightly greater frontal lobe planning capacities. Perhaps, most importantly, when practiced in areas of seasonally abundant food, such a life style would have permitted major population explosions and propelled social changes which encouraged the invention of art, jewellery, and quality tools. Thus, advances in technology and changes in social structure coincident to the seasonally migratory life style, rather than major advances in linguistic or cognitive capacity, may have initiated the Upper Palaeolithic cultural revolution and eventually allowed modern humans to out-compete the Neanderthals.

Acknowledgements

While working on this paper, the author was supported, in part, by the McDonald Institute for Archaeological Research, University of Cambridge.

Kathleen R. Gibson
Department of Basic Sciences
University of Texas at Houston, Dental Branch
P.O. Box 20068
Houston, TX 77225
USA

References

Allsworth-Jones, P., 1993. The archaeology of archaic and early modern *Homo sapiens*: an African perspective. *Cambridge Archaeological Journal* 3(1), 21–39.

Barber, E.W., 1994. *Women's Work: the First 20,000 Years*. New York (NY): W.W. Norton & Company.

Bar-Yosef, O., 1989. Geochronology of the Levantine Middle Palaeolithic, in Mellars & Stringer (eds.), 589–619.

Bar-Yosef, O., 1992. The role of Western Asia in modern human origins. *Philosophical Transactions of the Royal Society* Series B 337, 193–200.

Bednarik, R.G., 1995. Concept-mediated marking in the Lower Paleolithic. *Current Anthropology* 36, 605–34.

Binford, L., 1992. Cited By Joshua Fischman, in 'Hard evidence'. *Discover* 13, 44–51.

Binford, S.R., 1968. Early Upper Pleistocene adaptations in the Levant. *American Anthropologist* 70, 707–17.

Bräuer, G., 1989. The evolution of modern humans: a comparison of the African and non-African evidence, in Mellars & Stringer (eds.), 123–54.

Brooks, A., in press. Behaviour and human evolution, in *Issues in Human Evolution*, ed. F.C. Howell. San Francisco (CA): California Academy of Sciences.

Brooks, A. & J. Yellen, 1995. Morphological Archaics and Behavioural Moderns: a New Look at the Middle Stone Age. Unpublished paper presented at the fourth annual meeting of the Paleoanthropology Society, Oakland (CA).

Brooks, A., D. Helgren, J.S. Cramer, A. Franklin, W. Hornyak, J. Keating, R. Klein, W. Rink, H. Schwarcz, J.N.L. Smith, K. Stewart, N.E. Todd, J. Verniers & J. Yellen, 1995. Dating and context of three Middle Stone Age sites with bone points in the Upper Semliki Valley, Zaire. *Science* 268, 548–53.

Butterworth, G., J. Rutkowska & M. Scaife (eds.), 1985. *Evolution and Developmental Psychology*. Brighton: Harvester Press.

Byers, A.M., 1994. Symboling and the Middle–Upper Palaeolithic transition. *Current Anthropology* 15, 369–99.

Byrne, R. & A. Whiten, 1988. *Machiavellian Intelligence: Social Expertise and the Evolution of Intellect in Monkeys, Apes, and Humans*. Oxford: Oxford University Press.

Calvin, W.H., 1982. Did throwing stones shape hominid brain evolution? *Ethology and Sociobiology* 3, 115–24.

Calvin, W.H., 1993. The unitary hypothesis: a common neural circuitry for novel manipulations, language, plan-ahead, and throwing?, in Gibson & Ingold (eds.), 230–50.

Chase, P., 1989. How different was Middle Palaeolithic subsistence? A zooarchaeological perspective on the Middle to Upper Palaeolithic transition, in Mellars & Stringer (eds.), 321–37.

Churchill, S.E., 1993. Weapon technology, prey size selection, and hunting methods in modern hunter-gatherers: implications for hunting in the Palaeolithic and Mesolithic, in Peterkin *et al.* (eds.), 11–24.

Clark, J.D., 1988. The Middle Stone Age of East Africa and the beginnings of regional identity. *Journal of World Prehistory* 2, 235–303.

Conkey, M.W., 1980. The identification of prehistoric hunter-gatherer aggregation sites: the case of Altamira. *Current Anthropology* 21, 609–30.

Davidson, I. & W. Noble, 1993. Tools and language in human evolution, in Gibson & Ingold (eds.), 363–88.

Deacon, H.J., 1989. Late Pleistocene paleoecology and archaeology in the Southern Cape, South Africa, in Mellars & Stringer (eds.), 547–64.

de Beaune, S.A. & R. White, 1993. Ice Age lamps. *Scientific American* 268, 108–13.

de Grolier, E. (ed.), 1983. *Glossogenetics: the Origin and Evolution of Language*. New York (NY): Harwood Academic Publishers.

Diamond, M.C., 1988. *Enriching Heredity*. New York (NY): The Free Press.

Donald, M., 1991. *Origins of the Modern Mind: Three Stages in the Evolution of Culture and Cognition*. Cambridge (MA): Harvard University Press.

Falk, D., 1980. Language, handedness and primate brains: did the Australopithecines sign? *American Anthropologist* 82, 72–8.

Falk, D., 1983. Cerebral cortices of East African early hominids. *Science* 221, 1072.

Falk, D., 1992. *Braindance*. New York (NY): Henry Holt & Co..

Finger, S., 1991. Brain damage, development, and behaviour: early findings. *Developmental Neuropsychology* 7, 261–74.

Gibson, K.R., 1981. Comparative neuro-ontogeny; its implications for the development of intelligence, in *Infancy and Epistemology*, ed. G. Butterworth. Brighton: Harvester Press, 52–82.

Gibson, K.R., 1983. Comparative neurobehavioural ontogeny: the constructionist perspective in the evolution of language, object manipulation and the brain, in de Grolier (ed.), 52–82.

Gibson, K.R., 1985. Has the evolution of intelligence stagnated since Neanderthal man?, in Butterworth *et al.*

(eds.), 102–14.

Gibson, K.R., 1986. Cognition, brain size, and the extraction of embedded food resources, in *Primate Ontogeny, Cognition and Social Behaviour*, eds. J. Else & P.C. Lee. Cambridge: Cambridge University Press, 93–105.

Gibson, K.R., 1988. Brain size and the evolution of language, in *The Genesis of Language: a Different Judgment of Evidence*, ed. M. Landsberg. Berlin: Mouton de Gruyter, 149–72.

Gibson, K.R., 1990. New perspectives on instincts and intelligence: brain size and the emergence of hierarchical mental constructional skills, in Parker & Gibson (eds.), 97–128.

Gibson, K.R., 1991. Tools, language, and intelligence: evolutionary implications. *Man* 26, 602–19.

Gibson, K.R., 1993a. Animal minds, human minds, in Gibson & Ingold (eds.), 3–19.

Gibson, K.R., 1993b. Generative interplay between technical capacities, social relations, imitation and cognition, in Gibson & Ingold (eds.), 131–7.

Gibson, K.R., 1993c. Tool use, language and social behaviour in relationship to information processing capacities, in Gibson & Ingold (eds.), 251–69.

Gibson, K.R., 1994. Continuity theories of language origins versus the Lieberman model. *Language and Communication* 14, 97–114.

Gibson, K.R. & T. Ingold (eds.), 1993. *Tools, Language and Cognition in Human Evolution*. Cambridge: Cambridge University Press.

Gibson, K.R. & A.C. Petersen (eds.), 1991. *Brain Maturation and Cognitive Development: Comparative and Cross-cultural Perspectives*. Hawthorne (NY): Aldine de Gruyter.

Goody, J., 1977. *The Domestication of the Savage Mind*. Cambridge: Cambridge University Press.

Gowlett, J.A.J., 1984. Mental abilities of early man: a look at some hard evidence, in *Hominid Evolution and Community Ecology*, ed. R. Foley. London: Academic Press, 167–92.

Greenfield, P.M., 1991. Language, tools and the brain: the development and evolution of hierarchically organized sequential behaviour. *Behavioural and Brain Sciences* 14, 531–95.

Greenfield, P.M. & E.S. Savage-Rumbaugh, 1990. Grammatical combination in *Pan paniscus*: processes of learning and invention in the evolution and development of language, in Parker & Gibson (eds.), 540–78.

Harris, J.W.K. & S.D. Capaldo, 1993. The earliest stone tools: their implications for an understanding of the activities and behaviour of late Pliocene hominids, in *The Use of Tools by Human and Non-human Primates*, eds. A. Berthelet & J. Chavaillon. Oxford: Clarendon Press, 196–220.

Harrold, F.B., 1989. Mousterian, Châtelperronian, and early Aurignacian in Western Europe: continuity or discontinuity?, in Mellars & Stringer (eds.), 677–713.

Holloway, R., 1985. The poor brain of *Homo sapiens neanderthalensis*: see what you please, in *Ancestors: the Hard Evidence*, ed. E. Delson. New York (NY): Alan R. Liss, 319–24.

Jerison, H.J., 1982. Allometry, brain size, cortical surface, and convolutedness, in *Primate Brain Evolution*, eds. E. Armstrong & D. Falk. New York (NY): Plenum, 77–84.

Klein, R., 1989. Biological and behavioural perspectives on modern human origins in southern Africa, in Mellars & Stringer (eds.), 529–46.

Klein, R., 1992. The archeology of modern human origins. *Evolutionary Anthropology* 1, 5–14.

Knecht, H., 1993. Early Upper Palaeolithic approaches to bone and antler projectile technology, in Peterkin *et al.* (eds.), 33–47.

Kochetcova, V.I., 1978. *Paleoneurology*. Washington (DC): V.H. Winston.

Kuhn, S., 1993. Mousterian technology as an adaptive response, in Peterkin *et al.* (eds.), 25–31.

Kuhn, S., 1995. *Mousterian Lithic Technology*. Princeton (NJ): Princeton University Press.

Lieberman, D., 1993. Variability in hunter-gatherer seasonal mobility in the Southern Levant: from the Mousterian to the Natufian, in Peterkin *et al.* (eds.), 207–19.

Lieberman, D.E. & J.J. Shea, 1994. Behavioural differences between archaic and modern humans in the Levantine Mousterian. *American Anthropologist* 96, 300–332.

Marshack, A., 1989. Evolution of the human capacity: the symbolic evidence. *Yearbook of Physical Anthropology* 32, 1–34.

Marshack, A., 1991. The Taï plaque and calendrical notation in the Upper Palaeolithic. *Cambridge Archaeological Journal* 1(1), 25–61.

Marshack, A., 1994. Commentary. *Current Anthropology* 35, 387–8.

Marshall, L., 1976. *The !Kung of the Nyae Nyae*. Cambridge (MA): Harvard University Press.

Mellars, P.A., 1973. The character of the Middle–Upper Palaeolithic transition in southwest France, in *The Explanation of Culture Change*, ed. C. Renfrew. London: Duckworth, 255–76.

Mellars, P.A., 1989a. Major issues in the emergence of modern humans. *Current Anthropology* 30, 349–85.

Mellars, P.A., 1989b. Technological changes across the Middle–Upper Palaeolithic transition: economic, social, and cognitive perspectives, in Mellars & Stringer (eds.), 338–65.

Mellars, P.A., 1991. Cognitive changes and the emergence of modern humans in Europe. *Cambridge Archaeological Journal* 1(1), 63–76.

Mellars, P.A., 1992. Archaeology and the population dispersal hypothesis of modern human origins in Europe. *Philosophical Transactions of the Royal Society of London Series B* 337, 225–34.

Mellars, P.A. & C. Stringer (eds.), 1989. *The Human Revolution: Behavioural and Biological Perspectives on the Origins of Modern Humans*. Edinburgh: Edinburgh University Press.

Merzenich, M.M., & J.H. Kaas, 1982. Reorganization of mammalian somatosensory cortex following peripheral nerve injury. *Trends in Neurosciences* 5, 434–6.
Mishkin, M., B. Malamut & J. Bachavelier, 1984. Memories and habits: two neural systems, in *Neurobiology of Learning and Memory*, eds. G. Lynch, J.L. McGaugh & N.M. Weinberger. New York (NY): Guilford Press, 65–77.
Morell, V., 1995. Siberia: surprising home for early modern humans. *Science* 268, 1279.
Neville, H.J., 1991. Neurobiology of cognitive and language processing: effects of early experience, in Gibson & Petersen (eds.), 355–80.
Parker, S.T., 1985. Higher intelligence as an adaptation for social and technological strategies in early *Homo sapiens*, in Butterworth *et al.* (eds.), 83–100
Parker, S.T. & K.R. Gibson, 1979. A model of the evolution of language and intelligence in early hominids. *Behavioural and Brain Sciences* 2, 367–407.
Parker, S.T. & K.R. Gibson (eds.), 1990. *'Language' and Intelligence in Monkeys and Apes: Comparative Developmental Perspectives*. Cambridge: Cambridge University Press.
Parker, S.T., R.W. Mitchell & M.L. Boccia (eds.), 1994. *Self-awareness in Animals and Humans: Developmental Perspectives*. Cambridge: Cambridge University Press.
Peterkin, G.L., H.M. Bricker & P. Mellars (eds.), 1993. *Hunting and Animal Exploitation in the Later Palaeolithic and Mesolithic of Eurasia*. Washington (DC): American Anthropological Association.
Reynolds, P.C., 1983. Ape constructional ability and the origin of linguistic structure, in de Grolier (ed.), 185–200.
Reynolds, P.C., 1993. The complementation theory of language and tool use, in Gibson & Ingold (eds.), 407–45.
Rodseth, L., R.W. Wrangham, A.M. Harrigan & B. Smuts, 1991. The human community as a primate society. *Current Anthropology* 32, 221–49.
Roebrocks, W., NJ Conard & T. van Kolfschoten, 1992. Dense forests, cold steps, and the Palaeolithic settlement of Northern Europe. *Current Anthropology* 33, 551–86.
Savage-Rumbaugh, E.S. & D.M. Rumbaugh, 1993. The emergence of language, in Gibson & Ingold (eds.), 87–108.
Shea, J., 1989. A functional study of the lithic industries associated with hominid fossils in the Kebara and Qafzeh Caves, Israel, in Mellars & Stringer (eds.), 610–25.
Shea, J., 1993. Lithic use-wear evidence for hunting by Neanderthals and early modern humans from the Levantine Mousterian, in Peterkin *et al.* (eds.), 189–97.
Smith, E.A., 1981. The application of optimal foraging theory to the analysis of hunter-gatherer group size, in *Hunter-Gatherer Foraging Strategies*, eds. B. Winterhalder & E.A. Smith. Chicago (IL): University of Chicago Press, 36–65.
Soffer, O., 1989. The Middle to Upper Palaeolithic transition on the Russian plain, in Mellars & Stringer (eds.), 714–42.
Squire, L.R. & S. Zola-Morgan, 1988. Memory: brain systems and behaviour. *Trends in Neurosciences* 11, 170–75.
Stiner, M., 1994. *Honor Among Thieves*. Princeton (NJ): Princeton University Press.
Straus, L.G., 1993. Upper Palaeolithic hunting tactics and weapons in Western Europe, in Peterkin *et al.* (eds.), 83–93.
Tobias, P., 1987. The brain of *Homo habilis*: a new level of organization in cerebral evolution. *Journal of Human Evolution* 16, 741–62.
Tomasello, M., 1990. Questions regarding imitation, 'language', and cultural transmission in apes and monkeys, in Parker & Gibson (eds.), 274–311.
Tomasello, M., 1993. The Social Learning of Chimpanzees and Human Children. Unpublished paper presented American Anthropological Association Meetings, Washington, DC, December 1993.
Vandermeersch, B., 1989. The evolution of modern humans: recent evidence from Southwest Asia, in Mellars & Stringer (eds.), 156–64.
Visalberghi, E. & D. Fragaszy, 1990. Do monkeys ape?, in Parker & Gibson (eds.), 247–73.
Whallon, R., 1989. Elements of cultural change in the later Paleolithic, in Mellars & Stringer (eds.), 433–54.
White, R., 1989. Production complexity and standardization in early Aurignacian bead and pendant manufacture: evolutionary implications, in Mellars & Stringer (eds.), 367–90.
Whiten, A. (ed.), 1991. *Natural Theories of Mind*. Oxford: Basil Blackwell.
Wissler, C., 1957. *The American Indian*. 3rd edition. Gloucester (MA): Peter Smith.
Wreschner, E., 1980. Red ochre and human evolution: a case for discussion. *Current Anthropology* 21, 631–43.
Wynn, T.G., 1979. The intelligence of later Acheulean hominids. *Man* 14, 371–91.
Yellen, J., A.S. Brooks, E. Cornelissen, M.J. Mehlman & K. Stewart, 1995. A Middle Stone Age worked bone industry from Katanda, Upper Semliki Valley, Zaire. *Science* 268, 553–6.
Zubrow, E., 1989. The demographic modeling of Neanderthal extinction, in Mellars & Stringer (eds.), 212–31.

Part II

Brain Size and Structure in Earlier Homines

4
Relating Brain Size to Intelligence in Primates

Richard W. Byrne

Brain size is widely used to infer intelligence, but hidden assumptions underly the process. Scaling against body size, though commonly used, implies a 'switchboard model' of brain function, not an intelligent system with properties of a turing machine. Absolute brain size, or relative neocortical size, give more proper estimates of potential intelligence. Such estimates do correlate with social group size, and with tactical deception frequency, in haplorhine primates. This pattern is consistent with the Machiavellian intelligence hypothesis: that a trend to larger groups in haplorhines selected for greater intelligence to deal with social problems, shown in brain enlargement. The key to this increased intelligence is likely to be increased learning speed, allowing a richer social data base and acquisition of subtle tactics in only a few trials. Great apes' more insightful, representational intelligence may not be based on brain expansion, but on brain re-organization.

It is not surprising that discussion of brain size is central to reconstructing the early human mind. Skulls are robust, and tend to be well preserved as fossils. They can also readily be used to estimate the size, shape, and to some extent proportions, of the brains they once contained. Large brains are one of the defining characteristics of hominids, and it is seldom questioned that large hominid brains indicate specialization in intelligence as a biological adaptation; sometimes, indeed, the brain proportions are considered to show further specialization within intelligence, for instance in language (Holloway 1983).

The fact that exact anatomical measurements can be made on brains may sometimes lead to a spurious confidence in the measures' validity. (Analogously, the precise numerical nature of an IQ test score has historically had a seductive attraction, quite out of proportion with the tests' validity.) Just as behavioural tests that purport to assess intelligence in animals are only as good as their empirical support (see Warren 1973; Macphail 1982), so brain measurements require calibration and validation before they can be accepted as indications of intelligent function. I will argue that, until we possess a much clearer understanding of how animal intelligence varies (empirically), and what operations brains are performing (theoretically), we would be wise to take all claims with a good pinch of salt.

My purpose in this paper is therefore to argue that the available measures of brain size and intelligence in living primates (let alone extinct humans) are very shaky ones; and that neither is clearly sturdier as a measure than the other, despite superficial appearances to the contrary. Furthermore, I suggest that the logic used to deduce intelligent behaviour from fossil skulls is based on a series of hidden assumptions, some very reasonable, some less so. Certain of the underlying assumptions in much currently accepted work do not look reassuring in the cold light of day.

Underlying assumptions

1. Bigger is better

This assumption is fundamental to almost all comparative analyses of brain size. At first glance, the claim seems quite naïve; it is certainly difficult to support empirically, for several reasons. Unfortunately, the relation between cranial capacity and the skull's outside dimensions is not close. Many brains are available for examination in medical schools, but the IQ of the people when alive is not generally known. Passingham (1979) instead used socio-economic status as a rough and ready estimate of intelligence, and managed to show a small correlation in humans between this measure and brain size (but not head size, measured externally). It may be argued that only a small correlation between brain size and intelligence is all that is needed to produce a strong selective pressure for larger brains in hominids, given the presumed importance of intelligence for hominid adaptation. Lack of a striking relationship need not be fatal to the idea. More fundamentally, it may be questioned whether a present-day, within-species correlation in brain size and intelligence is to be expected at all, even on the strongest hypothesis relating brain volume to intelligence. The point is that given a strong and lasting selective pressure for any characteristic, the genetically-based variance that codes for the characteristic will reduce. In the case of brain size, genes coding for low brain sizes will be weeded out, as brain size increases under selection, as it did during later hominid evolution. Gene mutation rates are slow, so after such a period of very powerful and rapid selection, little remaining variance would be expected.

On the other hand, the whole idea of bigger-is-better has been challenged by pointing to the case of human hydroencephaly. Hydroencephalics include intelligent and efficient people, whose brain matter is nevertheless reduced to a thin 'rind' around a massive water-filled cranial cavity — a popular article reporting the discovery was titled 'Is your brain really necessary?'. These cases are a puzzle, certainly, but often the cranium is enlarged, accommodating the excessive water, so some hydroencephalics may well have a normal number of cerebral neurons.

And in general, it is surely reasonable on the whole to expect that an organ like the brain, that is repetitive in structure, does more or better work the more units it has. For humans, anyway, we have some cast-iron evidence for believing big brains have been worth paying for — because we do *pay*. Primates acquire adult brain weight by one of two routes: they may be born with relatively large brains, or show prolonged postnatal brain growth. Most species specialize in one or the other route, or a trade-off between the two; humans are unique in investing in both (Harvey *et al.* 1987). But despite extended extra-uterine growth in brain volume, human birth still causes problems from large head size. A human baby's skull at birth is slightly larger than the birth canal: the consequent extreme pain of many deliveries and occasional risk of death to mother and baby are familiar to us. In all other great apes, the birth canal is much larger than the baby's skull, and birth is unproblematic. Given the constraint of head size versus birth canal, it is clear that human birth could not readily occur after a longer gestation time; nevertheless, the human neonate is more vulnerable and fragile than that of any other ape or monkey, incurring further costs to the mother who must look after the baby. Natural selection inexorably works to optimize biological systems (locally: global optimization by natural selection may often be impossible, because each intermediate form on route must have a survival advantage, no adaptive 'troughs' can be passed). All the costs of this brain expansion *must* therefore be compensated by major advantage; the only convincing candidate is increased intelligence, in the sense of neural computing power of some sort.

2. Relative size is what matters

In animal biology, allometric scaling against body size has been a triumph: making sense of biological puzzles by allowing proper comparison across species. To give a familiar example, scaling shows that elephants do not have especially thick legs for a mammal of their weight. Their proportions are a simple consequence of the fact that cross-sectional area increases by the square of length but volume increases by the cube: relative thickness must increase to compensate. Unmanageable curvilinear relationships (provided they follow some geometric function) can be reduced to tractable straight lines. The underlying logic of animal sizes and shapes is thus exposed, and the changes of form with growth become comprehensible.

Bigger animals often do have bigger brains, so allometric methods have been extensively applied to comparisons of brain size (e.g. Jerison 1973; Gould 1975; Clutton-Brock & Harvey 1980). In allometric scaling, a double logarithmic plot of something, in this case brain size, is made against body size for a given group of animals. Provided the two are related by some sort of power relationship, this forces the species points onto a straight line. Plotting against body size on logarithmic co-ordinates gives a reasonably straight line for primates, as for other groups of animals; but see Deacon (1990) for a suggestion that the underlying relationship is a curvilinear function. Modern analyses suggest that a line with slope of 0.75 gives the best linear fit to the data for primates (Martin 1990). Then one can see whether any particular animal in the group lies above the line (has a relatively larger brain than one might expect); on the line (has average brain size); or below the line (has a relatively small brain). These deviations from the best-fit line, the residuals, are thus treated as 'real braininess', an indication of the species' (or taxon's) specialization for brain size and intelligence. The commonest version is Jerison's 'encephalization quotient', the ratio of the observed brain weight to the weight expected from the best-fit scaling function.

There exists, however, no accepted theory to underwrite this reasoning, unlike the case of leg thickness in elephants or indeed most other applications of allometric scaling. Using scaling of brain size against body size in this way is implicitly making a strong claim about the functioning of neural tissue. The implication of scaling against body weight is that an animal with an expected brain size of 2 g and a real brain weighing 2 g, and an animal with an expected brain size of 200 g and a real brain weighing 200 g, are 'really' equally brainy.

This amounts to a hidden assumption that brains function only for coping with sensorimotor inputs and outputs, like the relays in the switchboard of an automatic telephone exchange. Certainly, that is *one* of the things brain tissue is doing, but is that all? If one accepted this *switchboard model* of the brain, it would be inappropriate to use the term 'intelligence' at all. A larger telephone exchange, serving the same number of subscribers as a small one, is not treated as more intelligent; the exchange simply has some spare capacity. It could reasonably be argued that 'spare capacity' might permit computational intelligence to take place. However, residuals are only estimating spare capacity in relative terms, and it is anyway difficult to attach any sensible meaning to deviations *below* the best-fit line, on this sort of approach.

3. The brain is an on-board computer

The phrasing is modern, but the belief that the brain has an executive function, guiding and directing behaviour, is pervasive and long-standing. A popular childhood encyclopaedia of the 1920s, 'The World of Wonder', depicts the stomach as a furnace and the brain as a thoughtful manager sitting at his desk. Since those days, it has become accepted that thinking *can* be understood mechanically (Newell *et al.* 1958), as a process that a complicated physical organ like a brain might do. This view of mental process depends on the brain having the approximate properties of a universal *turing machine*. Alan Turing, in the 1930s, showed that a device which could read and store simple arithmetical instructions on an infinite storage medium, and which stored instructions (programs) that guided its read/write operations in the same medium, could in principle solve any problems that could be represented in definite terms. He called this (then) imaginary device a computer. The solid-state machines we now know as 'computers' are approximate turing machines, limited in that they do not have anywhere near to the infinite memory assumed in the theory. Note that the precise embodiment of a turing machine is irrelevant, whether it is made of silicon and constructed with a single central processing unit, like many conventional computers, or made of neurons and distributed in a network. The brain with its massively interconnected network of multiply branching neurons might well be a much better approximation to a universal turing machine. Equating thinking with intelligence, and viewing the brain as a complex active guidance system for the body, is implicit in any belief that the brain is the seat of intelligence. Of course, the computers on our desks are also limited by inefficient programming, but evolution has had the highly efficient mechanism of natural selection to optimize brain programs, so it is no major step to assume that brains are optimized (for their particular species in its environment of evolutionary adaptedness). This extra assumption is built into the 'on-board computer' metaphor, but it is a very reasonable one.

Given the view of the brain as a computer with optimized programs, bigger (neural) brains should be better computers: having more neurons and more interconnections, so more closely approximating a universal turing machine. In contrast to the underlying assumptions of the traditional allometry, the adoption of a *computer model* argues that absolute not relative size governs efficiency. As an on-board computer, a 200 g brain would work no better in a 3 kg animal than a 30 kg one. The implicit models — switchboard or computer — are in conflict.

In reality, of course, brains have dual roles: the switchboard model is partly right, and so is the computer one. Jerison (1973) realized this, and tried to separate the two, by calculating how many 'spare neurons' (the absolute number) a given brain had available for its computing, on the basis of the minimum necessary for vegetative, somatic, maintenance function (i.e. switchboard purposes) only. Bauchot & Stephan (1966) attempted the same separation in a more empirical way, comparing primate brain sizes against the size expected from an allometric scaling relationship for 'basal insectivores' alone — animals such as shrews with very small brains and little claim to intelligence. These attempts have attracted much criticism, but would I argue they are closer to the right track than conventional allometric scaling, in that at least they attempt to tackle the dual roles problem. Unfortunately, it is impossible to justify any assumption that a type of animal employs its brain only for somatic, vegetative functions. Without this baseline of animals with zero intelligence, the estimation of the proportion of neural tissue 'spare' for intelligent, computational purposes is rather problematic. One promising alternative is to try directly to measure switchboard size in each brain. Passingham did so, using medulla volume as the estimate of input/output traffic (Passingham 1982); arguably, spinal cord cross-sectional area would give a still better estimate. Passingham's approach is astute, because it makes no assumption that the switchboard and the computer functions are physically separate in the brain. Natural selection may well have optimized the efficiency of brain tissue, but this process need not have tidily separated the various logically distinct roles played by the organ. Some localization of function is certainly found in the brain, but the inevitable difficulty of studying non-localized processes mean that the impression of considerable localization given in elementary neuropsychology texts is likely to be an overestimate, if not actually a legacy of phrenology. Unfortunately Passingham's approach met a different problem, the practical difficulty that neither medulla volume nor spinal cord cross-sectional area are known for most species.

4. Intelligent function is localized in the neocortex
Dunbar (1992), attempting to solve the same problem of separating the brain's dual functions of body maintenance and intelligent action, has taken a more rough and ready approach which allows an extensive comparative data base to be used. He calculates the ratio of the volume of primate neocortex to that of the rest of the brain, making the point that brain expansion in the primate radiation has largely been of neocortex; the resulting 'neocortical ratio' can be obtained for many species. Using this measure makes the implicit assumption, not inconsistent with the teaching of most university neuropsychology courses, that intelligent function is not only localized, but largely a cortical business in mammals. Despite the obvious concern that this underestimates the contribution of the hindbrain and various subcortical structures to intelligence, the usefulness of Dunbar's index of brain enlargement has been validated by its correlations with indices of complexity in behaviour.

These measures themselves bring in further assumptions, however. In order to relate any measure of brain capacity to success in dealing with problems, the complexity of the problems has also to be judged. The problems confronting primates have been seen as being of two sorts: environmental and social. (Though the distinction is a somewhat artificial one, since an animal's society is clearly part of its environment, and physical problems of the environment can sometimes be solved in social ways.) Neither is easy to estimate.

5. Environmental complexity can be estimated
Where predation pressure is a serious problem for primates, it is generally believed that the evolutionary response has not been an individual one, but a matter of enlarging the typical social group (van Schaik 1983). Presumably the mechanism of this is the increased fitness of individual animals which tend to greater sociality under conditions of predation danger. The major individual problems confronting primates, potentially requiring intelligence, have therefore been argued to concern food acquisition (Parker & Gibson 1977; Mackinnon 1978; Milton 1981; Clutton-Brock & Harvey 1980; Gibson 1986). Typically, the main issue has been seen as the complexity of the 'cognitive map' (i.e. a mental representation of the spatial and temporal distribution of food) necessary to efficiently locate a patchy but predictable food source. As a measure of this task, range area or day range length have been used. Though clearly a gross oversimplification of environmental complexity, range size has nevertheless been shown to correlate with relative brain size (Clutton-Brock & Harvey 1980), an index similar to the encephalization quotient. However, this result may be an artefact of gut specialization (Byrne 1994). A large gut permits a coarser but simpler diet and hence smaller ranging area; but the large gut makes the brain appear relatively small, because the bigger gut is bound to need a bigger bodily frame to support it. Further, brain tissue and gut tissue compete for limited metabolic resources, since both are energetically costly to maintain; without switching from a plant diet, enlargement in both organs could not be supported, hence the same correlation could be produced, again artefactually (Aiello pers. comm.). More fundamentally, for the reasons outlined earlier, measures derived by traditional allometric methods, like encephalization quotient, are unlikely to properly assess brain capacity for problem solving, hence unlikely to produce meaningful correlations with measures of problem complexity. Turning to the more promising neocortical ratio measure, Dunbar (1992) has shown that this relates neither to primate range area or day range; neocortical expansion is not related to the memory problem involved in food finding.

6. Social complexity can be estimated

In recent years, there has been increasing interest in the idea that social complexity rather than environmental complexity selected for primate brain enlargement, the so-called 'Machiavellian Intelligence' hypothesis (Jolly 1966; Humphrey 1976; Byrne & Whiten 1988a). Estimating social complexity is no easier than estimating environmental complexity. The true social complexity an individual primate confronts in its group will depend not only on the number of other individuals, but on the degree of within-group competition (which will vary with mating system and diet), and on the amount and type of relevant information it can mentally represent about each other known individual. However, these details are difficult to quantify (and may never be known with any precision), whereas group size presents a straightforward measure that is certain to partly index social complexity, and has the great advantage of being known for most primate species. Dunbar (1992) also examined the relationship of group size to neocortical ratio, where he found a close correlation in haplorhine primates (monkeys and apes), but not strepsirhines (lemurs, galagos and lorises). This neatly fits with the way in which the Machiavellian intelligence theory explains the primate origin of human intelligence, since the lack of social sophistication in strepsirhines (Jolly 1966) was one of the original strands of evidence leading to the idea. Nevertheless, group size gives an index, not of primate intelligence or social sophistication, but of the extent of the social problem. Ideally, we need a direct measure of intelligent performance against which to validate and calibrate indirect measures like group size.

Extensive laboratory work in comparative psychology leads to little confidence in the experimental measurement of intelligence in animals (Warren 1973; Macphail 1985); in laboratory tests, given enough trials and variations of conditions, most animals seem to be able to learn much the same things. Passingham (1981) has reanalyzed some of this data, showing that species differences exist, not in what can ultimately be achieved, but in the speed of learning — giving an indication of what effect we might expect an enlarged neocortex to have. However, few species are kept in laboratories to be tested, so no broad evolutionary comparisons can be made.

In the field, different species of animals naturally do quite different things, so it is not generally possible to compare their abilities for underlying intelligence: is a vervet monkey more intelligent, because it knows the relative ranks of its fellow group members (Cheney & Seyfarth 1990), than a Clark's nutcracker which can remember where it stored 3000 of its hidden food caches (Balda & Turek 1984; Vander Wall 1982), or than an ant which can keep track by dead reckoning of its violently twisting and turning courses covering distances of over a kilometre and turn straight for home when it wants (Wehner & Srinivasan 1981)?

Comparison across species so different in phylogeny and life style may be impossible, but in the limited domain of social animals that are not unduly close relatives (and this includes most primates), progress towards a 'species-fair' measure of learning speed may be possible. An individual in one of these groups would benefit from gaining resources at the expense of its social group members from time to time, but it would suffer from leaving the group. This nice balancing act is what Humphrey (1976) suggested could result in the evolution of social intelligence. In particular, individuals would benefit from learning tactics that functioned to trick their fellows out of resources; 'tactical deception' (Byrne & Whiten 1985; Whiten & Byrne 1988). Having carried out extensive requests to the ethological and zoological community for any evidence of tactical deception (e.g. Byrne & Whiten 1988b; 1992), with little result, I think it is likely that most types of mammal seldom learn such tactics in the wild. Presumably, then, tactical deception is not easy to learn. Similar requests to primatologists, however, have produced abundant evidence of tactical deception (currently available in Byrne & Whiten 1990), showing that individuals in all haplorhine primate genera use tactical deception to gain resources at the expense of their fellow group members (Byrne & Whiten 1992). (Machiavellian intelligence could also involve more co-operative and superficially altruistic pieces of trickery; but these are much rarer even in primates, apparently, so they not useful for giving an overview of capabilities.) Learning tactical deception is probably difficult for animals because the learning has to be rapid (the chance circumstances that might serve to reinforce deceptive behaviour are not common, everyday ones); and yet the animal has to differentiate among quite subtle stimulus configurations — precise ranks, affiliations, and memories of past presence or absence, of individuals. Of course, if a primate could *plan* deception, as can humans, these limits would not apply. Current evidence is against this possibility, except in the great apes (Byrne & Whiten 1992) and even there evidence is sparse (Byrne & Whiten 1991); most primate tactical deception is learnt by trial and error.

Perhaps the observed frequency of tactical deception in haplorhine primates can be used to index their intelligence? The only assumption this makes is that the frequency of showing beneficial tactics that are hard to learn is dependent on learning ability. Whiten and I have already collated all available records of tactical deception, and independently rated each on whether we were convinced that tactical deception was involved, rather than some coincidence or simpler-to-learn process (Byrne & Whiten 1992). For those where we both agreed on the interpretation as tactical deception, we counted up within taxonomic categories (lumping together multiple cases of the same type by the same observer in the same species). For each taxonomic category where sufficient data existed (usually the genus, sometimes higher), we then compared deception frequency against the number of studies carried out over a period of recent years; the patterns across taxa were significantly different, by a chi-square test (Byrne & Whiten 1992). Knowing that tactical deception is not exhibited at random supports a further stage of analysis, in which the incidence is used as a metric of intelligence. For each genus, I have now calculated the deviation (above or below the frequency expected by chance), and expressed it as a proportion of the expected frequency: a rough-and-ready measure of the usage of deceptive manipulation in each primate genus. Plotting this 'deception index' against neocortical ratio (averaged over the genus), gives a significant linear relationship (Fig. 4.1).

Conclusion

Many implicit assumptions have been identified in the exercise of relating brain size to intelligence among primates. Despite this, the prognosis is overall a positive one. Taken alone, each one of the measures that need to be compared, in order to test theories of the function of primate brain enlargement, is imperfect. Neocortical ratio, though an improvement on measures that relate brain size to overall body size, depends on the belief that intelligence is a local property of the neocortex. Group size is only a first

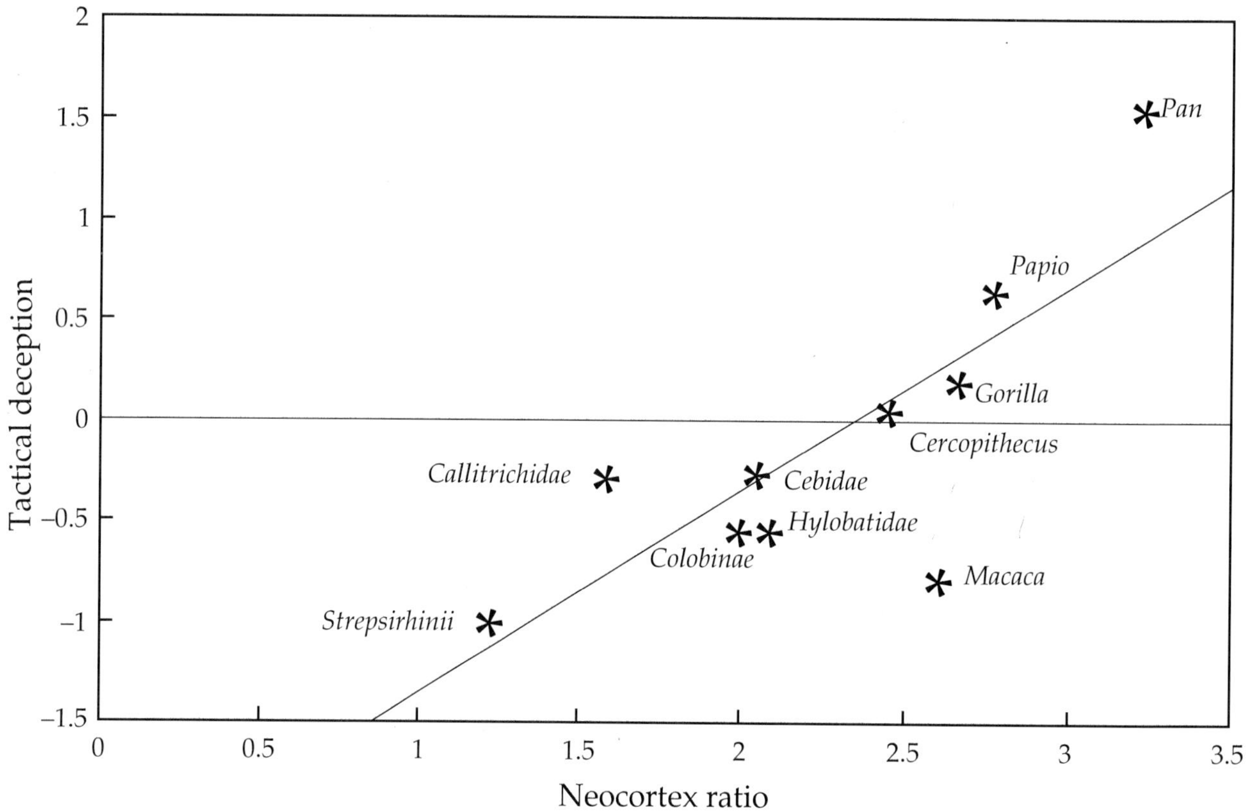

Figure 4.1. *Neocortex ratio predicts deception frequency. Here, neocortex ratio is plotted against the frequency with which tactical deception is observed (corrected for the number of field studies) in wild primates, for 10 taxa of primates. This relationship is significant on one-way ANOVA* ($F(1,8) = 11.89$, $P < 0.01$)*. Note that the great apes (*Pan *and* Gorilla*) overlap with monkeys (e.g.* Papio*) in neocortex ratio and index of deception.*

approximation to an adequate indication of the social problem faced by a primate. Range size, similarly, is only a first approximation to the cognitive load a primate incurs in foraging. And the deception index is based on data which were never more than fragmentary and incomplete. All are therefore rather speculative measures, but they are the best that currently exist for validating the relationship between brain size and intelligence. The consistent picture they present, when compared, goes some way to reassure that the exercise is meaningful. Neocortex size, group size and usage of deception all intercorrelate strongly, forming a strong argument that primate neocortex size is related to social intellect; range size and other measures of the demands of the cognitive map appear unrelated to cortical expansion.

Two caveats must be added to this conclusion. One is the obvious point that correlation does not indicate cause. The 'Machiavellian Intelligence' hypothesis would suggest that an increase in group size, caused by some external factor such as increasing predation in more open habitats, presented the need for increased intelligence in social manipulation. This in turn selected for an expansion in neocortex, to keep track of the sophistication and to speed up learning in social contexts. (It must be remembered that this teleological phrasing is a shorthand for the more ponderous, but strictly accurate one. Animals do not 'grow larger brains to achieve finer social manipulation'; individuals with gene mutations that code for larger brains are favoured in environments where learning social manipulations pays resource dividends, and thus they leave more descendants than those whose genes code for smaller brains.) Since increased neocortical development would potentially be at the metabolic expense of other body or brain parts, the more frugivorous diet of haplorhines presumably enabled the metabolic demands of brain expansion to be met. On this account, group size led to a pay-off for social sophistication, which caused neocortical enlargement (this enlargement itself allowing increased social sophistication, a positive feedback loop). Other versions are, however, possible. A shift in diet to the more frugivorous condition of haplorhines (in response some other, unknown, selective pressure) might have allowed increased brain size, by changing the cost-benefit equation — lowered metabolic costs selecting for larger brains than under the previously tighter constraints. Greater size of groups, with consequent increased social manipulation, might then have been a side-effect. Or, a need for greater social sophistication (in response to some undiscovered need) might have selected for larger brains, with an ability to live in larger groups produced as an incidental consequence. Neither of these scenarios, however, nor any other, has the parsimony of the Machiavellian intelligence hypothesis, since each requires more intervening variables. Why should primates switch to frugivory, with the attendant problems of a patchy and scarce resource, if not to enable larger brains or larger groups? Why should primates need greater social sophistication, if not to live in larger groups?

The second caveat concerns great apes. Neither neocortical ratio, nor the deception index, make a sharp separation between the great apes — orangutans, gorillas and the two species of chimpanzees — and the monkeys. The two taxonomic groups overlap, on both measures. (Absolute brain size is larger in all great apes, however.) Yet there is now increasing evidence that great apes possess a greater depth of understanding than any monkey, both in the social sphere (empathy, co-operation, theory of mind) and more technological areas (cause-and-effect relations, imitation, hierarchical structuring of skilled behaviour); see Byrne (1995a,b) for full references. While built on the capability for rapid learning, that is possessed by all haplorhines to some degree and enabled by their neocortical expansion, the quantum shift in intellectual understanding of the great apes is unlikely to be simply a matter of increased cortical size — unless there is a critical brain volume below which such a shift is impossible.

Richard W. Byrne
Scottish Primate Research Group
University of St Andrews
School of Psychology
St Andrews
Fife
KY16 9JU
Scotland

References

Balda, R.P. & R.J. Turek, 1984. The cache-memory system as an example of memory capabilities in Clark's nutcracker, in *Animal Cognition*, eds. H.L. Roitblat, T.G. Bever & H.S. Terrace. Hillsdale (NJ): Erlbaum, 89–105.

Bauchot, R. & H. Stephan, 1966. Données nouvelles sur l'encéphalisation des insectivores et des prosimians. *Mammalia* 30, 160–96.

Byrne, R.W., 1994. The evolution of intelligence, in *The Evolution of Behaviour*, eds. P.J.B. Slater & T.R. Halliday. Cambridge: Cambridge University Press, 223–65.

Byrne, R.W., 1995a. The ape legacy: the evolution of Machiavellian intelligence and interactive planning, in *Social Intelligence and Interaction*, ed. E. Goody. Cambridge: Cambridge University Press, 37–52.

Byrne, R.W., 1995b. *The Thinking Ape: Evolutionary Origins of Intelligence*. Oxford: Oxford University Press.

Byrne, R.W. & A. Whiten, 1985. Tactical deception of familiar individuals in baboons. *Animal Behaviour* 33, 669–73.

Byrne, R.W. & A. Whiten, 1988a. *Machiavellian Intelligence: Social Expertise and the Evolution of Intellect in Monkeys, Apes and Humans*. Oxford: Oxford University Press.

Byrne, R.W. & A. Whiten, 1988b. Towards the next generation in data quality: a new survey of primate tactical deception. *The Behavioral and Brain Sciences* 11, 267–73.

Byrne, R.W. & A. Whiten, 1990. Tactical deception in primates: the 1990 database. *Primate Report* 27, 1–101.

Byrne, R.W. & A. Whiten, 1991. Computation and mindreading in primate tactical deception, in *Natural Theories of Mind*, ed. A. Whiten. Oxford: Blackwells, 127–41.

Byrne, R.W. & A. Whiten, 1992. Cognitive evolution in primates: evidence from tactical deception. *Man* 27, 609–27.

Cheney, D.L. & R.M. Seyfarth, 1990. *How Monkeys See the World: Inside the Mind of Another Species*. Chicago (IL): University of Chicago Press.

Clutton-Brock, T.H. & P.H. Harvey, 1980. Primates, brains and ecology. *Journal of Zoology, London* 190, 309–23.

Deacon, T.W., 1990. Fallacies of progression in theories of brain-size evolution. *International Journal of Primatology*, 11, 193–236.

Dunbar, R.I.M., 1992. Neocortex size as a constraint on group size in primates. *Journal of Human Evolution* 20, 469–93.

Gibson, K.R., 1986. Cognition, brain size and the extraction of embedded food resources, in *Primate Ontogeny, Cognition and Social Behavior*, eds. J. Else & P.C. Lee. Cambridge, Cambridge University Press, 93–105.

Gould, S.J., 1975. Allometry in primates, with emphasis on scaling and the evolution of the brain. *Contributions to Primatology* 5, 244–92.

Harvey, P.H., R.D. Martin & T.H. Clutton-Brock, 1987. Life histories in comparative perspective, in *Primate Societies*, eds. B.B. Smuts, D.L. Cheney, R.M. Seyfarth, R.W. Wrangham & T.T. Struhsaker. Chicago (IL): University of Chicago Press, 181–96.

Holloway, R.L., 1983. Human paleontological evidence relevant to language behaviour. *Human Neurobiology* 2, 105–14.

Humphrey, N.K., 1976. The social function of intellect, in *Growing Points in Ethology*, eds. P.P.G. Bateson & R.A. Hinde. Cambridge: Cambridge University Press, 303–17.

Jerison, H.J., 1973. *Evolution of the Brain and Intelligence*. New York (NY): Academic Press.

Jolly, A., 1966. Lemur social behavior and primate intelligence. *Science* 153, 501–6.

Mackinnon, J., 1978. *The Ape within Us*. London: Collins.

Macphail, E.M., 1982. *Brain and Intelligence in Vertebrates*. Oxford: Oxford University Press.

Macphail, E.M., 1985. Vertebrate intelligence: the null hypothesis, in *Animal Intelligence*, ed. L. Weiskrantz. Oxford: Oxford University Press.

Martin, R.D., 1990. *Primate Origins and Evolution: a Phylogenetic Reconstruction*. London: Chapman & Hall.

Milton, K., 1981. Distribution patterns of tropical plant foods as a stimulus to primate mental development. *American Anthropologist* 83, 534–48.

Newell, A., J.C. Shaw & H.A. Simon, 1958. Elements of a theory of human problem solving. *Psychological Review* 65, 151–66.

Parker, S.T. & K.R. Gibson, 1977. Object manipulation, tool use and sensorimotor intelligence as feeding adaptations in cebus monkeys and great apes. *Journal of Human Evolution* 6, 623–41.

Passingham, R.E., 1979. Brain size and intelligence in man. *Brain and Behavioural Evolution* 16, 253–70.

Passingham, R.E., 1981. Primate specializations in brain and intelligence. *Symposia of the Zoological Society of London* 46, 361–88.

Passingham, R.E., 1982. *The Human Primate*. Oxford: W.H. Freeman.

van Schaik, C.P., 1983. Why are diurnal primates living in groups? *Behaviour* 87, 120–44.

Vander Wall, S.B., 1982. An experimental analysis of seed recovery in Clark's nutcracker. *Animal Behaviour* 30, 84–94.

Warren, J.M., 1973. Learning in vertebrates, in *Behaviour of Nonhuman Primates*, vol. 1, eds. A.M. Schrier, H.F. Harlow & F. Stollnitz. New York (NY): Academic Press, 249–81.

Wehner, R. & M.V. Srinivasan, 1981. Searching behaviour of desert ants, genus *Cataglyphis (Formicidae, Hymenoptera)*. *Journal of Comparative Physiology* 142, 315–38.

Whiten, A. & R.W. Byrne, 1988. Tactical deception in primates. *Behavioural and Brain Sciences* 11, 233–73.

5

Measuring Cognition in Extinct Hominids

Robert A. Foley

Any model of the evolution of the human mind should be testable against and consistent with the evidence drawn from the human palaeontological and the archaeological record. Such data are, however, notoriously sparse and incomplete. Here an attempt is made to measure a number of variables from the palaeobiological record that may provide evidence of the pattern of human cognitive evolution: relative brain size, neocortex ratio, duration of technological tradition, complexity of technology and rate of maturation. If these data do reflect cognitive states then they show a pattern of independent evolution, with maturation rate and neocortex ratio showing a relatively early and more continuous pattern than EQ and technology. Debates about the relative cognitive status of extinct hominid taxa should take into account that different skills may evolve independently. It is also argued that changes in life history strategy underpin cognitive evolution.

There are two necessary components of any successful explanation and description of the evolution of the human mind. The first of these must be an adequate understanding of the biological mechanics of cognition and a theory that can account for the range of variation we can see among animals today. The second is to place that theoretical framework into the specific context of the time and place of hominid evolution. Ultimately it is the way we put these two elements together that is crucial. The problem with such an integration, however, lies in the fact that these two components draw on radically different sets of data and theoretical frameworks.

Building a theoretical model of the way a mind works is the field of comparative psychology. The primary source of information here is the study of humans and other animals under laboratory and natural conditions, and an understanding of them in terms of the mechanics of neurobiology. The pre-human mind that emerges from such studies is seldom specific to the particular conditions of phylogeny or environment in which humans evolved. To the comparative psychologist the early human mind floats unanchored in the vast tracts of space and time between the development of the central nervous system and the evolution of human self-reflexivity.

The specifics of hominid evolution, on the other hand, which are the concern of palaeoanthropologists, do provide details of the context and timing of events in the evolution of human cognition. This evidence, though, tends to be lacking in detail, is sparse and unevenly distributed, and can seldom be directly linked to the cognitive skills of the extinct hominids that would be of concern to psychological models. For the most part palaeobiologically based interpretations of the early human mind are vast inferential arches built on one or two rather eroded foundation stones.

Solutions to these methodological problems will come from a number of directions. For example, rooting behavioural and anatomical characteristics onto phylogenetic structures through cladistics will allow cognitive skills to be assigned to taxa at various evolutionary nodes, thus helping to anchor the psychological and social capacities with the specific events of evolution (Foley 1987; 1989a; see Robson Brown this volume). Another approach would be to secure more firmly the skills involved in technology to broader cognitive models (Mithen this volume). In this paper no attempt is made to infer directly the actual cognitive skills and capacities of the pre-human hominids. Instead what I shall seek to do is to quantify the relative information processing and adaptive intelligence levels of extinct hominid taxa, specifically those of the genus *Homo*. The aim is to provide a more secure empirical basis for discussions of the relative capabilities of the hominids of the past, using the palaeobiological and archaeological evidence available. In simple terms, are modern humans twice as intelligent as *Homo habilis* as brain size data might imply; was *Homo erectus* two and a half times more intelligent than chimpanzees? The answer to such questions may help to solve some of the debates about the relative cognitive and intellectual capacities of neanderthals and *Homo sapiens*; whether the difference is a quantum leap, or a mere difference in style.

Which pre-human mind? The problem of hominid taxa

One of the primary discoveries of recent years is that the cognitive abilities of animals varies considerably between taxa (see Lee this volume; Byrne & Whiten 1988; Cheney & Seyfarth 1990; McGrew 1992; Parker & Gibson 1979; Savage Rumbaugh *et al.* 1983). Monkeys have very different abilities from apes, and within the Hominoidea there is also much variation. This can be seen in the differences between gorillas and chimpanzees which, despite a very close phylogenetic relationship, differ very considerably in both level of encephalization and cognitive abilities and styles. Even between the two species of chimpanzee, *Pan troglodytes*, and *P. paniscus*, there may be substantial differences in behavioural potential (Savage-Rumbaugh & Savage-Rumbaugh 1993). There may of course be still further levels of variation within species, such as between sexes and with age, but these will not be addressed here (see Lee this volume; Byrne this volume).

These results are significant for the way in which we approach the evolution of hominid cognition and the specifics of the pre-human mind. In particular, the key question is — which 'early human mind'?. It is clear from the fossil record, and recent interpretations of it that the pattern of human evolution is not one of a continuous and gradual anagenesis, but includes distinct species located at particular points in time and space (Fig. 5.1). The antecedents to the human mind therefore cannot be a general and unspecified precursor of that of modern humans, but is likely to have varied between taxa (Lee 1989). Furthermore, not all extinct hominids are ancestral to modern humans. Some, such as the robust Australopithecines, and possibly *Australopithecus africanus* and

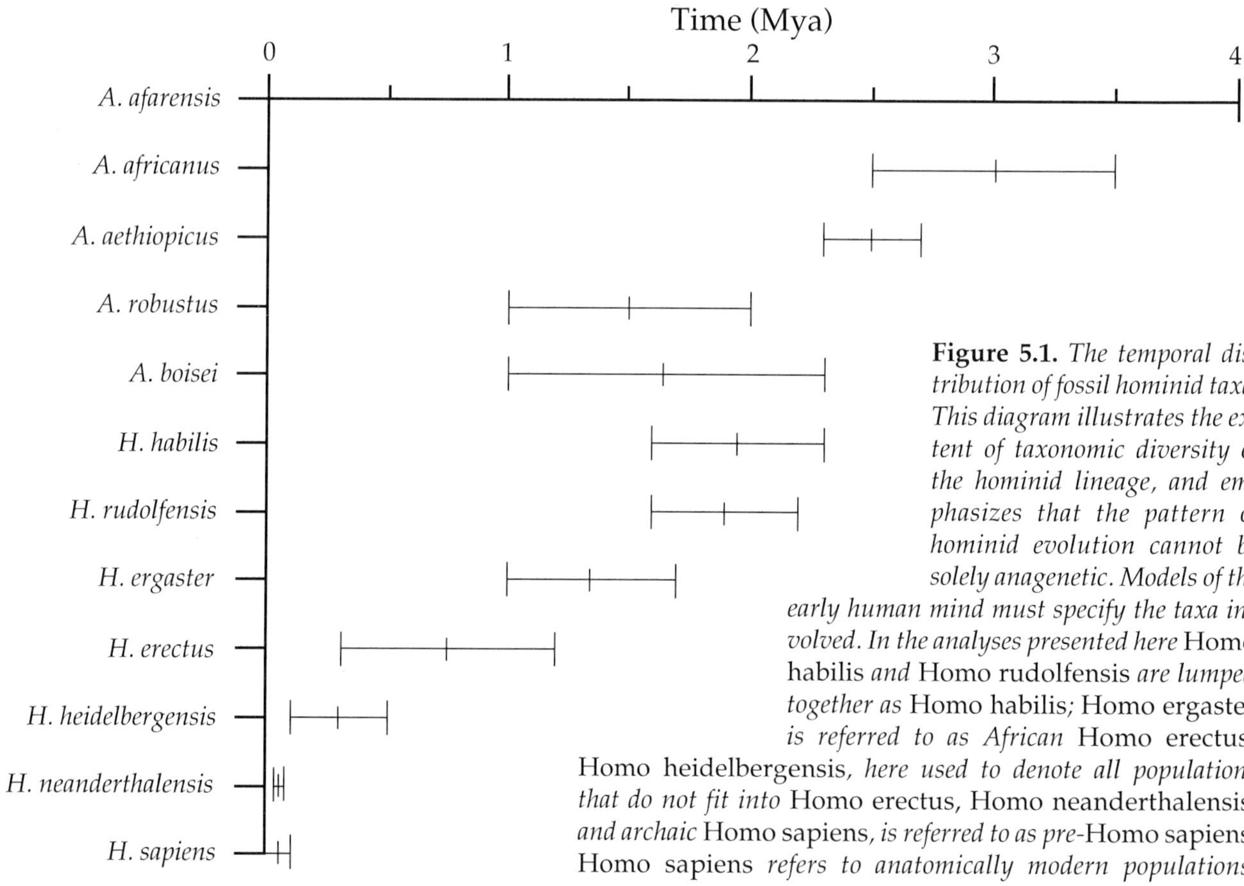

Figure 5.1. *The temporal distribution of fossil hominid taxa. This diagram illustrates the extent of taxonomic diversity of the hominid lineage, and emphasizes that the pattern of hominid evolution cannot be solely anagenetic. Models of the early human mind must specify the taxa involved. In the analyses presented here* Homo habilis *and* Homo rudolfensis *are lumped together as* Homo habilis; Homo ergaster *is referred to as African* Homo erectus; Homo heidelbergensis, *here used to denote all populations that do not fit into* Homo erectus, Homo neanderthalensis, *and archaic* Homo sapiens, *is referred to as pre-*Homo sapiens. Homo sapiens *refers to anatomically modern populations.*

Homo neanderthalensis, are distinct evolutionary developments, and cognitively as much as anatomically they are likely to have their own unique traits. It is no longer possible to adopt a progressivist approach to the evolution of human cognition any more than it is to other hominid features.

The aim here will be to establish the relative cognitive capacities of specific hominid taxa. The definitions of these are set out below. Owing to our inability to establish data on very specific taxonomic groups these are relatively generalized, and it is not necessarily the case that the units used here represent real species, nor that they comprise the totality of the hominid record. From the range of taxa shown in Figure 5.1 a number of key units are employed in the analyses presented here. In particular, only the genus *Homo* is included in these analyses as there is no clear evidence for stone technology among the Australopithecines (Foley 1987; Susman 1988) and this limits their use for comparative purposes.

1. Early *Homo*: this refers to *Homo habilis* in the sense used by Tobias (1991) to include all the early African hominids not specifically assigned to *Homo erectus* or the australopithecines (i.e. with enlarged cranial capacity, reduced face and dentition, but lacking the cranial specializations of *Homo erectus*). Many authors, myself included, would prefer to split this material into two species, *Homo habilis* (e.g. the Olduvai hominids + KNM-ER1813) and *Homo rudolfensis* (e.g. KNM-ER 1470), a larger and more robust species (Wood 1991). This hominid occurs in sub-Saharan Africa between 2.3 and 1.6 Myr (Hill *et al.* 1992).
2. Early (African) *Homo erectus*: in the generalized sense the taxon *Homo erectus* occurs in Asia and Africa, from 1.6 Myr to 0.3 Myr. It is possible to distinguish various attributes that are unique to the later Asian forms, and so some authorities recognize an early African taxon (*Homo ergaster*) (Andrews 1984; Wood 1991) and true, later, Asian *Homo erectus*. This distinction is made here, and 'early African *Homo erectus*' refers to the material, such as KNM-ER3733, occurring in Africa between 1.6 and <1.0 Myr.
3. Later (Asian) *Homo erectus* represents the remainder of '*Homo erectus*', and is generally later (<1.0 Mya) and confined to Asia.
4. Pre-*Homo sapiens*: *Homo sapiens* is conventionally used to include all material that post-dates *Homo erectus* and shows cranial expansion (>1000 cc) and a reduction in facial height. This makes the taxon extremely polytypic. Recently it has been argued that the taxon *Homo sapiens* should be confined to anatomically modern humans (Stringer & Andrews 1988; Tattersall 1986; Foley 1989b). This leaves various problems in assigning the later Middle Pleistocene and early Upper Pleistocene material consistently. Here 'pre-*Homo sapiens*' refers to the, primarily European, later–Middle Pleistocene fossil hominids.
5. *Homo neanderthalensis*: that part of the 'archaic *Homo sapiens*' category of the conventional classification that has distinct mid-facial prognathism and a number of other unique features, and occurs in Europe during the upper Pleistocene until approximately 35 Kyr, and contemporaneously in the Middle East. It represents the best known non modern hominid, and to many authorities constitutes a distinct species evolving in parallel to modern humans (see Tattersall 1986).
6. *Homo sapiens*: refers to anatomically modern humans, occurring from approximately 100 Kyr in Africa and the Levant. Where associated data relating to *Homo sapiens* is used (see below), late Pleistocene and hunter-gathering modern humans are used.

In the subsequent analyses these taxa constitute the units used.

Quantitative measures of cognitive states in extinct hominids

The fossil and archaeological record contains a number of lines of evidence that have been used to infer the cognitive states of the hominids. These are both anatomical and behavioural. Here only those that are susceptible to quantitative analysis and can be assigned to several of the taxa are used. The following variables may be defined:

1. *Brain size and encephalization quotient*. Overall brain size, which can be calculated from cranial capacity with considerable accuracy, is generally thought to be a good indicator of the overall information-processing capacity of a species. However, the primary determinant of brain size is body size. In assessing the significance of brain size it is therefore necessary to factor out the effect of brain size. The relationship between brain size and body size is allometric and a number of models have been developed to describe the relationship. These may be used to calculate the encephalization quotient (EQ), which is the size of the brain relative to expected size for a particular body weight. Two exponents are generally used — two thirds (Jerison 1973) and three quarters (Martin 1983). The

former scales with surface area, the latter with metabolic rate. Both are presented where available. Use of EQ in the fossil record depends upon the availability of good estimates of body size. The data used here are from Aiello & Dunbar (1993).

2. *Neocortex ratio*. Dunbar (1992) has recently argued that neocortex ratio, rather than EQ is the best estimate of 'significant higher intelligence' and therefore it is the ratio of this to the rest of the brain that needs to be measured. This has been calculated for a number of extinct hominids (Aiello & Dunbar 1993).
3. *Technocomplex duration*. Once stone tools are made then there is greater access to behavioural information about hominids, and by deduction, their cognitive capacities (Gowlett 1984). Quantifying such information, however, is no easy task. The length of time technological traditions persist, defined here as the length of time from first to last appearance, is one such variable that may be assessed, if assumptions are made about the use of specific artefacts as 'technological apomorphies' (see Foley 1987; 1989c). Pronounced persistence may provide insights into cognition either in terms of limited scope for innovation and introducing variation and novelty, or else in terms of the maintenance of social traditions (McGrew & Tutin 1978). Isaac (1972) produced a similar set of data, and this has been updated here (Table 5.1). The assumption is made here that *Homo habilis* is responsible for the Oldowan and that this technology persists throughout the duration of the species; that the duration of the Acheulean, associated with African and (by inference only) European *Homo erectus* is over 1 million years; that the pebble tool tradition of Asia is associated with *Homo erectus*, with similar levels of longevity. Pre-*Homo sapiens* has been linked here with the technologies associated with the African MSA, while *Homo neanderthalensis* is associated with the Levallois tradition lasting over 200,000 years. For *Homo sapiens* a number of archaeological hunter-gatherer populations were used.
4. *Technocomplex complexity*. The number of identifiable stone tool types is basic information that can be derived from archaeological data, and from which inferences about cognition may be derived. Clearly these data are subject to a certain amount of subjectivity, but the differences between taxa are often an order of magnitude and therefore the effect of classificatory bias may be insignificant. The data used here are drawn from Isaac (1972), updated from Schick & Toth (1993) for the earlier phases.
5. *Age of eruption of mandibular M1*. There is growing acceptance that levels of encephalization are related to other life history parameters (gestation length, longevity, age of first reproduction) (Harvey *et al.* 1987). It is also known that the rate of growth and hence age of maturation changes very substantially during the course of hominid evolution (Smith 1992). Recent developments in studies of dental microstructure (Bromage & Dean 1985; Beynon & Dean 1988; Beynon & Wood 1987) now allow the age at which teeth erupt to be determined independently, and as the age of eruption of the first molar correlates with brain size among primates (Smith 1989), this can be used to indicate evolutionary changes in life history strategy that are likely to be associated with cognitive and social traits. Data here are drawn from sources described in Foley & Lee (1991).

Patterns in the evolution of hominid cognition

Table 5.1 shows the values obtained for these variables for the fossil hominid taxa. For a number of data points it was not possible to calculate a figure; for EQs this is due to an absence of good body size estimates, and for age of eruption of M1 no material has yet been analyzed for one taxon. It should also be stressed that there may be considerable variance involved and these are taxon means based on single data points or small samples. A further problem lies in the fact that the data for each cognitive variable varies considerably in range, and it is difficult to establish equivalence. Table 5:2 therefore shows the same data expressed as a ratio relative to *Homo habilis*, where *Homo habilis* always equals zero; in Table 3 a ratio relative to *Homo sapiens*, where *Homo sapiens* always equals 1. In Table 5.4 the ratio are calculated in such a way as to root *Homo habilis* at zero and *Homo sapiens* at 1. These data give a better impression of the relative positions of the hominid taxa and the amount of change between them.

The information that can be derived from these data do not relate to the actual cognitive or behavioural skills of the hominids. What they can do is provide a scale of the relative abilities of the hominids, and thus when and in what general areas of cognition evolutionary changes are occurring.

The most obvious point to emerge is that whatever scale is used the various traits change independently, or at least the association is not such that all change is simultaneous. Figures 5.2 and 5.3 show for the two ratios used, the pattern of change for

Table 5.1. *Anatomical, behavioural and life history variables of extinct hominid taxa and early* Homo sapiens *from which cognitive states may be inferred. See text for discussion and definition of variables and taxa. See text for details and sources.*

Taxon	Encephalization Quotient ('EQ') (Jerison)	Encephalization Quotient ('EQ') (Martin)	Neocortex Ratio	Duration of archaeological technocomplex (Myr)	Archaeological complexity (number of tool types)	Age of eruption of mandibular M1 (years)
Homo habilis	4.31	3.38	3.44	1.20	6.00	4.00
Homo erectus (African)	4.40	3.34	3.66	1.20	11.00	4.60
Homo erectus (Asian)	4.40	3.34	3.79	1.00	7.00	5.40
Pre-*Homo sapiens*	–	–	3.95	0.10	19.00	–
Homo neanderthalensis	–	–	4.06	0.21	42.00	6.60
Homo sapiens	8.07	6.28	4.12	0.02	69.00	6.10

Table 5.2. *Ratio of behavioural and neurobiological characteristics of* Homo *species. In all cases* Homo habilis *= 0.*

Taxon	Jerison's Encephalization Quotient	Neocortex Ratio	Archaeological Longevity	Technological Complexity	Age of Maturation
Homo habilis	0.00	0.00	0.00	0.00	0.00
Homo erectus (African)	2.00	6.00	0.00	7.00	9.00
Homo erectus(Asian)	2.00	9.00	0.00	1.00	23.00
Pre-*Homo sapiens*	–	13.00	17.00	19.00	–
Homo neanderthalensis	–	16.00	7.00	52.00	42.00
Homo sapiens	47.00	17.00	98.00	91.00	34.00

Table 5.3. *Ratio of behavioural and neurobiological characteristics of* Homo *species. In all cases* Homo sapiens *= 1.*

Taxon	Jerison's Encephalization Quotient	Neocortex Ratio	Archaeological Longevity	Technological Complexity	Age of Maturation
Homo habilis	0.53	0.83	0.02	0.09	0.66
Homo erectus (African)	0.55	0.89	0.02	0.16	0.75
Homo erectus (Asian)	0.55	0.92	0.02	0.10	0.89
Pre-*Homo sapiens*	–	0.96	0.19	0.28	–
Homo neanderthalensis	–	0.99	0.09	0.61	1.08
Homo sapiens	1.00	1.00	1.00	1.00	1.00

Table 5.4. *Ratio of behavioural and neurobiological characteristics of* Homo *species. In all cases* Homo habilis *= 0 and* Homo sapiens *= 1.*

Taxon	Jerison's Encephalization Quotient	Neocortex Ratio	Archaeological Longevity	Technological Complexity	Age of Maturation
Homo habilis	0.00	0.00	0.00	0.00	0.00
Homo erectus (African)	0.04	0.35	0.00	0.08	0.26
Homo erectus (Asian)	0.04	0.53	0.00	0.01	0.68
Pre-*Homo sapiens*	–	0.76	0.17	0.21	–
Homo neanderthalensis	–	0.94	0.07	0.57	1.24
Homo sapiens	1.00	1.00	1.00	1.00	1.00

each of the variables, and it can be seen clearly that, for example, technological parameters alter very little for most of the non-modern hominids. It is only with *Homo sapiens* that there is much change, and then it is very dramatic. In contrast, neocortex ratio shifts gradually and relatively continuously across the taxa. An attempt can be made to assess the degree of association statistically, although the small samples and unknown error levels rule against using these measures to assess statistical significance. However, Table 5.5 shows the correlation coefficients for the relationship between the cognitive variables. The strongest associations are between archaeological complexity and EQ ($r = 1$); archaeological duration and EQ ($r = -.99$); maturation rate and neocortex ratio ($r = .96$); and archaeological duration and neocortex ratio ($r = -.91$). It should be remembered for the first two associations there are no EQ data for two of the taxa. If the ratios are used to calculate the correlation coefficients (in order to remove scaling effects) the level of association drops although the rank order remains the same.

A related point is that the baseline of change is very different for the different taxa. *Homo habilis* possesses over four-fifths of the neocortex ratio of *Homo sapiens*, about half of the EQ, and matures at about two-thirds of the rate found in modern humans. In contrast, the archaeological/technological measures show less than 10 per cent of the capabilities found in early modern humans.

Furthermore, change is not always unidirectional towards greater complexity, reaching a peak in modern humans. For example, the Asian *Homo erectus* display less technological complexity compared to earlier African *Homo erectus*. In the case of Neanderthals and modern humans there is considerable complexity. Neanderthals share some technological characteristics with early *Homo sapiens* (Skhūl

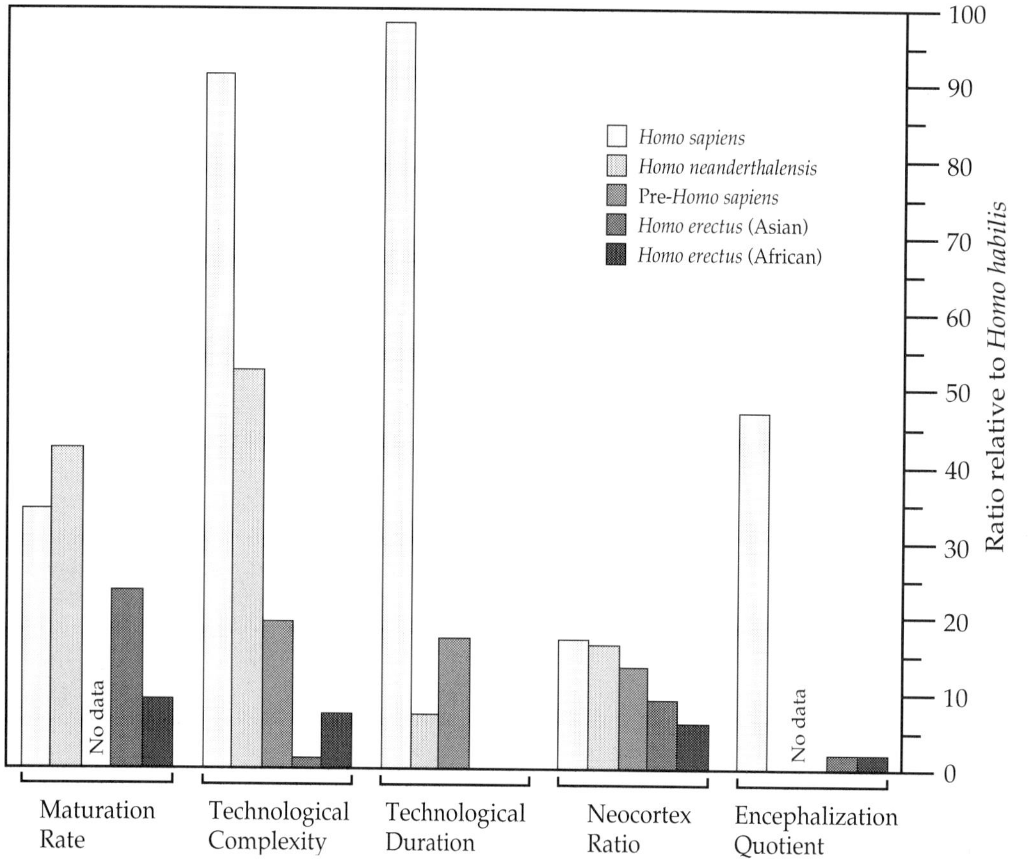

Figure 5.2. *Histograms representing the 'cognitive variables' for the hominid taxa. To ease comparison all data are expressed as a ratio relative to* Homo habilis *which equals zero (and hence is not shown).*

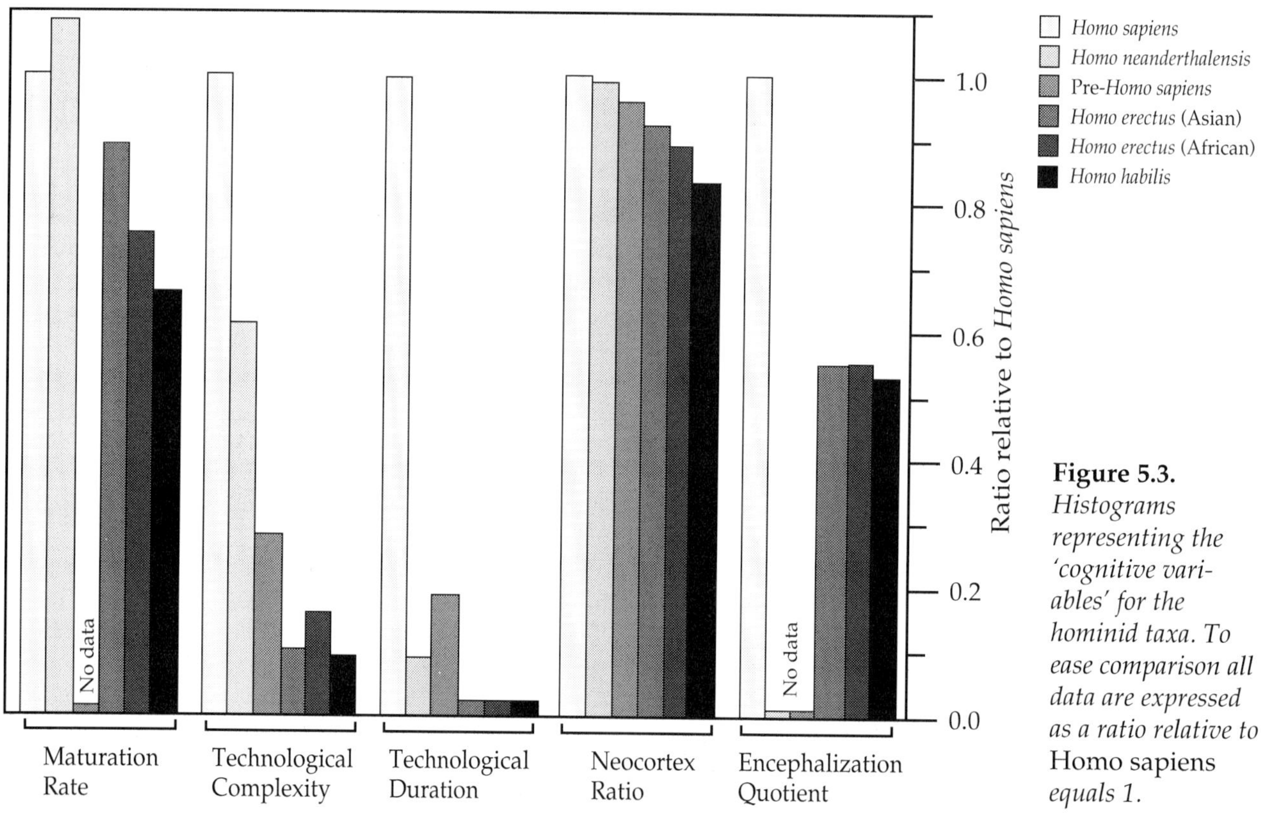

Figure 5.3. *Histograms representing the 'cognitive variables' for the hominid taxa. To ease comparison all data are expressed as a ratio relative to* Homo sapiens *equals 1.*

Table 5.5. *Correlation coefficients between behavioural and neurobiological variables.*

	Encephalization Quotient	Neocortex Ratio	Archaeological Longevity	Maturation Rate	Technological Complexity
Encephalization Quotient	1.00				
Neocortex Ratio	0.87	1.00			
Archaeological Longevity	–0.99	–0.91	1.00		
Maturation	0.79	0.96	–0.89	1.00	
Technological Complexity	1.00	0.82	–0.79	0.76	1.00

and Qafzeh), and also show increasing complexity at the end of their time span (Châtelperronean, Szeletian, etc.). Conversely complex and 'modern' technology (*cf.* Upper Palaeolithic) is not always associated with modern *Homo sapiens*. It is the case, however, that in Europe at the transition between Neanderthals and *Homo sapiens* there is an increase in technological complexity (Mellars this volume).

These three observations together have important implications for debates about the abilities and cognitive states of extinct hominids relative to modern humans. The level of difference between modern humans and other species depends upon the measure used. Technology will exaggerate the differences, while neocortex ratio will emphasize continuity. EQ and maturation rate indicate an intermediate position. The simplest conclusion to be drawn is that different skills and capacities evolved at different rates at different times, and therefore that the various extinct hominids may have had genuinely different combinations of behavioural and intellectual traits.

From the relative patterns of these various traits it is perhaps more interesting to see whether any biological insights can be gained from the pattern of change, rather than being concerned with the relative rankings of the taxa. These may provide a better understanding of the evolutionary processes that led to the evolution of modern humans.

For the earliest part of the hominid evolutionary history, the rate of maturation appears to change most significantly. This pattern would be seen even more strongly if comparisons with Australopithecines were made. This implies that changes in life history strategies and their underlying patterns of parental effort and investment are the first shifts to occur. These are associated with a gradual change in neocortex ratio. Dunbar (1992) has argued that this change reflects increasing group size, and it may be argued here that the conjunction of these two variables implies an ecological and energetic change in the earlier phases of hominid evolution, resulting in a change in social strategy and hence both the conditions under which encephalization may occur and the selective pressures leading to such a change. Subsequent to these life history and neurological changes there are associated, possibly threshold, effects in EQ and technology. These, however, occur later (with pre-*Homo sapiens*) and also very rapidly.

It should be remembered that the analyses presented here have been based on comparisons between taxa, and these do not necessarily relate to the actual rates of change through time. Figure 5.4 shows the extrapolation from this data of the possible rates of change for these variables. Such extrapolations would support the notion that for neocortex ratio and maturation rate there is an earlier, more extended and gradual pattern of change than is found in the technological record.

Conclusions

The primary purpose of this paper has been to argue that any attempt to examine the evolution of human cognition must be based on the palaeobiologically observable pattern of hominid evolution. Models that seek to map and explain the evolutionary pathways from the pre-human mind to the abilities of modern humans must be consistent with the only direct evidence of the timing of this evolution and the conditions under which it occurred. Such an argument in turn places the onus on palaeobiologists to provide data on the cognitive abilities of extinct hominids that allows comparison between taxa. Some steps in this direction are made here, but it needs to be stressed that the data presented are based on sparse and small samples, and analysis has only been possible using pooled taxonomic information that ignores any within-group variance or time-transgressive trends.

Such data, however, do provide some tentative insights into the pattern of human cognitive evolution. The first of these is that no single variable can represent the character of cognition, and when several variables are used a level of independence can be observed. It is possible to show that different abilities evolve at different rates at different times.

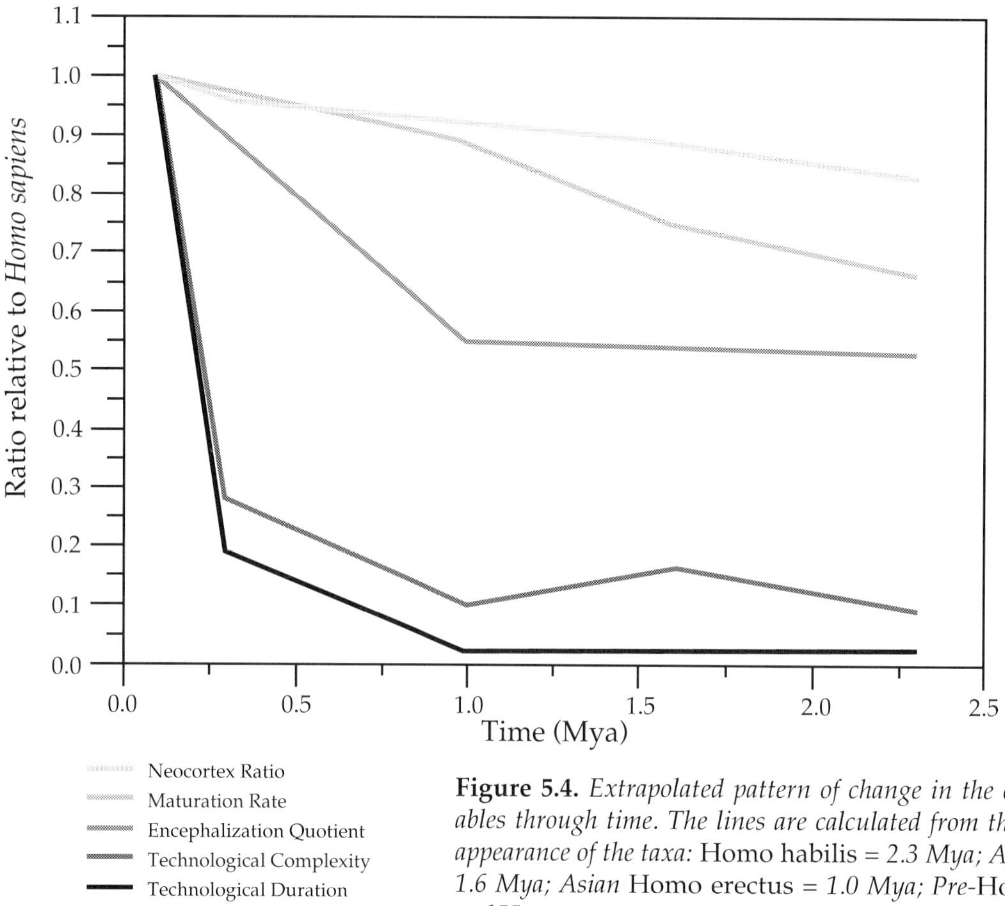

Figure 5.4. *Extrapolated pattern of change in the cognitively related variables through time. The lines are calculated from the estimated date of first appearance of the taxa:* Homo habilis = *2.3 Mya; African* Homo erectus = *1.6 Mya; Asian* Homo erectus = *1.0 Mya; Pre-*Homo sapiens = *.5 Mya; and* Homo sapiens = *.1 Mya. Neanderthals are excluded from the extrapolation on the grounds that they do not appear to be part of the evolutionary lineage leading to* Homo sapiens, *but appear later. The values associated with them are shown by the letters M (Maturation Rate), N (Neocortex Ratio), C (Technological Complexity), and D (Technological Duration). Data used are ratio data where* Homo sapiens = *1.*

An important implication that would require further investigation is whether this means that the cognitive characteristics of the modern human mind evolved as a response to a number of different selective pressures, or whether there is a greater degree of integration both neurobiologically and in terms of selection. Answering this question may provide some insights into the debate about specific versus general intelligence (see various papers, this volume). It may also help to resolve some of the debates about the relative levels of cognitive skills of extinct hominids and the question of whether there is a 'revolution' with the appearance of modern humans (Mellars & Stringer 1989) or whether neanderthals are only trivially different to *Homo sapiens*. The answer to this question may well depend upon what is being compared.

Perhaps the most provocative suggestion that may be made on the basis of the limited data presented here is the apparent precedence in time of changes in the rate of maturation. If correct this underpins arguments (Humphrey 1976) that even if social complexity provides the key selective pressure leading to greater levels of encephalization in humans in particular and mammals in general, nonetheless this social complexity is predicated upon the energetic feasibility of specific reproductive strategies, and hence upon the pivotal links between social strategies and ecological ones. It also means that the evolution of the human mind cannot be divorced from the evolution of our more basic biological traits.

Robert A. Foley
Human Evolutionary Biology Research Group
Department of Biological Anthropology
University of Cambridge
Downing Street
Cambridge
CB2 3DZ

References

Andrews, P.J., 1984. An alternative interpretation of the characters used to define *Homo erectus*. *Courier Forschungsinstitut Seckenberg*, 167–75.

Aiello, L.C. & R.I.M. Dunbar, 1993. Neocortex size, group size and the evolution of language. *Current Anthropology* 34, 184–93.

Beynon, A.D. & M.C. Dean, 1988. Distinct dental development patterns in early fossil hominids. *Nature* 335, 509–14.

Beynon, A.D. & B.A. Wood, 1987. Patterns and rates of molar crown formation times in East African hominids. *Nature* 326, 493–6.

Bromage, T.G. & M.C. Dean, 1985. Re-evaluation of the age at death of Plio-Pleistocene fossil hominids. *Nature* 317, 525–8.

Byrne, R. & A. Whiten (eds.), 1988. *Machiavellian Intelligence*. Oxford: Clarendon Press.

Cheyney, D.L. & R.M. Seyfarth, 1990. *How Monkeys See the World*. Chicago (IL): University of Chicago Press.

Dunbar, R.I.M., 1992. Neocortex size as a constraint on group size in primates. *Journal of Human Evolution* 22, 469–93.

Foley, R.A., 1987. Hominid species and stone tool assemblages: how are they related? *Antiquity* 61, 380–92.

Foley, R.A., 1989a. The evolution of hominid social behaviour, in *Comparative Socioecology*, eds. V. Standen & R.A. Foley. Oxford: Blackwell Scientific Publications, 473–94.

Foley, R.A., 1989b. The ecology of speciation: comparative perspectives on the origins of modern humans, in Mellars & Stringer (eds.), 298–320.

Foley, R.A., 1989c. Alternative models of Pleistocene cultural evolution. *Antiquity* 63, 159–62.

Foley, R.A. & P.C. Lee, 1991. Ecology and energetics of encephalization in hominid evolution. *Philosophical Transactions of the Royal Society*, London Series B 334, 223–32.

Harvey, P.H., R.D. Martin & T.H. Clutton-Brock, 1987. Life histories in comparative perspective, in *Primate Societies*, eds. B.B. Smuts, D.L. Cheney, R.M. Seyfarth, R.W. Wrangham & T.T. Struhsaker. Chicago (IL): University of Chicago Press, 181–96.

Hill, A., S. Ward, A. Delno, G. Curtis & R. Drake, 1992. Earliest *Homo*. *Nature* 355, 719–22.

Humphrey, N.K., 1976. The social function of intellect, in *Growing Points in Ethology*, eds. P.P.G. Bateson & R.A. Hinde. Cambridge: Cambridge University Press, 303–17.

Jerison, H.J., 1973. *Evolution of the Brain and Intelligence*. New York (NY): Academic Press.

Lee, P.C., 1989. Comparative ethological approaches in modelling hominid behaviour. *Ossa* 14, 113–26.

McGrew, W.C., 1992. *Chimpanzee Material Culture*. Cambridge: Cambridge University Press.

McGrew, W.C. & C.E.G. Tutin, 1978. Evidence for a social custom in wild chimpanzees. *Man* 13, 234–51.

Martin, R.D., 1983. *Human Brain Evolution in an Ecological Context. 52nd James Arthur Lecture on the Evolution of the Brain*. New York (NY): American Museum of Natural History.

Mellars, P.A. & C.B. Stringer (eds.), 1989. *The Human Revolution: Behavioural and Biological Perspectives on the Origins of Modern Humans*. Edinburgh: Edinburgh University Press.

Parker, S. & K. Gibson, 1979. A developmental model for the evolution of language and intelligence in early hominids. *Behavioural and Brain Sciences* 2, 367–408.

Savage-Rumbaugh, E.S. & D.M. Savage-Rumbaugh, 1993. The emergence of language, in *Tools, Language and Cognition in Human Evolution*, eds. K.R. Gibson & T. Ingold. Cambridge: Cambridge University Press, 86–108.

Savage-Rumbaugh, E.S., J.L. Pate, J. Lawson, S.T. Smith & S. Rosenbaum, 1983. Can a chimpanzee make a statement? *Journal of Experimental Psychology* 112, 457–92.

Smith, B.H., 1989. Dental development as a measure of life history in primates. *Evolution* 43, 683–8.

Smith, B.H., 1992. Life history and the evolution of human maturation. *Evolutionary Anthropology* 1(4), 134–42.

Stringer, C.B. & P.J. Andrews, 1988. Genetic and fossil evidence for the origin of modern humans. *Science* 1263–8.

Susman, R.L., 1988. Hand of *Paranthropus robustus* from Swartkrans Member 1, fossil evidence for tool behavior. *Science* 240, 781–4.

Tattersall, I., 1986. Species recognition in human palaeontology. *Journal of Human Evolution* 15, 165–75.

Tobias, P.V., 1991. *Olduvai Gorge*, vol. 4: Homo habilis — *Skulls, Endocasts and Teeth*. Cambridge: Cambridge University Press.

Trinkaus, E. (ed.), 1989. *The Emergence of Modern Humans*. Cambridge: Cambridge University Press.

Wood, B.A., 1991. *Koobi Fora Research Project*, vol. 4: *Hominid Cranial Remains*. Oxford: Oxford University Press.

6

The Functions of the Prefrontal Cortex in Humans and Other Animals

Angela C. Roberts, Peter Collins & Trevor W. Robbins

This paper reviews data from both phylogenetic and ontogenetic studies to illustrate the importance of the prefrontal cortex in higher cognitive function. The thesis proposed is that some of these higher cognitive abilities considered unique to man and which are thought to depend on the prefrontal cortex may well have evolved from more elementary cognitive processes already mediated by the prefrontal cortex of lower primates. Similarly, the earliest cognitive abilities to be expressed by a maturing prefrontal cortex during human development may act as the building blocks for the future development of more complex functions. Thus, a comparative neuropsychological as well as comparative neurobiological approach to the study of the prefrontal cortex of monkeys and man will provide us some insight into the evolution of mind.

The importance of the prefrontal cortex for the integration of man into his social environment can be seen when studying those patients with damage to this structure, sustained either as a consequence of neurosurgery for trauma or epilepsy or following neurodegeneration, as in dementia of the frontal type. There is often gross disturbance of their everyday life, apart from the most routine activities. They are extremely poor organizers and thus even the simplest of tasks, such as going out to dinner, can take hours to plan, and even then a decision may not be made. They seldom plan ahead and show little spontaneous behaviour, relying instead on others to direct and organize their activities. Apart from these informal descriptions of their behaviour, the cognitive abilities of subjects with damage to their frontal lobes have also been characterized formally by numerous neuropsychological tests. This has resulted in the discovery of a range of tests that are especially sensitive to prefrontal damage including the Wisconsin Card Sort test (WCST), verbal fluency, self-ordered search and the 'Tower of London'. These demonstrate respectively, impairments in set-shifting, i.e. the ability to shift the focus of attention (Milner 1963), the spontaneous generation of words from defined categories such as animals or letters (Miller 1984), working memory, i.e. the capacity to hold information in mind while performing a task (Milner & Petrides 1984) and finally planning, i.e. the ability to formulate and evaluate strategies (Shallice 1982). The importance of the prefrontal cortex for this range of abilities has been substantiated by brain-imaging studies using positron emission tomography (PET) that have identified specific areas of prefrontal cortex in normal subjects that are active selectively during the performance of some of these tests (Berman *et al.* 1986; Jonides *et al.* 1993; Morris *et al.* 1993; Petrides *et al.* 1993).

The pattern of impairment following prefrontal cortex damage is often heterogeneous, with different patients being impaired on different subsets of tests. Indeed, in some cases patients perform the same as control subjects on most of these tests and yet are clearly impaired in their daily activities (Penfield & Evans 1935; Shallice & Burgess 1991). This suggests that the prefrontal cortex has multiple functions and that the pattern of impairment is dependent upon the particular region of damage. However, as no two patients have exactly the same locus of damage, it has proven difficult to correlate the region of damage with impairment on particular tests. Moreover, given that many of the tests are not only measuring different cognitive abilities, e.g. set-shifting

a) Computerized Tower of London

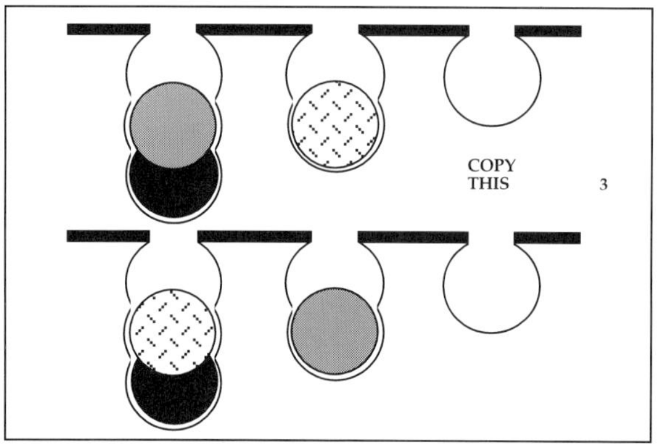

b) Spatial Short-term Memory Task

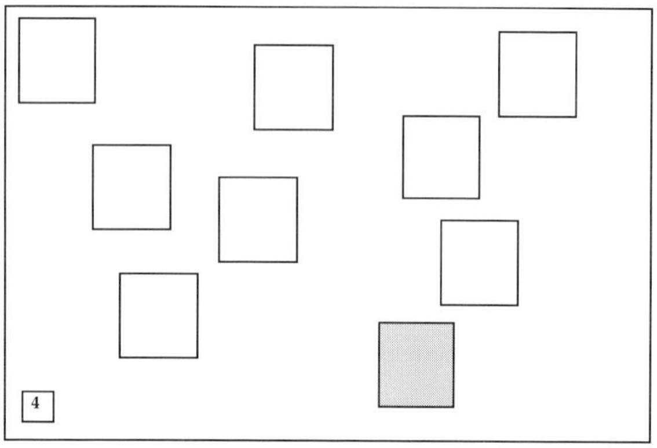

c) Spatial Working Memory Task

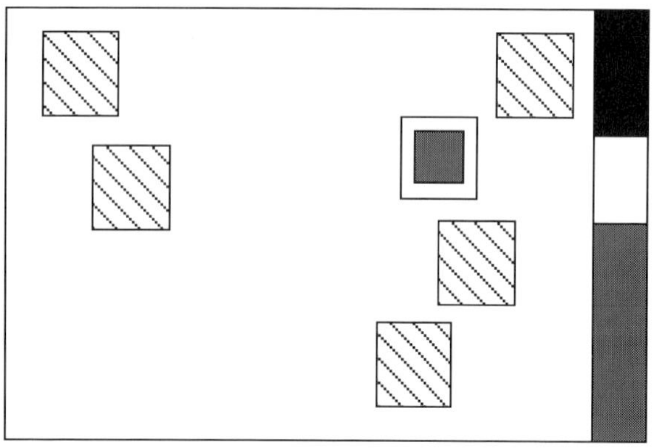

Figure 6.1. *(a) This task is a variation on the Tower of London developed by Shallice and McCarthy (Shallice 1982). Two sets of three coloured balls are presented, one in the top half of the screen and one in the bottom half. These are described to the subject as snooker (or 'pool') balls since they appear to be hanging in 'pockets'. There are three pockets in each half of the screen, one that can hold three balls, one that can hold two balls and finally one that can hold only one ball. On each trial a red ball, a green ball and a blue ball are placed into the pockets of each of the two displays, in predetermined positions. The subject is asked to re-arrange the balls in the bottom display, such that their positions match the goal arrangement in the top display. A ball can be moved by first touching it and then by touching an empty position in one of the other pockets. The subject is told at the outset that they cannot place a ball high in a pocket if there is no other ball beneath it and they cannot remove a ball from a pocket if there is another ball sitting above it in the same pocket.*

There are two blocks of six problems each. The first six test trials comprise two problems each of two, three and four moves, followed by a further six trials comprising two at four moves and four at five moves. The subjects are instructed to examine the arrangement of the balls at the beginning of each problem and to try and solve it in the minimum possible number of moves.

(b & c) These tests measure spatial short-term memory capacity and spatial working memory respectively, two component cognitive processes required in the Tower of London task. In the computerized version of the Corsi Block Tapping test (b), spatial short-term memory capacity is determined from the ability of subjects to remember a sequence of squares on the screen. Each trial begins with the same arrangement of nine white squares, presented in a pseudo-random pattern. Subjects are instructed to observe the boxes, as some will change colour, one after the other. Their task is to remember the location and the sequential order of the boxes which change and to then repeat the observed sequence by touching the boxes in the same order. The test begins at the simplest level of difficulty with a two box problem. After each successful trial the number of boxes changing in the next sequence is increased by one to a maximum of nine boxes. After an incorrect attempt at any level an alternative sequence of the same length is presented until the subject fails three consecutive trials at any one level. A number in the bottom left-hand corner indicates the length of the current sequence on each trial. Spatial short-term memory capacity is defined as the final level at which the subject recalls at least one sequence of boxes.

In the Spatial working memory task (c), the subject is required to 'search through' a number of boxes presented on the screen by touching each one such that it 'opens up' revealing its contents. The object of the task is to collect 'blue tokens' hidden inside the boxes and once found, to use them to fill an empty column at the side of the screen. The subjects are instructed that at any one time there will be a single token hidden inside one of the boxes. Their task is to search until they find it at which point the next token will be hidden. The key instruction is that for each trial once a blue token has been found within a particular box then that box will never be used again to hide a blue token. On every trial each box is used once so that the total number of tokens to be found corresponds to the total number of boxes on the screen. Subjects are presented with four test trials each of two, three, four, six and finally eight boxes. (For a detailed description of all three of these tasks see Owen et al. *1990.)*

vs working memory, but also depend upon the processing of different *types* of information, e.g. visual *vs* spatial, it is unclear which of these dimensions is responsible for the differential pattern of impairments in frontal lobe damaged patients. Finally, performance on some tests depends upon more than one cognitive function and indeed some cognitive abilities depend upon a number of identifiable cognitive processes. For example, planning ability is measured in the Tower of London by requiring subjects to move coloured blocks from an initial position to a target position in a minimum number of moves (see computerized version in Fig. 6.1a). Not only does this depend upon the formulation and evaluation of strategies, but also upon mental imagery, the ability to shift from one possible strategy to another, spatial working memory to hold all of the relevant information in mind while performing the task and finally the expression and sequencing of the selected action plan — many of which functions have been attributed to the prefrontal cortex. In our own studies (Owen *et al.* 1990) we attempt to assess the relative contribution of some of these components including spatial span (Fig. 6.1b) and spatial working memory (Fig. 6.1c).

When investigating the parallel and hierarchical organization of higher cognitive functions within the prefrontal cortex, it is important to consider evidence from both phylogenetic and ontogenetic studies, as both of these approaches may afford us some insight into the cognitive capabilities of early man. The greatest increase in the size of the prefrontal cortex has occurred within the primate order and by comparing cognitive functioning across different species of primate it may be possible to identify an *evolutionary* development in their cognitive abilities in parallel with the enlargement of the prefrontal cortex. Thus, some of the higher-order cognitive abilities considered unique to man and which are thought to reside in the prefrontal cortex may well have evolved from more elementary component cognitive processes already mediated by the prefrontal cortex of lower primates. In addition, by studying the cognitive effects of localized lesions within the prefrontal cortex of monkeys it should be possible to determine not only *which* cognitive abilities are mediated by this area of the brain but also *how* it is organized. For example, are different areas specialized to perform different cognitive operations, e.g. set-shifting versus working memory or, alternatively, to perform these operations on different types of information, e.g. visual *vs* spatial? Insights into the hierarchical organization of higher cognitive functions are also provided by studying cognition during *human* development. It is likely that the earliest cognitive abilities to be expressed by a maturing prefrontal cortex act as the building blocks for the future development of more complex functions. The remainder of this chapter will consider evidence from these two approaches that together support a central role of the prefrontal cortex in the development of higher order cognitive control.

Anatomical development

Phylogenetic studies

The increase in size of the prefrontal cortex with phylogenetic development is most apparent in the primate order where, according to Brodmann (1912), the prefrontal cortex constitutes 29 per cent of the total cortex in humans, 17 per cent in chimpanzees, 11.5 per cent in macaques and 8.5 per cent in lemurs (Fig. 6.2a). Studies of neocortical development have revealed that the neocortex of mammals is made of two separate components or moieties, one from cingular (parahippocampal) and the other from insular (parapiriform) regions, that develop in opposite directions around the hemisphere and meet on its dorsolateral aspect. Around the rostral pole of the hemisphere these two moieties, or 'trends', i.e. archicortical (dorsal) and paleocortical (ventral), form the prefrontal neopallium (see Fig. 6.2b) (Barbas & Pandya 1989). However, in primates, Sanides (Fuster 1989) has suggested that the prefrontal cortex has resulted from three, rather than two differentiating fields, the two primordial fields and a third, more recent field from the motor cortex which proceeds forward through premotor areas to dorsolateral prefrontal cortex (area 46 on Fig. 2b). The entire dorsal prefrontal region, along with the parvocellular portion of the mediodorsal (MD) nucleus of the thalamus to which it is connected, undergo phylogenetically greater enlargement up the primate scale than do the orbital prefrontal region and the corresponding magnocellular portion of the MD nucleus (reviewed in Fuster 1989). Thus, it has been speculated that the more recent growth of the dorsal region of prefrontal cortex may reflect the increasing importance, in higher species, of the cognitive functions that this region supports. This is in general agreement with the findings of Barbas & Pandya (1989) which suggest that the evolutionary older areas within the dorsal and ventral 'trends' are located on the orbital and medial surfaces of the prefrontal cortex, whilst the more recent areas within the two 'trends' are located on the dorsolateral surface (Fig. 6.2b).

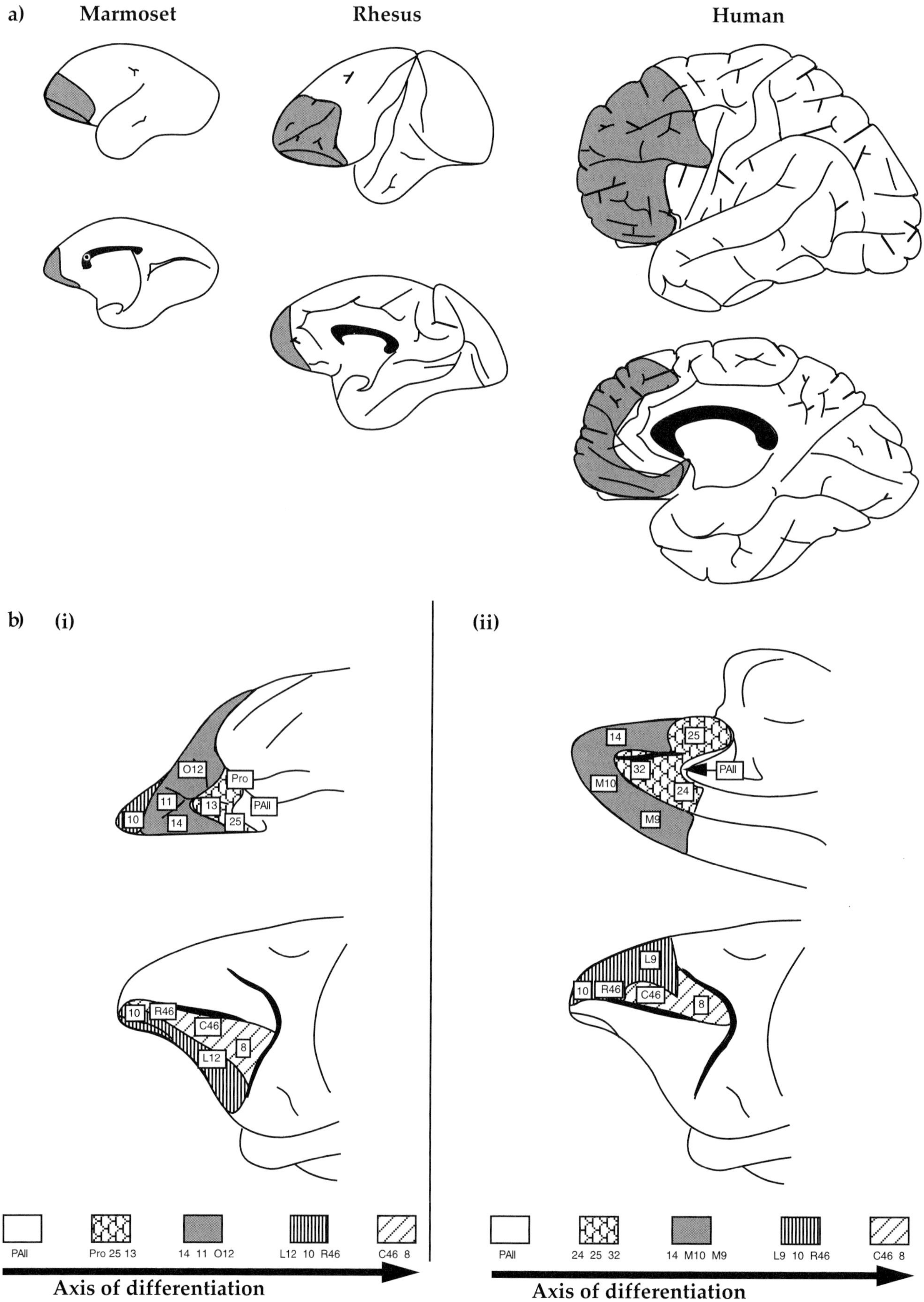
a)
Marmoset
Rhesus
Human
b)
(i)
O12
Pro
11
13
PAII
10
14
25
10
R46
C46
8
L12
PAII
Pro 25 13
14 11 O12
L12 10 R46
C46 8
Axis of differentiation
(ii)
14
25
32
PAII
M10
24
M9
L9
10
R46
C46
8
PAII
24 25 32
14 M10 M9
L9 10 R46
C46 8
Axis of differentiation

Ontogenetic studies

At birth, the laminar arrangement and cortico-cortico and cortico-caudate efferent projections of the prefrontal cortex are already present. However, dendritic organization is rudimentary and cell volume relatively small when compared to the adult. During the first two years of post-natal development there is considerable re-organization of the prefrontal cortex including a rapid increase in dendritic branching, particularly of cortico-cortical fibres, co-incident with increases in synaptic density along the sprouting dendrites (Kostovic 1990; Mrzljak *et al.* 1990). Myelination of intrinsic and extrinsic nerve fibres of cortical association areas also occurs during this period, the process starting later compared to other cortical areas and continuing for many years. The discovery of myelogenetic stages led Fleschig (reviewed in Fuster 1989) to suggest that the development of function follows the same sequence as myelination and may be partly dependent on it. Certainly, cognitive functions considered frontal in nature, begin to develop during the latter half of the first year when there is a spurt of growth within prefrontal areas (see later). Moreover, in parallel with phylogenetic studies there is evidence that orbital areas mature earlier than do areas on the dorsolateral surface which again may be reflected in the time-course of the cognitive development of these two regions.

Cognition

In the past the paradigms used to study cognitive function in non-human primates and other mammals were different from those used in human studies. Thus, even though similarities were reported between the impairments in human and non-human primates following damage to the prefrontal cortex, the comparability of these deficits was unclear. More recently, a number of studies have taken classical tests of prefrontal cognitive function in monkeys and adapted them for use in humans. An example is the spatial delayed response task (Freedman & Oscar-Berman 1986; Verin *et al.* 1993) which investigates the ability of monkeys to remember the location of food that they had seen hidden in one of two identical food wells a few seconds earlier, and a variant of this task, spatial delayed alternation. Alternatively, clinical neuropsychological tests such as the WCST have been rendered suitable for use in monkeys (Roberts *et al.* 1988). By comparing directly the effects of damage to the prefrontal cortex on the performance of both humans and monkeys on the *same* cognitive tests, it will be possible to establish with more certainty the functional homology of prefrontal cortex between monkey and man — until recently this was based primarily upon the pattern of connectivity with the rest of the brain. Moreover, the finding of qualitatively similar impairments in monkeys and man will also help to demonstrate possible homology of the cognitive functions under investigation. Finally, the study of discrete lesions in monkeys can address the issue of functional heterogeneity within the prefrontal cortex and whether it reflects functional differentiation of the distinct regions.

Figure 6.2. *(a) Lateral and medial views of the prefrontal cortex in three primate species. (After Brodmann, 1912: not to scale.) The prefrontal cortex (shaded areas) can be differentiated from adjacent areas of the frontal lobes by the presence of a well developed internal granular layer (iv), which is absent or less prominent in premotor or motor cortex. Although not a definitive criterion, the prefrontal cortex also receives a* larger *proportion of its thalamic input from the mediodorsal nucleus, than from other thalamic nuclei, relative to other cortical areas.*
(b) Diagrams showing the architectonic stages within the prefrontal cortex of the rhesus monkey. Within both the ventral (i) and dorsal (ii) trends the axis of cytoarchitectonic differentiation proceeds in a direction from the least differentiated periallocortex (PAII) to area 8, which represents the most differentiated cortex. (Redrawn from Barbas & Pandya 1989.)

Attentional set-shifting

One of the most common tests used to assess prefrontal function in humans is the WCST which requires subjects to sort a pack of cards along one of three perceptual dimensions, colour, form or number, according to feedback from the person conducting the test. Occasionally, the tester will change the rule and the subject has to learn to switch from sorting the cards according to one category to sorting them according to another, for example, from colour to number. Patients with damage to the prefrontal cortex are particularly impaired at switching between categories, tending instead to perseverate along the original sorting category. Although similar impairments in shifting 'attentional set' have been purported to occur in monkeys with prefrontal damage, the tests used have been various types of discrimination reversal paradigms (Mishkin 1964; Passingham 1972), which do *not* require shifts of attention between perceptual dimensions. Commonly these tests have required the monkey to learn to respond to one of two stimuli in order to obtain reward and then merely to reverse their responding to the other stimulus. Therefore, at the reversal stage the monkeys are required to reverse a specific stimulus-response or stimulus-reward association *within* a particular perceptual dimension rather than to shift their attention to *another* perceptual dimension. Recently, however, a computerized paradigm has been developed that can measure attentional set-shifting ability in both human and non-human primates alike. Essentially the test compares performance on compound visual discriminations that either require the continued maintenance of attention, i.e. 'set', towards a previously relevant perceptual dimension, an *intra-dimensional shift* (IDS), or require a shift of attention from one perceptual dimension to another, an *extra-dimensional shift*. (EDS). The slower the acquisition of the discrimination requiring an EDS, the greater the difficulty in shifting an attentional set.

The various stages of the test are illustrated in Figure 6.3. Initially, subjects are trained to discriminate stimuli in one dimension, such as shapes, presented on the computer screen. The subject chooses a stimulus simply by touching the screen and learns if their response is correct or not either from verbal feedback presented on the screen, in the case of humans, or by the presence or absence of reward, in the case of monkeys (Fig. 6.3a). After subjects have successfully reached a criterion of learning a second, irrelevant dimension is introduced, e.g. lines, but the subject must continue responding to the previously correct stimulus, ignoring the presence of the new, irrelevant dimension (Fig. 6.3b). There then follows two transfer tests in which completely novel sets of shapes and lines are presented and the subject must either maintain attention towards the previously relevant dimension (IDS, Fig. 6.3c) or switch attention towards the previously irrelevant dimension (EDS, Fig. 6.3d).

Comparison of performance on these two types of shift has been studied in representative species of primates at three different levels of phylogenetic development; the common marmoset (New World monkey) (Roberts *et al.* 1988), the rhesus macaque (an

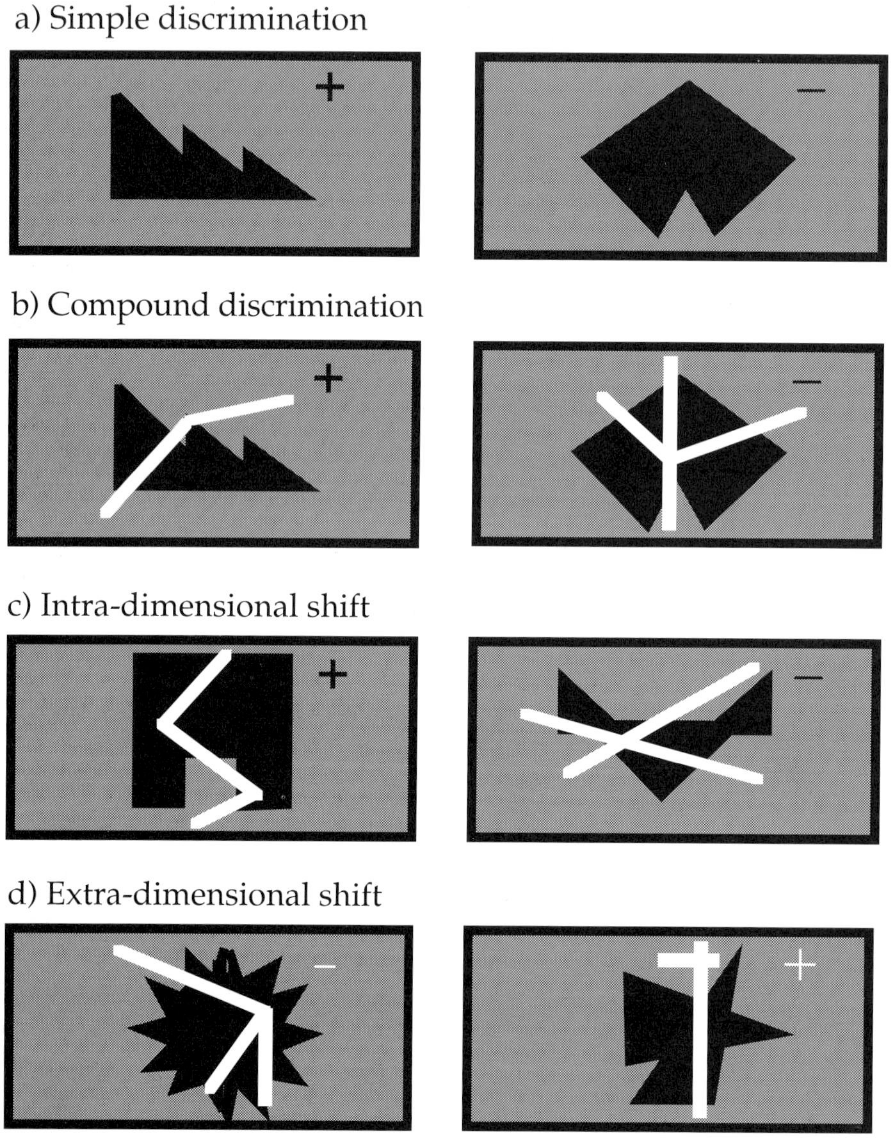

Figure 6.3. *The 'shape' and 'line' exemplars and the combinations used for the various stages of the attentional set-shifting paradigm. In this example the dimension of 'shape' is relevant to the first three discriminations whilst the 'line' dimension is relevant to the final discrimination. In any one trial of a compound discrimination a shape exemplar may be paired with one or other of the line exemplars. The correct and incorrect choice for each discrimination is indicated by the '+' and '–' respectively. Black typeface specifies that 'shape' is the relevant dimension whilst white typeface specifies that 'line' is the relevant dimension.*

Old World monkey) (Gold, Roberts, Robbins and Koob, unpublished findings) and man (Roberts *et al.* 1988). In all cases acquisition of the discrimination requiring an intra-dimensional shift was superior to that requiring an extra-dimensional shift, confirming that in the early stages of the test all three species had learned to attend selectively to the relevant dimension. However, it is clear from Figure 6.4a that there was a progressive reduction in the number of errors made before acquiring the two transfer tests as one moved from marmoset, to rhesus, to man. This may reflect an increase in the degree of abstraction, i.e. the ability to attend to the dimensional properties of a stimulus and to use just one of these to govern responding, exhibited by primate species as one ascends the phylogenetic scale.

The qualitatively similar behaviour between monkey and man on the attentional set-shifting paradigm substantiates the claim that the paradigm is tapping the *same* cognitive function across species. The question can then be addressed as to whether the prefrontal cortex in both monkey and man supports this set-shifting ability. A significant proportion of humans with frontal lobe damage (Owen *et al.* 1991) fail to complete the EDS stage of the test and even those that do pass take many more trials to reach criterion than do control subjects (Fig. 6.4b). A similar impairment at the EDS stage of the test has been demonstrated in marmosets (Fig. 6.4b) following excitotoxic lesions of the prefrontal cortex (Roberts in press). Therefore, these effects of lesions together with the comparative behavioural studies, suggest that the ability to shift an attentional set in both monkey and man is dependent upon the prefrontal cortex. This is further substantiated by a recent study investigating attentional set-shifting in children of three different age groups, 4–5, 6–8, and 9–11 years (Hughes 1993; Hughes *et al.* 1994). Whilst 40 per cent of the children in the two lower age groups failed to complete the EDS stage of the test, all children aged between 9 and 10 years passed. Moreover, the younger children that did complete the EDS stage were disproportionately slower compared to their preceding IDS, than the oldest age group (Fig. 6.4c). This may reflect a progressive development in the proficiency of shifting an attentional set.

Working memory: spatial delayed response task and A not B

There is overwhelming evidence that ablation of the prefrontal cortex in monkeys impairs performance on the spatial delayed response task (SDR) (Jacobsen 1935; Rosvold *et al.* 1961; Divac & Warren 1971). On each trial, monkeys watch as a reward (e.g. peanut) is hidden randomly in one of two food wells positioned to the left and right, in front of them. After a delay, during which a screen is lowered to prevent the monkey seeing the food wells, the screen is removed and the monkeys can retrieve the reward if they choose the well in which the reward was hidden. Whilst monkeys with removal of the prefrontal cortex, specifically the dorsolateral region (area 46 on Fig. 6.2b), succeed at 0 sec. delay, they fail the test with delays as short as 2–5 secs. These results have been replicated many times in a variety of experimental conditions, including localized cooling of the prefrontal cortex through an implanted probe, which produces a 'reversible lesion' (Fuster & Alexander 1970). Moreover, using a sophisticated version of the task, the delayed saccade procedure, Goldman-Rakic and colleagues have identified a map of visual space within the principal sulcus, such that permanent or temporary manipulations to localized areas within the principal sulcus disrupt memory guided eye movements to selective visual field locations (Funahashi *et al.* 1993). Goldman-Rakic has interpreted this deficit as one of working memory, in which the monkey is unable to hold 'on-line' a representation of the location of a goal such as a food morsel. With a task as simple as SDR it was somewhat of a surprise to find that humans with damage to the prefrontal cortex are also impaired on this test at delays of 10 secs. or more (Freedman & Oscar-Berman 1986).

In developmental studies Diamond (Diamond & Goldman-Rakic 1989) was one of the first to point out the similarities between SDR and the A not B test, the latter having been used by Piaget (1954) to demonstrate one of the classic markers of human developmental change during the second half of the first year of life. The only difference between the two tests is that in A not B the toy is not hidden randomly on every trial but is consistently hidden on one side until the infant chooses correctly and then the side of hiding is reversed and the procedure repeated. Between 7$^1/_2$ to 11 months, infants can retrieve the toy from the first well but when the side of hiding is reversed to B, they reach back to A (hence 'A not B'). The expression of the A not B error is dependent upon the length of delay between hiding and retrieval (Diamond 1990). Whilst infants at 7$^1/_2$ months produce the A not B error even at the shortest delay of 1 second, infants between 8 and 11$^1/_2$ months of age produce the error only after progressively increasing delays. By 12 months, infants can reach correctly at delays as long as 10 secs. (Fig. 6.5a). A similar delay dependent developmental

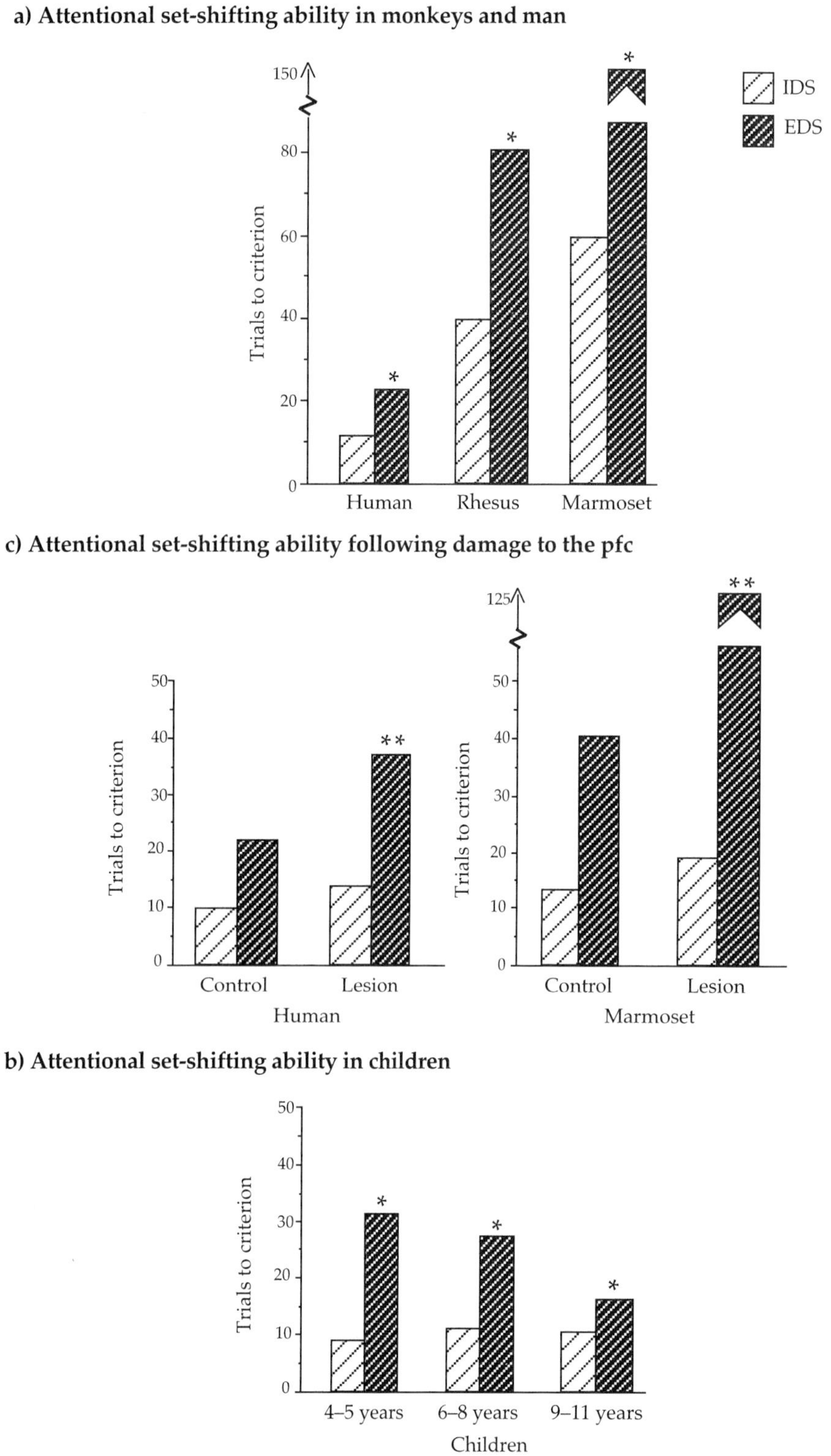

Figure 6.4. *Phylogenetic and ontogenetic studies of attentional set-shifting ability. Each graph plots the number of trials taken to reach criterion on a visual compound discrimination that either requires an intra-dimensional shift (IDS) or an extra-dimensional shift (EDS). (a) A comparison of the performance of New World monkeys (marmosets), Old World monkeys (rhesus) and humans. (b) A comparison of the effects of ablation of the prefrontal cortex (pfc) in humans and excitotoxic lesions of the pfc in marmosets. The marmosets in this study were given a 'hand-testing', i.e. more concrete (or less abstract), version of the attentional set-shifting paradigm rather than the 'computerized' version, resulting in fewer errors being made overall on both the IDS and EDS discriminations compared to the marmosets performing the 'computerized' version in (a). Although identical stimuli were used in these two versions, in the former the stimuli were presented on the front-face of two wooden boxes and the monkey was required to choose one of the boxes. If he chose the correct box then he was able to retrieve food reward from inside the box. In contrast, in the latter, the monkey was required to touch one of the two stimuli presented on the screen and if he touched the correct stimulus he received liquid reward from a licking spout. By making the response and the reward spatially compatible, as in the 'hand-testing' version, it is far easier for animals to learn the association between a response and reward than if the response and reward are spatially separate, as in the 'computerized' version. * indicates that EDS performance is significantly different from IDS performance. ** indicates that the lesion group is significantly different from the control group on the EDS. (c) A comparison of the performance of children at three different age groups. Drawn from data from the dissertation of Hughes (1993).*

progression in performance of A not B has also been reported in infant rhesus monkeys between $1^1/_2$ and 4 months of age (Fig. 6.5b). The improvement in the performance of both human and rhesus infants on this test has been suggested to reflect the development of the ability to inhibit a prepotent response tendency. However, given that this ability is delay dependent, it has been proposed that successful performance must also depend upon the ability to control action by information in memory, and that the length of time that information can be held 'on-line' increases progressively during development.

The similarities between the failure of infants to perform the A not B task and the failure of both monkeys and humans with prefrontal damage to perform SDR has led to the suggestion that the developmental changes in the performance of infants on the A not B task may be related to the maturation of prefrontal cortex. Certainly, the period over which improvement occurs corresponds to a rapid period of growth and myelination within the prefrontal cortex of both rhesus monkeys and humans. This has been substantiated by the finding that monkeys, not only display the characteristic A not B error following ablation of dorsolateral prefrontal cortex (Fig. 6.5c), but also produce a similar pattern of errors on SDR (Diamond 1990).

Social behaviour

One of the most characteristic features of damage restricted to orbitofrontal cortex in both monkey and man is reduced emotional responding and social withdrawal. The latter, in part, may be self-imposed or alternatively, inappropriate social behaviour on the part of the subject may lead to their being ostracized by the community. For example, the submissive and solicitation behaviour shown by intact female monkeys in response to the advances of a male is replaced either by indifference or aggression in females with orbitofrontal damage (Franzen & Myers 1973). Tragically, these females also show indifference towards their own offspring. A striking resemblance to these reports can be seen in the description of a patient, EVR, with damage restricted to ventromedial prefrontal cortex, who was characterized as 'incapable of either maintaining an enduring attachment to a sexual partner or functioning as a responsible parent' (Eslinger & Damasio 1985). These overt alterations in the patient's behaviour were found to be accompanied by abnormal responses in autonomic activity as shown by a complete abolition of his skin conductance response (SCR) to socially meaningful stimuli, e.g. slides depicting social disaster, nudity and mutilation (Damasio *et al.* 1990). The specificity of this effect was shown by his normal SCR response to unconditioned, 'orienting' stimuli, e.g. hand claps. Although such comparisons have not been made in monkeys with orbitofrontal damage, an early study did demonstrate a general reduction in the galvanic skin response (GSR) of monkeys in response to novel auditory stimuli, following ablation of the lateral orbital region (area 12 on Fig. 6.2b) (Kimble *et al.* 1965). These findings are consistent with the known connections of the orbital region of the prefrontal cortex with autonomic control structures including the amygdala as well as effector sites in the brainstem (Neafsey 1990).

Hierarchical *vs* hetararchical organization

Evidence has been provided in this chapter that two cognitive functions, attentional set-shifting and spatial working memory, as measured by SDR, are common to monkeys and man. Both these cognitive abilities are qualitatively similar across different primate species and lesions of structurally homologous regions of the frontal lobes have been shown to produce equivalent deficits. Monkeys and man have also been shown to exhibit a similar profile of abnormal responses in their social behaviour following damage to homologous regions of the prefrontal cortex, in this case, the ventromedial aspects. It remains then to consider the question originally posed at the beginning of this chapter as to how higher cognitive functioning is organized, both anatomically and psychologically, within the prefrontal cortex and whether this follows a *fixed* hierarchical arrangement, whereby there is a fixed *superordinate–subordinate* relationship between functional units, or, a *labile* hierarchical or hetararchical arrangement whereby the *superordinate–subordinate* relationship between functional units varies depending upon the context (see Gallistel 1980). Evolutionary studies of the neuronal organization and intrinsic connectivity of the prefrontal cortex would support the case for both types of arrangement. Within the archicortical (dorsal) and paleocortical (ventral) trends (Barbas & Pandya 1989), older areas are less differentiated than more recent areas and there is considerable reciprocity of connections between adjacent areas, suggesting that hierarchical organization exists *within* a trend. In contrast, connections between these two trends only tend to link areas that are at a *similar* stage of differentiation, suggestive of hetararchical organization. Whether it is possible to map cognitive functioning onto this

a) Performance of human infants on the A not B task

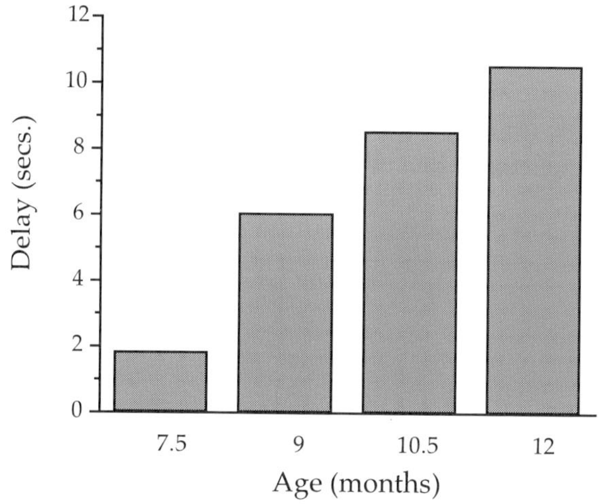

b) Performance of infant rhesus monkeys on the A not B task

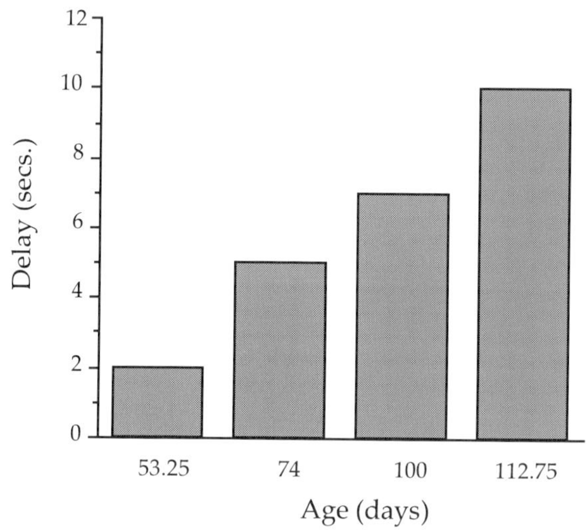

c) Performance of control and lesioned adult rhesus monkeys on the A not B task

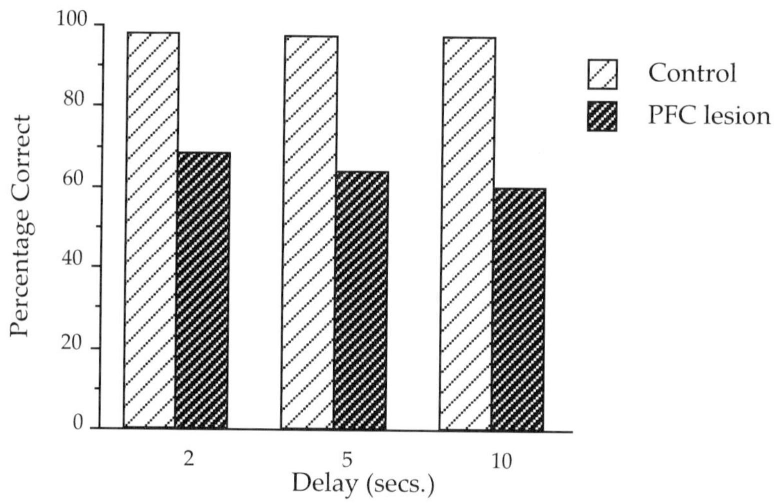

Figure 6.5. *Ontogenetic and phylogenetic studies of performance of the A not B task. (a) A comparison of the performance of children from seven and a half to twelve months of age indicating the length of the delay at which the A not B error is committed. (b) The same as in (a) except that the comparison is between the performance of infant rhesus monkeys from one and a half to four months of age. (c) The effects of ablation to the dorsolateral prefrontal cortex of rhesus monkeys on their performance of the A not B task. All graphs were redrawn from Diamond (1990).*

anatomical arrangement within the prefrontal cortex will be discussed below.

One of the clearest behavioural distinctions to arise from the studies of monkeys is that between damage to dorsolateral (area 46) *vs* lateral orbital (area 12) regions of prefrontal cortex. Whilst the latter is associated with perseverative responding on discrimination reversal paradigms (McEnaney & Butter 1969; Iversen & Mishkin 1970) the former impairs spatial delayed response performance. This has led to the proposal that these regions are mediating separate cognitive processes, namely, mnemonic processing in the case of dorsolateral prefrontal cortex and the inhibition of prepotent response tendencies, in the case of lateral orbital prefrontal cortex. Whilst impairments in attentional set-shifting have not, as yet, been localized in non-human primate studies, given the perseverative nature of this impairment it would be predicted that damage to lateral orbital rather than dorsolateral regions would be the critical factor. However, the cognitive processing required to shift between 'attentional sets' is of a higher order than that required to reverse specific stimulus-response or stimulus-reward associations in a discrimination reversal task. (As mentioned earlier, the latter do not require shifts of attention between perceptual dimensions as both stimuli remain relevant to the discrimination.) Thus, if attentional set-shifting, like discrimination reversal learning, is impaired by damage to lateral orbital prefrontal cortex, this would provide evidence of hierarchical processing within this ventral trend. Limited evidence from clinical studies, however, has attributed the impaired performance of patients with frontal lobe damage on the WCST to a loss of functioning in dorsolateral regions of prefrontal cortex (Milner 1963) and, in agreement with these data, recent studies using PET have shown increased metabolic activity within this area during performance of this test (Berman *et al.* 1986). Moreover, preliminary findings in the marmoset have suggested that deficits in set-shifting and reversal learning can be dissociated following selective lesions of the prefrontal cortex (Dias, Robbins and Roberts, unpublished data). If substantiated, then these findings suggest that the emphasis placed on the rigidity and perseveration of responding following damage to lateral orbital, but not dorsolateral regions, may be misplaced. Although definitive conclusions on the localization of attentional set-shifting impairments are not yet possible it is clear from the studies in monkeys with selective impairments to dorsolateral prefrontal cortex, that they too fail to inhibit a prepotent response on the A not B and SDR paradigms. However, impairment on these tests is also critically dependent upon the delay between hiding and retrieval. Indeed, in human and monkey developmental studies it has been shown that the impairment is observed *only* at progressively longer and longer delays (see Fig. 6.5a & b). Thus, it would appear that it is the development of the ability to control action by spatial information in memory that is responsible for the progressive improvement on these tests and that inhibiting a prepotent response is a *necessary* but *coincidental* factor. If this is the case then perhaps the perseverative responding that has been described following lateral orbital damage should also be considered in this light and another explanation of the impairments sought.

Recently, Damasio has proposed a theory of orbitofrontal function derived from the study of social behaviour in humans following damage to this region. He suggests that autonomic changes that usually accompany social interactions may be used as 'somatic markers' which at a later date, are re-activated under similar social circumstances and provide an advance signal of the affective outcome of particular responses (Damasio *et al.* 1991). As a consequence, certain actions are more or less likely to be chosen depending upon whether they induce a negative or positive somatic state. It is proposed that damage to the orbitofrontal cortex results in the failure of these somatic markers to be reactivated, thus depriving the individual of an automatic device to signal the longer term consequences of their actions. Thus their responses are more likely to be governed by immediate goals, resulting in impulsivity as well as deficits in longer term decision-making and planning, as seen in patients like EVR. Such an interpretation may also explain the perseverative responding of monkeys with lateral orbital damage on a variety of discrimination reversal paradigms. For example, at the time of a reversal the expected outcome, i.e. delivery of reinforcement, fails to occur and this may normally result in the animal experiencing a negative somatic state which may act to trigger a rapid change in their responding. If the event is not accompanied by this negative somatic state, as may be the case in monkeys with lateral orbital damage, then the reduced negative impact of the reversal will slow down the process responsible for changing the current response pattern, resulting in prolonged perseverative responding. Consistent with this theory is the discovery of 'error-related' cells in the orbitofrontal cortex of monkeys that fire to the unexpected negative outcome following the reversal of reinforcement contingencies in a go–no go paradigm (Thorpe *et al.* 1983). Together with another subset of cells in this region that fire selectively to specific stimuli when they are associated with reinforcement, the orbitofrontal cortex is capable of *predicting*, as well as *inducing*, the affective outcome of a particular response in advance of it being made and as a consequence is also well placed to detect the occurrence of an unexpected or unpredicted event, resulting in the termination of the current response plan and the initiation of another.

The differential effects then that follow damage to dorsolateral and lateral orbital regions of prefrontal cortex may be best understood in terms of the information that these two regions process. Both of these areas are involved in the formulation of action plans, but whereas this is guided by spatial information in memory within dorsolateral prefrontal cortex, it is guided by affective information in memory in orbital prefrontal cortex. Other areas of prefrontal cortex, such as the dorsomedial region may be involved in controlling action plans that depend upon kinaesthetic (Passingham 1978) information. All these regions may then work either in parallel, or in an integrated fashion to select and execute the most appropriate action plan. This may take the form of a single response, as required in a discrimination reversal paradigm or a complex sequence of responses that require temporally continuous control directed towards a single goal, as in the Tower of London task. The action plan may also be prospective in nature, involving the scheduling of multiple goals, as for example in the planning and implementation of a dinner party. Clearly these more complex action plans will require multiple cognitive operations using multiple types of information and as a consequence will also require integrative control. Thus the complexities of social behaviour can be governed by both emotional (affective) and cognitive processes that are mediated by the ventral and dorsal trends within the prefrontal cortex respectively. Although these two regions may often act together to *facilitate* the selection of an outcome, under certain circumstances they may also *compete* for the selection of an outcome. Thus the final selection may well be the result of a mixture of competition and integration between both regions. Of relevance to this issue are the studies of Damasio *et al.* (1990) described above, which measured the skin conductance response of subjects with ventromedial prefrontal damage (areas 10, 11 & 13) during the presentation of socially meaningful stimuli. Although these subjects failed to show a rise in their SCR whilst they were passively viewing the slides, this abnormal response could be partially reversed if the subjects were asked to report the content of the slides and provide an impression of their feelings towards them. This 'cognitive' condition illustrates that the ability to generate autonomic responses has not been completely compromised by the lesion but their mechanism of induction has been altered. The contribution made by cognition to social behaviour may also explain why EVR is able to access appropriate social knowledge when questioned about it verbally. However, EVR did report that he had not experienced the kind of feeling that he thought he ought to have in relation to the stimuli on the slides (even in the 'cognitive' condition), suggesting that without the orbitofrontal cortex, there is a blunting of affect or emotion independent of changes in autonomic activity.

In summary, from phylogenetic and developmental studies it would appear that the prefrontal cortex is made up of at least two separate moieties or trends. Whilst emotional processing has developed along the ventral 'trend', cognitive processing has developed along the dorsal 'trend'. During development this has probably resulted in the elaboration of cognitive and emotional abilities within each trend. For example, whilst dorsolateral cortex enables a monkey to hold spatial information in mind over a considerable time period as required in the spatial delayed response task, in humans it mediates the complex spatial planning requirements of the Tower of London. Similarly, whilst orbitofrontal cortex can support the behavioural interactions required in a social group of monkeys, it mediates the refined emotional responses that are seen in humans. As these two moieties have developed, so too has the need for integration of their functions and this is probably reflected in the connectivity between trends of areas at a similar stage of differentiation. Thus, this chapter has illustrated how by taking a comparative neuropsychological, as well as comparative neurobiological approach to the study of prefrontal cortex some of the boundary conditions may be defined for the evolution of mind in early hominids.

Angela C. Roberts, Peter Collins & Trevor W. Robbins
Department of Experimental Psychology
University of Cambridge
Downing Site
Cambridge
CB2 3EB

References

Barbas, H. & D.N. Pandya, 1989. Architecture and intrinsic connections of the prefrontal cortex in the rhesus monkey. *Journal of Comparative Neurology* 286, 353–75.

Berman, K.F., M.D. Zec & D.R. Weinberger, 1986. Physiologic dysfunction of dorsolateral prefrontal cortex in Schizophrenia. *Archives of General Psychiatry* 43, 126–35.

Brodmann, K., 1912. Neue ergebneisse über die vergleichende histologische lokalisation der grosshirnrinde mit besonderer berücksichtigung des stirnhirns. *Anatomischer Anzeiger* (Suppl.) 41, 157–216.

Damasio, A.R., D. Tranel & H.C. Damasio, 1990. Individuals with sociopathic behavior caused by frontal damage fail to respond autonomically to social stimuli. *Behavioural Brain Research* 41, 81–94.

Damasio, A.R., D. Tranel & H.C. Damasio, 1991. Somatic markers and the guidance of behavior: theory and preliminary testing, in *Frontal Lobe Function and Dysfunction*, eds. H.S. Levin., H.M. Eisenberg & A.L. Benton. Oxford: Oxford University Press, 217–29.

Diamond, A., 1990. The development and neural bases of memory functions as indexed by the AB and delayed response tasks in human infants and infant monkeys, in 'The development and neural bases of higher cognitive functions'. *Annals of the New York Academy of Sciences* 608, 266–317.

Diamond, A. & P.S. Goldman-Rakic, 1989. Comparison of human infants and rhesus monkeys on Piaget's AB task: evidence for dependence on dorsolateral prefrontal cortex. *Experimental Brain Research* 74, 24–40.

Divac, I. & J.M. Warren, 1971. Delayed response by frontal monkeys in the Nencki testing situation. *Neuropsychologia* 9, 209–17.

Eslinger, P.J. & A.R. Damasio, 1985. Severe disturbance of higher cognition after bilateral frontal ablation : patient EVR. *Neurology* 35, 1731–41.

Franzen, E.A. & R.E. Myers, 1973. Neural control of social behaviour: prefrontal and anterior temporal cortex. *Neuropsychologia* 11, 141–57.

Freedman, M. & M. Oscar-Berman, 1986. Bilateral frontal lobe disease and selective delayed esponse deficits in humans. *Behavioural Neuroscience* 100, 337–42.

Funahashi, S., C.J. Bruce & P.S. Goldman-Rakic, 1993. Dorsolateral prefrontal lesions and oculomotor delayed-response performance: evidence for mnemonic 'scotomas'. *Journal of Neuroscience* 13, 1479–97.

Fuster, J.M., 1989. *The Prefrontal Cortex*. New York (NY): Raven Press.

Fuster, J.M. & G.E. Alexander, 1970. Delayed response deficit by cryogenic depression of frontal cortex. *Brain Research* 20, 85–90.

Gallistel, C.R., 1980. *The Organisation of Action: a New Synthesis*. Hillsdale (NJ): Lawrence Erlbaum Associates.

Hughes, C., 1993. Executive Dysfunction in Autism. Unpublished Ph.D. dissertation, University of Cambridge.

Hughes, C., J. Russell & T.W. Robbins, 1994. Evidence for executive dysfunction in autism. *Neuropsychologia* 32, 477–92.

Iversen, S.D. & M. Mishkin, 1970. Perseverative interference in monkeys following selective lesions of the inferior prefrontal convexity. *Experimental Brain Research* 11, 376–86.

Jacobsen, C.F., 1935. Functions of the frontal association area in primates. *Archives of Neurology and Psychiatry* 33, 558–69.

Jonides J., E.E. Smith, R.E. Koeppe, E. Awh, S. Minoshima & M.A. Mintum, 1993. Spatial working memory in humans as revealed by PET. *Nature* 363, 623–5.

Kimble, D.P., M.H. Bagshaw & K.H. Pribram, 1965. The GSR of monkeys during orientation and habituation after selective partial ablations of the cingulate and frontal cortex. *Neuropsychologia* 3, 121–8.

Kostovic, I., 1990. Structural and histochemical reorganisation of the human prefrontal cortex during perinatal and postnatal life. *Progress in Brain Research* 85, 313–24.

McEnaney, K.W. & C.M. Butter, 1969. Perseveration of responding and non-responding in monkeys with orbital frontal ablations. *Journal of Comparative and Physiological Psychology* 68, 558–61.

Miller, E., 1984. Verbal fluency as a function of a measure of verbal intelligence and in relation to different types of cerebral pathology. *British Journal of Clinical Psychology* 23, 53–7.

Milner, B., 1963. Effects of different brain lesions on card sorting. *Archives of Neurology* 9, 90–100.

Milner, B. & M. Petrides, 1984. Behavioural effects of frontal-lobe lesions in man. *Trends in Neurosciences* 7, 403–7.

Mishkin, M., 1964. Perseveration of central sets after frontal lesions in monkeys, in *The Frontal Granular Cortex and Behaviour*, eds. J.M. Warren & K. Akert. New York (NY): McGraw-Hill, 219–41.

Mishkin, M., E.S. Prockop & H.E. Rosvold, 1962. One-trial object-discrimination learning in monkeys with frontal lesions. *Journal of Comparative and Physiological Psychology* 55, 172–81.

Morris, R.G., S. Ahmed, G.M. Syed & B.K. Toone, 1993. Neural correlates of planning ability: frontal lobe activation during the Tower of London test. *Neuropsychologia* 13, 67–78.

Mrzljak, L., H.B.M. Uylings, C.G. Van Eden & M. Judas, 1990. Neuronal development in human prefrontal cortex in prenatal and postnatal stages. *Progress in Brain Research* 85, 313–24.

Neafsey, E.J., 1990. Prefrontal cortical control of the autonomic nervous sytem:anatomical and physiological observations. *Progress in Brain Research* 85, 147–66.

Owen, A.M., J.D. Downes, B.J. Sahakian, C.E. Polkey & T.W. Robbins, 1990. Planning and spatial working memory following frontal lobe lesions in man. *Neuropsychologia* 28, 1021–34.

Owen, A.M., A.C. Roberts, C.E. Polkey, B.J. Sahakian & T.W. Robbins, 1991. Extra-dimensional versus intra-dimensional set-shifting performance following frontal lobe excisions, temporal lobe excisions or amygdalo-hippocampectomy in man. *Neuropsychologia* 29, 993–1006.

Passingham, R.E., 1972. Non-reversal shifts after selective prefrontal ablations in monkeys (*Macaca mulatta*). *Neuropsychologia* 10, 41–6.

Passingham, R.E., 1978. Information about movements in monkeys (*Macaca mulatta*) with lesions of dorsal prefrontal cortex. *Brain Research* 152, 313–28.

Penfield W. & J. Evans, 1935. The frontal lobes in man: a clinical study of maximum removals. *Brain* 58, 115–33.

Petrides, M., B. Alivisatos, A.C. Evans & E. Meyer, 1993. Dissociation of human mid-dorsolateral from posterior dorsolateral frontal cortex in memory processing. *Proceedings of the National Academy of Sciences* 90, 873–7.

Piaget, J., 1954. *The Construction of Reality in the Child*. New York (NY): Basic Books.

Roberts, A.C., in press. Comparison of cognitive function in human and non-human primates, in *Models in Cognitive Behavioural Pharmacology*, Special issue of *Cognitive Brain Research*, eds. T. Steckler & G.D. D'Mello. Amsterdam: Elsevier.

Roberts, A.C., T.W. Robbins & B.J. Everitt, 1988. The effects of intradimensional and extradimensional shifts on visual discrimination learning in humans and non-human primates. *Quarterly Journal of Experimental Psychology* 40, 321–41.

Roberts, A.C., T.W. Robbins, B.J. Everitt & J.L. Muir, 1992. A specific form of cognitive rigidity following excitotoxic lesions of the basal forebrain in marmosets. *Neuroscience* 47, 251–64.

Roberts, A.C., M.A. De Salvia, L.S. Wilkinson, P.C. Collins, J.L. Muir, B.J. Everitt & T.W. Robbins, 1994. 6-Hydroxydopamine lesions of the prefrontal cortex in monkeys enhance performance on an analog of the wisconsin card sort test: possible interactions with subcortical dopamine. *Journal of Neuroscience* 14, 2531–44.

Rosvold, H.E., M.K. Szwarcbart, A.F. Mirsky & M. Mishkin, 1961. The effect of frontal-lobe damage on delayed response performance in chimpanzees. *Journal of Comparative Physiological Psychology* 54, 368–74.

Shallice, T., 1982. Specific impairments of planning. *Philosophical Transactions of the Royal Society*, London (Biology) 298, 199–209.

Shallice, T. & P.W. Burgess, 1991. Deficits in strategy application following frontal lobe damage in man. *Brain* 114, 727–41.

Thorpe, S.J., E.T. Rolls & S. Maddison, 1983. The orbitofrontal cortex: neuronal activity in the behaving monkey. *Experimental Brain Research* 49, 93–115.

Verin, M., A. Partiot, B. Pillon, C. Malapani, Y. Agid & B. Dubois, 1993. Delayed-response tasks and prefrontal lesions in man-evidence for self-generated patterns of behaviour with poor environmental modulation. *Neuropsychologia* 31, 1379–96.

7

The Anatomical Foundations of Cognition: Suggestions for a Reinterpretation

Elizabeth Whitcombe

Anatomical models of the human mind are conventionally drawn from the reference maps of higher cerebral functions currently in use in neuroscience. These maps are based on a conception of cortical topography which arises from the doctrine of aphasia: the localization of higher functions of speech or language by lesion to specific higher motor and sensory speech areas, Broca's and Wernicke's, in the human cortex. This paper suggests that Broca's and Wernicke's hypotheses, their premises and the methods used to demonstrate them, are anachronistic and that the history of aphasia to date has failed to substantiate these nineteenth-century assumptions. An alternative interpretation is proposed, in which speech may be considered in its biological context, as a form of 'animal signalling' exploiting, by functional adaptation, the distinctive properties of human physiological anatomy, peripheral structures together with their corresponding neural organization which serve the fundamental biological functions of breathing, feeding, hearing and seeing.

The anatomy of mind, of cognition, is commonly represented by a map of the human brain, or rather the cortex, on which certain cognitive functions are localized as follows: (1) areas of higher — volitional or intentional — motor function and motor-programming in the frontal cortex, distinguished from the four motor areas acknowledged at present, viz. primary ('excitable') motor, premotor and prefrontal cortex and supplementary motor area; (2) areas of higher sensory function — visual, auditory, somatosensory perception distributed over the temporo-parieto-occipital cortex, adjacent to the primary sensory areas; (3) higher speech, or language, areas, demarcated on the left hemisphere: a Broca's, or anterior cortical speech area, including but not necessarily confined to the posterior part of the inferior frontal gyrus anterior to, but excluding the inferior primary motor cortex; and a Wernicke's, or posterior cortical speech area, located, variously, in the posterior superior temporal gyrus and/or supramarginal and angular gyri.

While these essential features are found in all versions of the map of cortical functions, no precise demarcation has yet been agreed on. Nonetheless, the features themselves, however they may be represented, are commonly taken, together with the relative size of the human brain in relation to that of other primates, as the indicators of human cognitive status. Steps in the evolution of human cognition are traced through the fossil record by calculating brain size from volumetric measurement of the neurocranium, and by examining endocasts for the presence or absence of features to which some cognitive significance is conventionally assigned. These methods have certain shortcomings. The interpretation of endocasts, despite advances in technical sophistication, remains hampered by the sketchy and incomplete record of the cerebral surface on the inner surface of the calvarium: 'it must be remembered that detail of the sulci and gyri of the brain is not reflected in the brain capsule' (DuBrul 1988, 8). To date, the neuroanatomy of the human cerebral cortex is quite unknown (Crick & Jones 1993). The functional map is a temporary expedient, a compilation of two sets of data: extrapolations from animal experiments, and the 'topographical pathology' of aphasia and associated disorders of higher cerebral function. These two sets of data have not yet been consistently reconciled. How sensorimotor and higher sensorimotor functional anatomy are related has not been explained; neither the anatomical connectivity, intracortical or cortico-subcortical, has been elucidated nor the precise neurophysiological operations described; the core of the cognitive map, viz. the anterior and posterior speech areas, are derived from the doctrine of aphasia; the neuroanatomical organization of speech is not known (Lecours 1975).

The history of this map is instructive. Its origins lie, with the doctrine of aphasia, in phrenology, in the 'principle of localization' propounded at the turn of the nineteenth century by F.-J. Gall to explain the anatomical basis of human mind and human emotion. This early essay in materialism purported to dispose of Cartesian dualism. To eradicate the distinction between an incorporeal 'mind' and 'body', the materialist 'mind' was incorporated, not into body as such, but into brain, the uniquely human brain, abstracted from the nervous system and identified as an autonomous organ, analogous to the liver, or the heart. The distinguishing structural features of the brain, its 'convolutions' (gyri), were identified as 'cerebral organs', the seat, according to Gall's version of this doctrine, of 27 fundamental faculties. Emotional faculties were distributed over the posterior convolutions, intellectual faculties over the anterior. Faculty XV of Gall's system, the 'faculty of articulate language', or speech, the pinnacle of human intellect, was localized to the inferior frontal lobes (Gall & Spurzheim 1818; 1810).

Gall's methods of localization, by skull bumps and quirks of physiognomy, while fashionable in his time, brought phrenology into scientific disrepute. Contrary to common supposition, the essence of Gall's doctrine, his 'principle of localization' survived intact, refined by his contemporary and disciple, Bouillaud: intellectual faculties were localized in the anterior lobes; proof of the localization of the 'faculty of articulate language' alone would suffice to validate the principle. 'Articulate language', as Bouillaud conceived of it, was organized on two discrete levels: a lower, executive faculty served by the nerves and muscles of speech presided over by a higher, governing 'nervous principle', the 'faculty of articulate language'. Pathology, in cases of loss of speech from trauma, infection or vascular accident, was to provide the precise localization: the seat of function localized to the site of lesion (Bouillaud 1825). Bouillaud himself failed in his attempts to identify the exact localization of this higher faculty. In 1861, Broca, a surgeon and distinguished craniologist expressly committed to the phrenological 'principle of localization', in Bouillaud's version, provided the requisite demonstration: a patient, Leborgne, with a history of loss of speech, except for a monosyllable tan, a progressive right hemiplegia and terminal gangrene. Broca defined Leborgne's loss of speech, in the presence of an intact general intelligence, as the loss of an isolated intellectual faculty, the 'faculty of articulate language'. This disorder was so extraordinary as to warrant a special name, aphemia (changed in 1863 to aphasia), which distinguished it sufficiently from the 'mere' motor symptoms of hemiplegia. Broca interpreted the extensive perisylvian lesion in the left hemisphere demonstrated postmortem as having progressed in two distinct phases: confined, in the early phase to the posterior inferior frontal convolutions, the organs of intellect, it had subsequently progressed deep to the convolutions to involve the corpus striatum, that part of the basal ganglia identified, incontemporary functional anatomy, as 'the highest motor organ'. Broca then matched the phases of his pathology and symptomatology: early, 'frontal' with loss of speech, fixing a focus of origin for the lesion in the posterior one-third of the inferior frontal convolutions, 'probably the third'; late, involvement of the *corpus striatum*, with right hemiplegia. Broca was explicit in his method and his premises: the convolutions were not formed by the mere working of natural forces, but so as to function as organs with specific functions, the functions of the intellect; 'there is no one who does not know that the convolutions are not motor organs' (Broca 1861). Broca's description of the appearances of Leborgne's brain, preserved at the Salpetriere, has now been verified, and supplemented, by CT-scannography (Castaigne *et al.* 1980). These findings do not of themselves, however, substantiate his interpretation.

In 1870, the functional anatomy on which Broca had relied was displaced by new experimental findings. Fritsch & Hitzig (1870) demonstrated by stimulation in the dog the presence of an 'excitable motor cortex'. It could no longer be said, with Broca, that the convolutions were 'not motor organs'. Wernicke, however, who created the classical doctrine of aphasia, retained Bouillaud's principle from conviction, accepted Broca's localization of it and his anatomical thinking, and translated the old phrenological conceptions of lower and higher functional departments into Meynert's new anatomical theory of projection and association centres and pathways. In the light of the new experimental findings of reflex circuitry which included the cortex, Wernicke devised a scheme, a 'mechanism', for the 'the elementary psychological function of speech' as a 'form of volitional movement': not one but two reflex arcs, a 'physiological reflex arc', such as the experimentalists demonstrated, by which the mechanical, sensory and motor functions of speech were transacted, and a 'psychological reflex arc' for the higher, volitional element in speech. Broca's area — Wernicke accepted Broca's probable localization, left posterior F3, as definitive — constituted the psychomotor word-image centre, adjacent to but distinct from the inferior motor cortex serving the oro-facial 'speech organs'. The scheme required a psychosensory word-image centre as counterpart; Wernicke placed it, *ex hypothesi*, adjacent to the auditory cortex, with no formal demarcation since unlike the primary motor cortex, the primary auditory cortex had yet to be described. How the scheme of psychological and physiological reflex arcs operated, and co-operated, in the act of speaking and of hearing speech, was not explained. Wernicke concluded his demonstration with a series of 10 cases, not one of which substantiates his scheme. Only cases 2, 5, 8 and 10 came to autopsy. In case 10, the lesion was a diffuse abscess, which ruled out localization. In case 5, the history of both left and right strokes was corroborated by bilateral lesions; the patient was described as completely speechless, but the left lesion involved the first and second temporal gyri and 'Broca's area' was reported as intact. In case 8, described as 'sensory aphasia', comprehension was lost but bilateral deafness was suspected, while the patient's speech was limited to 'ja' and movements of the face and tongue were impaired; both hemispheres were atrophied, with extensive softening throughout the perisylvian region (Wernicke 1874, 45–67).

Confirmation from clinical observation was still confidently expected when Wernicke produced his last definitive statement on aphasia, in 1903, illustrated by an apparently definitive anatomy. This consisted of a diagram of the lateral aspect of the left hemisphere with 'language areas' marked on it, Broca's, Wernicke's and a 'visual word-image centre' over the supramarginal and angular gyri, together with idealized coronal and axial sections of language areas and pathways as they might appear if they existed, from the work of von Monakow and Dejerine (Wernicke 1903). Wernicke's scheme was not physiology, or psycho-physiology but consisted, as Adolf Meyer recognized, of verbal categories described in physiological terms (Meyer 1905). The anatomy with which he had correlated it was not a functional anatomy of speech but a topographical pathology, idealized, of aphasia. The two were, and have remained, confused. Wernicke's pupil Liepmann extended the scheme anteriorly, to embrace all higher volitional movement, other than speech. Liepmann did as Broca had done; in a celebrated case of progressive paralysis, 'Herr T.', he separated and distinguished early symptoms from late. The first, loss of fine, skilled, learned movements he called *apraxia*, loss of the higher, ideational faculty of volitional movement (Liepmann 1900). Herr T. went on to demonstrate, over five years, many symptoms, amongst which a progressive hemiplegia figured prominently. The lesions at postmortem were diffuse and defy any attempt to establish a convenient sequence (Liepmann 1905). Meanwhile Freud had given a collective term, *agnosia*, to the higher sensory disorders, of perception (other than the perception of speech) as distinct from sensation (Freud 1891), conceived of by the extension of aphasia, posteriorly, through the temporo-parieto-occipital region.

Cartesian dualism had not been abandoned, merely translated into the new cerebral physiology, psychosensory-motor and sensory-motor centres. The localization of higher speech, and higher functions in general according to the classical doctrine of aphasia, with apraxia and agnosia was almost universally accepted, and sustained for more than 50 years without any agreement being reached as to where the actual anatomical boundaries were to be drawn. Developments in histology and cytoarchitectonics complicated the issue further, as did the corticography of Penfield and Foerster in the 1920s and 1930s. There were now many maps, without corresponding demarcations.

Wernicke's scheme, in its simplicity and ambiguity, had in practice necessitated endless qualifications and amendments. Geschwind, however, saw in the scheme itself the solution to the confusion it had caused (Geschwind 1967). He revived it, and modernized its anatomical description: Broca's area, in the left inferior frontal gyrus, somewhat posteriorly; Wernicke's area in the posterior superior temporal gyrus; a cross-modality association centre, the seat of a mental function Geschwind saw as uniquely human, and the foundation of language, located to gyral regions found exclusively in the human brain, the left *planum temporale* and adjacent parts of the angular and supramarginal gyri (Geschwind 1965). The asymmetry of the left planum temporale was taken as evidence of its functional significance vis-à-vis language (Geschwind & Levitsky 1968). Wernicke's area was connected in this scheme with the cross-modality association region and, by a specific fibre-bundle, the arcuate fasciculus, with Broca's area (Geschwind 1970). Clinical and pathological evidence has again proved equivocal. A case of carbon monoxide poisoning, in which the patient could not speak spontaneously but repeated 'echolalically', could complete stereotyped phrases and learn the words and music of songs new to her, was interpreted as demonstrating the 'isolation of the speech area', since, it was argued, from findings *postmortem*, that 'Wernicke's area (i.e. the left posterior superior temporal gyrus)' was preserved, along with primary auditory cortex, and arcuate fasciculus despite much parietal damage, and that 'Broca's area appears nearly normal' (Geschwind *et al.* 1968). But there is no agreement as to where, exactly, Broca's area is, still less Wernicke's, it would seem difficult to argue for their isolation by the many and diffuse anoxic lesions of this case. The 'speech-area' has continued to prove elusive. Mohr's solution, reviewing hundreds of cases of 'Broca's aphasia', argued for a revised description, for an 'anterior' aphasia, and an 'anterior' speech area (Mohr *et al.* 1978). The Bogens listed many of the versions of Wernicke's area current in the 1970s (Bogen & Bogen 1976). Their solution was to go back to the perisylvian 'confluent language zone' of the classical school; back, that is, to the 'primordial convolution surrounding the Sylvian fissure' of the pre-experimental era of cortical anatomy.

The description of cerebral lesions has undoubtedly been improved by advances in imaging, but, in the search for the anatomy of cognitive functions by way of pathology, it is lesion itself which is the limiting factor, an unreliable means, *a priori*, of localization. While a focal lesion may be the direct cause of a disorder, by damaging a specific functional area, the disorder, as clinically observed, may be the result of a lesion's blocking of neural transmission into, or out of the area, from a site up- or downstream, or, specifically, the blocking of excitatory impulses to cortical functional units, or the cutting off of their vascular supply; or the disorder observed might be abnormal 'release' phenomena, from the blocking by lesion of inhibitory impulses (Carpenter 1990, 17).

New techniques of encephalography and dynamic imaging have raised expectations that the problems which have beset the localization of cognitive functions might now be solved. This would seem premature. Despite the remarkable sophistication of the new machinery, its powers of resolution are as yet insufficiently precise for the interpreter to determine the exact significance of the activity demonstrated. As Raichle acknowledges in a recent review (1994), PET and MRI scanning are based on regional cerebral blood flow and oxygenation respectively, the rates of which vary between some hundreds of milliseconds and several seconds; the rate of neural transmission in cortical networks may be 0.01 msecs. or less. Spatial resolution also presents difficulties; the relation of the schematic hemisphere, as imaged, to actual connectivity is obscure. It is unclear, therefore, what the patterns of cortical activity recorded during the performance of experimental tasks actually represent. In the case of cognitive tasks, further difficulties arise from the variety of interpretations of cognitive processing current in neuropsychology. No agreement has yet been reached as to which model is to be used in the design of the experiment, nor which version of the map of cerebral 'language-areas' in the interpretation of the results. Recent experiments in which normal subjects spoke, and silently 'generated' nouns and verbs respectively demonstrated different patterns of 'cortical' activity associated with each task (Raichle 1994). Claims that these patterns provide evidence of anatomical correlates of grammatical categories would seem questionable. Such categories are notoriously unstable. Linguistic indeterminacy must be allowed for (Matthews 1991, 54–9); the shifting of lexical and morphological boundaries over historical time would suggest any strict neuroanatomical correlation is at best doubtful. Neural activity recorded by PET or MRI in association with the spoken word may be related to specific speech-events, reflecting tonal complexity or prosodic contour, and these may be markers of grammatical class, but this cannot be determined owing to the lack of synchrony between recording and performance. It is well recognized that patterns of activity — 'patterns of spots' — elicited by volitional motor tasks and cognitive tasks overlap and that consequently neither can be determined with confidence (Passingham 1991). We may ask therefore whether, and how far, the 'spots' recorded for tasks on spoken language may represent the physical, sensorimotor activity of the neural apparatus controlling the structures of the head, neck and trunk employed in speech; and, by way of corollary, how much of the activity recorded in association with silent cognitive tasks represents the activity of that same controlling apparatus when not engaged in speech but maintaining respiratory and digestive functions and the vigilance of special visual and auditory sensing.

It may be useful to follow the lead given by the first experimentalists in central motor function,

Fritsch, Hitzig, Goltz and others. Rather than abstracting the cerebral hemispheres, or the cortex, from their context in the nervous system and the body it regulates, may we not restore them to continuity not only with the brainstem and cord but also with the peripheral end-organs which determine the specific character of sensory and motor modalities. Structural features of the cortex of cerebral, and cerebellar hemispheres, flexures and gyrification, inexplicable in terms of conventional localization, may be more profitably interpreted in the light of regional growth and development within the central nervous system (Welker 1990). Function may be seen as the property not of some discrete focal area, or some fraction of connectivity of cortex, brain or central nervous system; but rather as a property of anatomical systems, which consist of specialized peripheral end-organs, receptors and effectors, together with their *corresponding* neural organization, peripheral and central. Cajal recognized the determining character of peripheral receptors in functional neuroanatomy, and how their highly specialized properties could explain specific features of central neural organization: the partial decussation of the visual pathway at the chiasm found in most mammals could be explained by the image-reversing property of the lens, the function of the decussation being to 'rectify' the reversed image for correct integration (Guillery 1994). Experimental findings suggest that the degree of correspondence between peripheral receptors (and, we may assume, effectors) and their corresponding functional units in the central nervous system varies, amongst species, and in relation to the habits of the animal. In star-nosed mole, for example, tactile receptors are aligned in rays around the nostrils, each ray corresponding to a specific stripe formed by a multicellular functional unit in the somatosensory cortex (Catania *et al.* 1993). Similar findings of a 1:1 correspondence were established for the vibrissae in certain rodents; in oppossum, however, the correspondence is much more approximate (Woolsey *et al.* 1975).

It would seem not unreasonable to see in functional anatomy, in each of any creature's systems of specialized peripheral end-organs and corresponding neural organization, a 'series of anatomical possibilities', as Hughlings Jackson put it, ranging from 'narrow' to 'broad', or 'cultivated'. The 'narrow' range of the series, or spectrum, consists of sensori-motor systems specialized in structure, with species-specific modifications, to serve physiological functions essential to life: posture and locomotion, breathing, feeding, excretion, reproduction, sensing, for example. The 'broad' range represents the *functional adaptations* of these structural arrangements, skilled, learned routines devised to exploit their properties. We see how the structural properties of the human foot, for example, exquisitely specialized for bipedal locomotion can be exploited, by systematic training for the extraordinary manoeuvres of ballet-dancing (Alexander 1992, 57). It would surely be perverse to search for a ballet-dancing centre in the cerebral cortex; or for swimming or bicycling centres and therefore, as Pierre Marie argued (1908), why look for cortical centres of that most highly trained sensori-motor routine, speech? A *Virtuosität*, was how Wernicke described it, common to all normal humans, but acquired only after years of laborious training.

What is commonly called the peripheral 'vocal tract' is a misnomer. The structures of the human head, neck and upper trunk utilized in the production of voice and speech are primarily specialized for the fundamental functions of breathing and feeding, each peripheral receptor and effector component with its corresponding neural organization, from periphery to cerebral cortex. The valvular larynx originates in a protective mechanism for the airway, much modified in primates where it serves as a mechanism for fixing the thorax by the momentary arrest of respiration as a stable base for the act of prehension (Negus 1929). The hyoid bone is part of the stabilizing apparatus of the cranio-vertebral junction, and of the tongue (DuBrul 1988). The sensorimotor organization of the oro-naso-pharyngeal cavities, the 'aero-digestive crossroads' (Kelemen 1938) serve the functions of feeding — chewing and swallowing — in co-ordination with breathing. Exploration of the cortical connectivity active in respiration is in its infancy. The cortical circuitry in the control of automatic respiratory function, through chemoreceptors, is thought, from experimental evidence in non-human primates, to be located in the inferior frontal, supraorbital, region (Wall *et al.* 1951). Recent experiments in human volitional respiration with PET imaging have demonstrated activity in the middle and inferior reaches of premotor and primary motor cortex and supplementary motor areas; activity in the expiratory phase was recorded in Penfield's vocalization areas adjacent to and overlapping with motor cortex (Ramsay *et al.* 1993). This is readily explicable: human speech utilizes only the expiratory phase. The cortical connectivity associated with chewing and swallowing has yet to be defined. Foerster, however, in early experiments in corticography in epileptic patients elicited rhythmic chewing movements and swallowing sequences from the inferior frontal cortex, roughly where the classical, circumscribed version of Broca's area might lie (Foerster 1936).

Phonation, the use of purposive sound, may thus be seen as an adventitious functional adaptation of respiratory apparatus, peripheral organs co-ordinated under central neural control, monitored by the auditory system; resonance and amplification the adaptation of specializations or swallowing in the oro-naso-pharynx; articulation, similarly, as functional adaptation of mandibular and labial, palatal and above all lingual structures primarily specialized for feeding. When and where the peripheral structural specializations of the head and neck characteristic of the modern human, together with their neural controls, evolved we may say speech became possible. It may well have been developed in rudimentary form in many versions, many now lost, over many millennia, as the facial compartment took its distinctively modern shape: strikingly retruded, with shortened jaw, the modification in masseter insertion securing maximum efficiency, and thereby speed, in elevation and depression, while the corresponding modification in tongue insertion permitted a greatly increased range of movement within an enlarged oral cavity (DuBrul 1988; Duchin 1990). The characteristic modern modifications of the functional anatomy of the facial compartment are associated with an enlargement of the neurocranium and, in turn, with the development of erect, bipedal posture and locomotion; the nature and timing of the associations remains open to speculation. It would not be illogical to interpret the increase in brain-size, that is, the enlargement of the anterior compartment of the central nervous system, and of fronto-parietal regions in particular deduced from changes in the the neurocranium, in the context of these developments in sensori-motor functional anatomy of the head and neck, the independent upper limbs and hands and lower limbs specialized, with most complex and extensive postural arrangments, for bipedalism.

It could well be that speech and spoken language has developed from routines devised to exploit the vocal and articulatory possibilities of breathing and feeding arrangements peculiar to the human species, rather than arising mysteriously as some 'faculty' with which the human brain is uniquely endowed. If the organizing principles of functional anatomy, in a continuum from periphery to centre and *vice versa* can be clarified, we are then in a stronger position to interpret the effects of lesion, and particularly the relation of lesion to disorder. At the cerebral, cortical-subcortical level, of sensori-motor integration, lesion would appear to bring cause disintegration. Take, for example, the classic cases where patients cannot speak or perceive speech, as speech, from lesion of the middle cerebral artery. In speech, timing is crucial. The fine, fast sequences of speech — the normal rate for spoken English is 8 syllables a second — are characterized by rapid and abrupt alternation of sounds from the extremes of the vocal range; up to 100 neuromuscular co-ordinations must be accomplished in producing a phrase. A patient has only to lose the ability to time the phases of breathing correctly with phonation, resonance, amplification and the complex manoeuvres of articulation, or the ability to control the specific muscular actions required for specific speech-sounds for speech-sequences to be disrupted, and the correct prosodic performance to be unattainable. Where the impairment is sufficiently intrusive, the structural features of language itself are distorted. A patient has only to mistime the hearing of certain sounds or prosodic cues for the familiar sequences of his language to become those of a foreigner's.

Examined in the light of a verifiable functional anatomy and basic physiological observations, speech, speech-disorders and their associated lesions may well help clarify the relation between sensorimotor and cognitive experience.

Note

Arguments discussed briefly in this paper are set out in detail in *The Power of Speech* (Whitcombe in prep.).

Elizabeth Whitcombe
Spinal Injuries Unit
Royal National Orthopaedic Hospital Trust
Stanmore
Middlesex
HA7 4LP

References

Alexander, R.McN., 1992. *The Human Machine*. London: Natural History Museum.

Bogen, J. & G.M. Bogen, 1976. Wernicke's region — where is it? *Annals of the New York Academy of Sciences* 280, 834–43.

Bouillaud, J., 1825. Recherches cliniques propres à démontrer que la perte de la parole correspond à la lésion des lobules antérieurs du cerveau, et a confirmer l'opinion de M. Gall, sur le siège de l'organe du langage articulé'. *Archives Générales de Medicin* 8, 25–45.

Broca, P., 1861. Remarques sur le siège de la faculté du langage articulé, suivies d'une observation d'aphémie (perte de la parole)'. *Bulletins de la Société anatomique de Paris*, 36e année, 2e Série 6, 330–57.

Carpenter, R.H.S., 1990. *Neurophysiology*. 2nd edition. London: Edward Arnold.

Castaigne, P., F. Lhermitte, J.-L. Signoret & R. Abelanet, 1980. Description et étude scannographique du cerveau de Leborgne. La découverte de Broca. *Revue Neurologique* 136, 563–83.

Catania, K.C., R.G. Northcutt, J.H. Kaas & P.D. Beck, 1993. Nose stars and brain stripes. *Nature* 364, 493.

Chadwick, D.J. & J. Whelan (eds.), 1991. *Exploring Brain Functional Anatomy with Positron Tomography*. (CIBA Foundation Symposium 163.) Chichester: John Wiley.

Crick, F. & E. Jones, 1993. Backwardness of human neuroanatomy. *Nature* 361, 109–10.

DuBrul, E.L. (ed.), 1988. *Sicher and DuBrul's* Oral Anatomy. 8th edition. St Louis (MI): Ishiyaku EuroAmerica.

Duchin, L.E., 1990. The evolution of articulate speech: comparative anatomy of the oral cavity in *Pan* and *Homo*. *Journal of Human Evolution* 19, 687–97.

Foerster, O., 1936. Motorische Felder und Bahnen, in *Handbuch de Neurologie*, 6: *Allgemeine Neurologie*, eds. O. Bumke & O. Foerster. Berlin: Springer, 1–357.

Freud, S., 1891. *Zur Auffassung der Aphasien. Eine kritische Studie*. Leipzig: Deuticke.

Fritsch, G. & E. Hitzig, 1870. Uber die elektrische Erregbarkeit des Grosshirns. *Archiv für Anatomie, Physiologie und wissenschaftliche Medicin* 37, 300–332.

Gall, F-J. & G. Spurzheim, 1818. *Anatomie et Physiologie du Système nerveux en general, et du cerveau en particulier*, vol. 3. Paris: Schuell.

Geschwind, N., 1965. Disconnexion syndromes in animals and man. *Brain* 88, 237–94, 585–644.

Geschwind, N., 1967. Carl Wernicke's contribution to the study of aphasia. *Cortex* 3, 449–63.

Geschwind, N., 1970. The organization of language and the brain. *Science* 170, 940–44.

Geschwind, N. & W. Levitsky, 1968. Human brain: left-right asymmetries in temporal speech-region. *Science* 161, 186–7.

Geschwind, N., F.A. Quadfasel & J.M. Segarra, 1968. Isolation of the speech area. *Neuropsychologia* 6, 327–40.

Guillery, R.W., 1994. No crossing at the crossing. *Nature* 367, 597–8.

Kelemen, G., 1938. Comparative anatomical studies on the junction of larynx and resonant tube. *Acta Otolaryngologica* 26, 276–83.

Lecours, A.R., 1975. Myelogenetic correlates of the development of speech and language, in *Foundations of Language Development*, ed. E.H. Lenneberg. New York (NY): Academic Press, 121–35.

Liepmann, H., 1900. Das Krankheitsbild der Apraxie ('motorischen Asymbolie') auf grund eines Falles von einseitiger Apraxie. *Monatsschrift für Psychiatrie und Neurologie* 8, 15–44, 102–32, 182–97.

Liepmann, H., 1905. Die linke Hemisphäre und das Handeln. *Münchener medizinische Wochenschrift* 52, 2322–6, 2375–8.

Marie, P., 1908. [Discussion], Société de neurologie, 3e Discussion sur l'aphasie. *Revue neurologique* 16, 1042.

Matthews, P.H., 1991. *Morphology*. 2nd edition. Cambridge: Cambridge University Press.

Meyer, A., 1905. Aphasia. *The Psychological Bulletin* 2, 261–77.

Mohr, J.P. *et al.*, 1978. Broca aphasia: pathologic and clinical. *Neurology* 28, 311–24.

Negus, V.E., 1929. *The Mechanism of the Larynx*. London: William Heinemann.

Passingham, R.E., 1991. Discussion, in Chadwick & Whelan (eds.), 273.

Raichle, M., 1994. Visualizing the mind. *Scientific American* 270, 58–64.

Ramsay, S.C. *et al.*, 1993. Regional cerebral blood flow during volitional expiration in man: a comparison with volitional inspiration. *Journal of Physiology* 461, 85–101.

Wall, P.D., P. Glees & J.F. Fulton, 1951. Corticofugal connections of posterior orbital surface in rhesus monkey. *Brain* 74, 66–71.

Welker, W., 1990. Why does cerebral cortex fissure and fold?, in *Cerebral Cortex*, vol. 8, eds. E.G. Jones & A. Peters. New York (NY): Plenum Press, 3–136.

Wernicke, C., 1874. *Der aphasische Symptomencomplex. Eine psychologische Studie auf anatomischer Basis*. Breslau: Cohn und Weigert.

Wernicke, C., 1903. Der aphasische Symptomencomplex. *Der Deutsche Klinik* 6, 487–556.

Whitcombe, E., in prep.. *The Power of Speech. An Essay in Anatomy*. Oxford: Oxford University Press.

Wise, R., U. Hadar, D. Howard & K. Patterson, 1991. Language activation studies with positron emission tomography, in Chadwick & Whelan (eds.), 218–34.

Woolsey, T.A., C. Welker & R.H. Schwartz, 1975. Comparative anatomical studies of the Sm1 face cortex with special reference to the occurrence of 'barrels' in layer IV1,2. *Journal of Comparative Neurology* 164, 79–94.

8
Hominine Preadaptations for Language and Cognition

Leslie C. Aiello

The pattern of hominine brain size increase over the past 4 mya shows that there are two periods of relatively rapid change. The first of these marks the transition between the australopithecines and early members of the genus Homo, *while the second occurs about 250,000 years ago. Arguments are presented which suggest that the first period of increase is most probably an inevitable consequence of the transition to a more open, hot and dry environment by a bipedal hominine that had skeletal preadaptations stemming from an earlier arboreal climbing niche. Other features found at this stage of human evolution can also be considered to be an inevitable consequence of this transition. These features include the incorporation of a higher proportion of animal food in the diet, the reduction in the level of sexual dimorphism and the development of specific anatomical preadaptation to modern human language. This line of reasoning suggests that many features of* Homo erectus, *including its brain size and inferred cognitive ability, were correlates of selective pressures other than those specifically related to human symbolic cognition.*

The human brain is one of our most distinctive anatomical features. The size of the human brain, together with human cognitive and linguistic abilities, are arguably the primary features that distinguish humans from other animals. Current evidence supports the idea that there were three major stages in the evolution of human brain size (Fig. 8.1). The first of these was the australopithecine stage where the relative brain size was within the range found in the non-human primates. The second stage was the early *Homo* (or *Homo erectus*) stage. This stage lasted for a relatively long period of time from the first appearance of early *Homo erectus* (*Homo ergaster*) at approximately 1.7 million years ago (Wood 1992) to the middle part of the Middle Pleistocene period, one and a half million years later. During this stage there was little or no increase in the size of the hominine brain (Rightmire 1981; Leigh 1992). The final stage got underway approximately 250,000 to 300,000 years ago and saw an escalation of brain size to its current levels. This pattern of brain size increase implies that features that might have been instrumental in the earlier phases of the evolution of our brain size and cognitive ability may not have been the same factors that were important in the later evolution of the human brain (Aiello & Dunbar 1993).

This contribution will consider the anatomical background for the evolution of language and cognition by assessing some of the possible causes and consequences of the increase in hominine brain size. The purpose of this exercise is to offer some ideas pertinent to reconstructing the cognitive ability of *Homo erectus* who occupied the intermediate stage between the more ape-like australopithecines and modern humans. By arguing directly from the anatomy that is preserved in the fossil record, it emerges that the brain size increase characteristic of the *Homo erectus* level of organization need not have had anything whatsoever to do with direct selection for a human-like symbolic or cognitive capacity. Rather, the early expansion of the brain is explicable largely in the context of motor and behavioural modifications that were associated with the change from a more woodland habitat characteristic of the australopithecines to a more open habitat that is associated with *Homo erectus*. Any advances in cognitive or technical capacity in these early hominines over those of our living hominoid relatives can be viewed as emergent properties in the context of neural expansion and reorganization that occurred primarily in response to other factors.

The arguments presented here are based on the premise that the brain and the rest of the nervous system are one integrated unit and that the size and complexity of this integrated system reflects the functional complexity of the organism (Whitcombe 1995; this volume). From the point of view of human evolution, anthropologists have tended to explain increase in the size of the human brain by selection for language, intelligence and/or symbolic function (Falk 1987). But it could equally well be explained in the context of other aspects of functional complexity. If we approach the evolution of the human brain from a broad comparative point of view, there are at least two areas that would appear to be particularly relevant to the earlier phases of human brain evolution. The first of these is functional complexity having to do with sensory-motor complexity and integration. The second is cognitive complexity resulting from social integration and control.

In relation to sensory-motor integration, the larger the brain, the more complex can be the sensory-motor control of the body. There can be more nerve tracts and more integration between those tracts which premit separate firing of smaller muscle groups and more complex coordination of function. The relevance of this can be seen in the context of bipedal locomotion where the forelimbs have been freed from their ancestral locomotor role and function independently from the hindlimbs. Standing or walking on two legs also requires a much more complexly integrated system of balance than does quadrupedal locomotion. The position of the centre of gravity has to be constantly monitored and small groups of muscles need to be continuously recruited to correct its position and bring it into alignment. In bipedal walking or running, the movements of the hindlimbs also have to be integrated with the movements of the forelimbs and trunk which help to maintain the dynamic balance.

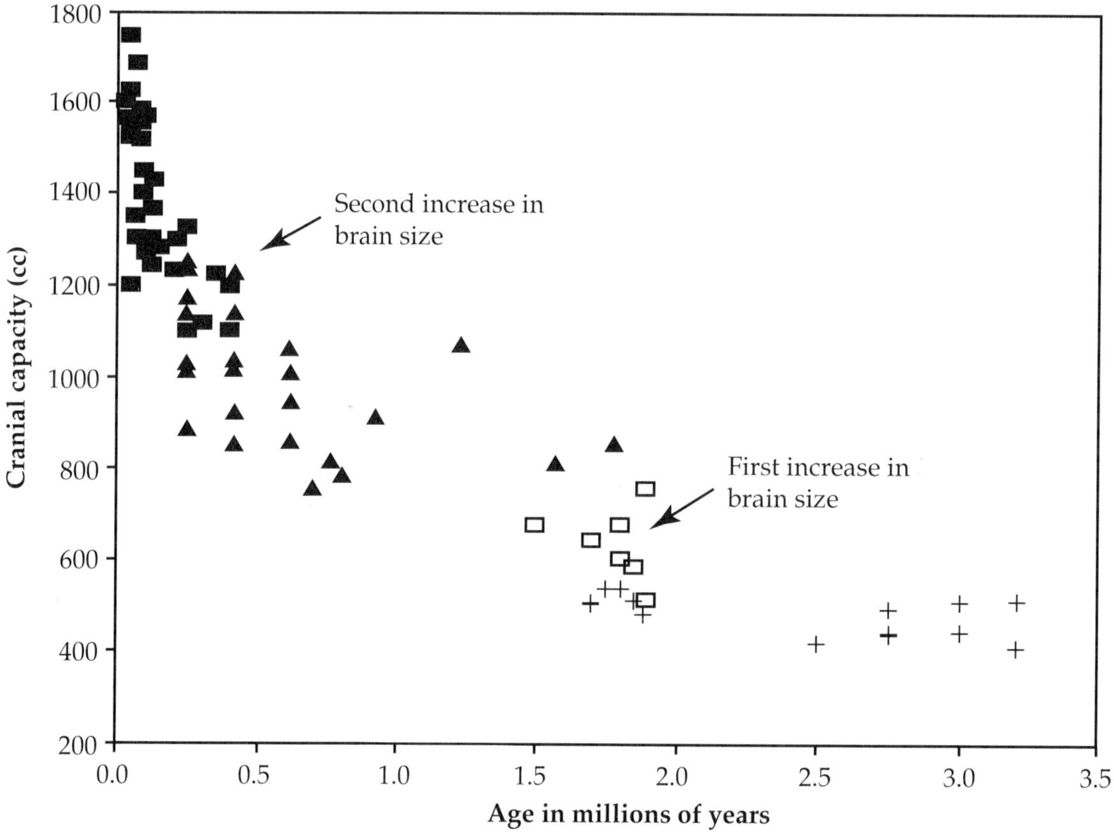

Figure 8.1. *The relationship between cranial capacity (cubic centimetres) and age (millions of years) for the hominines. Plus = australopithecines, open box =* Homo habilis *and* Homo rudolfensis, *triangle =* Homo erectus, *filled box = archaic and modern* Homo sapiens *including the Neanderthals. Note the two periods of rapid increase in cranial capacity and the relative stability in cranial capacity during the* Homo erectus *period.*

Object manipulation, which is assoicated with prehensile hands in the primates, also demands enhanced hand–eye coordination and an associated increase in neural circuitry. Furthermore, it has been argued that speed and coordination of arm movement, particularly in the context of aimed throwing would require increased neural capacity and integration (Calvin 1983; 1993). Bipedalism, and the associated dedicated use of the forelimb for non-locomotory functions would, therefore, presuppose a complex and highly fractionated nervous system and this would in turn presuppose a larger brain (Aiello 1995).

If dedicated, habitual bipedalism does require a larger and more complex nervous system it is perhaps no coincidence that those early hominines with the largest brain sizes (*Homo rudolfensis* and early *Homo erectus*) also have the most modern, bipedally adapted postcrania (Wood 1992; Walker & Leakey 1993). Over the past decade, it has been well established that the australopithecines had skeletons that are very different from those of modern humans (Stern & Susman 1983; Susman *et al.* 1984; Grausz *et al.* 1988). It is true that they show evidence of terrestrial bipedalism, particularly in the pelvis, but they also show clear evidence of arboreal climbing (Hunt 1994). Although the efficiency of australopithecine bipedalism has been debated over the years (Lovejoy 1974; 1988; Stern & Susman 1983), it is evident that these hominines did spend sufficient time in the trees for that arboreal adaptation to be clearly expressed in the skeleton (Hunt 1994). The suggestion that the australopithecines may not have yet developed the balance and coordination associated with human bipedalism is perhaps best supported by recent CT-based analyses of the bony labyrinth of the inner ear in the hominines (Spoor 1993; Spoor *et al.* 1994). The bony labyrinth, or semi-circular canals, are the structures in the ear that are directly involved with the sense of balance. The australopithecines have distinctly ape-like semi-circular canals while *Homo erectus* is the first of the hominines yet analyzed that has semi-circular canals that are of human size, proportion and orientation (Spoor 1993; Spoor *et al.* 1994).

In relation to the second function, social cognition, it has recently been demonstrated that the size of the social group in primates is correlated with the relative size of the neocortex (Dunbar 1992; 1993; Aiello & Dunbar 1993). In much the same fashion as the sensory-motor integration argued for bipedalism, a relatively large neocortex (in relation to the size of the rest of the brain) appears to facilitate the integration and coordination of greater numbers of social relationships needed to maintain the integrity of larger social groups. In simian primates, the relative size of the neocortex (the neocortex ratio) is highly correlated with overall brain size ($r^2 = 0.86$; Aiello & Dunbar 1993). Therefore the larger the brain size, the larger the neocortex ratio and the larger the social group in which the hominines would be expected to have lived. It is, therefore, conceivable that one of the factors behind the early evolution of the hominine brain may well have been pressure to live in larger and larger social groupings.

This hypothesis is attractive in that it also offers an explanation for the early origin of language, in the sense of enhanced vocal communication. Social cohesion is maintained in primates by grooming behaviour and the larger the social group the more time an individual would have to devote from its daily time-budget to 'servicing' its social relationships. Vocal communication is an alternative system of social bonding which would allow an individual to reduce its social time budget by simultaneously servicing its social relationships with multiple individuals. In the context of this hypothesis, the origin of human language would be in vocal grooming behaviour which would need not necessarily have a symbolic content (Aiello & Dunbar 1993). At this stage vocal grooming may have resembled group chattering such as is found in the living gelada baboons (Richman 1976; 1978; 1987; Dunbar 1993) which would express mutual interest and commitment and thereby instill a sense of social cohesion and well being.

It is a relatively easy matter to suggest when this might have happened in human evolution (Aiello & Dunbar 1993). One can predict the neocortex ratio from cranial capacity, group size from neocortex ratio and percentage grooming time from group size. On this basis it has been shown that the australopithecines would have lived in groups of unremarkable size. With the expanded brain of early *Homo*, the inferred group sizes would have been larger and these hominines would have been under pressure to have enhanced vocal communication for the purposes of social cohesion. It would have been at the stage of the plateau in brain evolution characteristic of early *Homo*, and particularly *Homo erectus*, that vocal grooming would most probably have been in the repertoire of our early ancestors.

Viewed from the perspective of these two examples, sensory-motor integration related to bipedal locomotion and object manipulation as well as to social cohesion, the first period of expansion of the hominine brain need not have had anything fundamentally to do with selection for a human type of symbolic cognition. Rather it may have had everything to do with factors that were associated with the need for larger group sizes and the adoption of bipedal locomotion in our early ancestors (Aiello 1995). These factors would involve not only the environmental changes that were occurring in Africa during the Pliocene and Early Pleistocene but also the anatomical preadaptions of the early hominines that were experiencing these changes.

In relation to the environment, palaeoecological reconstructions of the habitats of the early hominines suggest that the earlier australopithecines were adapted to more forested environments while the later hominines, both the robust australopithecines and early *Homo*, were adapting to increasingly open and more arid environments (Behrensmeyer & Cooke 1984; Foley 1987; Menocal 1995). This transition appears to have occurred both in southern and eastern Africa (Bromage & Schrenk 1995). Because group size in primates tends to correlate with terrestrial adaption (Foley 1987), one would expect that a transition to a terrestrial lifestyle by the early hominines would also have been associated with a significant increase in the effective group size and this would have provided a pressure for the increase in brain size. This pressure could have been operating in parallel in the robust australopithecines and the contemporaneous hominines belonging to early *Homo*. The robust australopithecines not only seem to be adapted to more open environments than the earlier, more gracile, australopithecines (Foley 1987), but they also have the largest absolute and relative brain sizes of their genus (Aiello & Dean 1990; Aiello & Wheeler 1995). The even larger relative and absolute brain size of early *Homo erectus* could conceivably be related to larger group sizes than found in the robust australopithecines, but, along the lines argued here, it could also relate to the more perfected bipedal locomotion in *Homo*.

It may be surprising, particularly in the context of the uniqueness of bipedalism as a locomotor adaptation, that bipedal locomotion can be viewed as an almost inevitable consequence of adaptation of the early hominines to a more terrestrial, hot, dry habitat. Other terrestrially adapted primates such as the baboons are not bipedal. This is most probably because they do not have the preadaptions for bipedalism. Baboons are Old World monkeys and all Old World monkeys have a limited component of below branch, or suspensory, feeding in their repertoires (Rose 1973). As a result they have a skeleton that is very different from that of the apes in features associated with mobility and truncal erectness (Andrews & Aiello 1984). Comparative studies have shown that suspensory feeders, particularly vertical climbers, are very similar to bipedal humans in joint excursion, muscle usage and kinematics (Kimura *et al.* 1979; Prost 1980; Fleagle *et al.* 1981; Ishida *et al.* 1985; Kimura 1985; Okada 1985). Below branch feeding, which is found in chimpanzees and is inferred for the immediate proto-hominine, would have made bipedalism a distinct possibility for the early hominines where it was not an option for the Old World monkeys entering the same habitat.

Hominine bipedalism can be related to the terrestrial environment through the thermoregulatory costs to the hominines in occupying a hot, open country habitat (Wheeler 1984; 1985; 1991a,b; 1992a,b; 1993; 1994). It is essential to maintain the temperature of the brain within about $\pm 2°$ of its normal temperature. Wheeler (1991a,b) has shown that bipedalism would reduce the thermoregulatory demands on, and improve the water budget of, a hominine opting for an open country adaptation by reducing the heat load on that hominine at midday by about 60 per cent. This may or may not have been one of the fundamental reasons for the initial evolution of australopithecine bipedalism (Wood 1993; Hunt 1994) however it would have provided the australopithecines with a thermoregulatory preadaption to the hotter and drier open savannah habitat. Wheeler (1992a; 1993) has also shown that bipedalism would have been an even more significant thermoregulatory adaptation for early *Homo*, and particularly early *Homo erectus*, which has a significantly different body build than do the australopithecines (Walker & Leakey 1993). For a hominine coming from a climbing heritage it would seem that bipedalism would be one of the most effective ways to adapt to the heat loads of the savannah.

If this line of argument is accurate, then the initial expansion of the human brain would have been a direct consequence of an adaptation to hotter, more open environments by a hominine with the preadaptions to bipedal locomotion that were acquired in an arboreal forested habitat (Aiello 1995). Together with a larger brain and bipedal locomotion, the transition to a terrestrial environment by such a hominine may also be related to other human features such as the incorporation of higher proportions of animal foods in the diet, the reduction in the level of sexual dimorphism and the development of specific anatomical preadaptions to the evolution of modern human phonemic language.

In relation to diet, it has recently been suggested that any increase in brain size in hominines in particular, and primates in general, would have been inevitably related to an increase in dietary quality (implying for the hominines an incorporation in the diet of larger amount of animal derived food) and a concomitant reduction in the size of the digestive tract (Aiello & Wheeler 1995). The connection between brain size, diet and the size of the digestive tract lies in the relationship between the metabolic cost of the brain and the gut and the overall basal metabolic rate of the body. Brain tissue is metabolically very expensive tissue, its mass specific metabolic rate being over 20 times that of skeletal muscle. The human brain is just over a kilogram heavier than would be expected for an average mammal of our body size (Aiello & Wheeler 1995) but there is no elevation in our BMR to account for this large and expensive brain. This would suggest that we have compensated for this by reducing the metabolic cost of one or more other tissues in the body. The metabolically expensive tissues are, in addition to the brain, the heart, the liver, the kidneys and the gastro-intestinal tract. All of these are highly constrained by body size except for the gut whose size can vary significantly in relation to diet. The higher quality and more easy to digest the diet, the smaller the gut can be.

It follows from this that as the brain increased in size in the early hominines, the diet would have had to have been of higher quality. One of the only other alternatives would have been to increase the BMR. However this would have had considerable cost in relation to increased intraspecific competition for food and also to increased foraging time which would have seriously effected the time budgets of the early hominines. A logical conclusion is that the incorporation of more animal-derived food in the diet would be expected. This has recently been confirmed by strontium-calcium and stable carbon isotope ratios which have shown that robust australopithecines were likely omnivorous rather than strictly herbivorous in their diet (Sillen 1992; Lee-Thorpe *et al.* 1994). Furthermore, the diet of early *Homo* most probably differed from that of the robust australopithecines in either the incorporation of more underground storage organs (soft bulbs, etc.) or the preferential consumption of animals having relatively elevated Sr/Ca ratios such as hydraxs (Sillen *et al.* 1995). The necessity for a high quality diet could also have fed back on itself. A higher quality diet would require a more sophisticated foraging strategy, and this in turn may have provided an additional selection pressure for a larger brain size and increased cognitive ability (Parker & Gibson 1977; 1979; Gibson 1986).

Another unique feature of *Homo* in relation to the earlier australopithecines is an increase in body size of both males and females and a marked reduction in the level of sexual dimorphism (McHenry 1994). Inferred species average body weights for the australopithecines range between 35 and 44 kg while body weight dimorphism is between 64 per cent and 79 per cent. In early *Homo erectus* inferred species average body weight has increased to 58 kg while percentage dimorphism is 83 per cent (McHenry 1994). This change could be interpreted as the result of change in mating strategy which would have relaxed the intrasexual competition among the males for access to the females. This explanation is, however, unconvincing in that the females are apparently getting bigger rather than the males becoming smaller (McHenry 1994) as would be expected if relaxation in selection for larger males, as the result of change in mating pattern, was the cause of the reduced sexual dimorphism. There are, however, environmental correlates of increased female size that are relevant to the transition to an open country environment and the arguments that have been presented above. The first of these is thermoregulation. In the context of the thermoregulatory model for the origin of bipedalism, smaller sized hominines would have been at a distinct disadvantage in a hot open environment in relation to larger bipedal hominines (Wheeler 1992a). This would result in the smaller females being under much more severe thermoregulatory stress in an open environment than the males. If both the males and the females were adopting similar foraging strategies in the open savannah, the females would have been under quite strong selection pressure for increase in body size. Furthermore, the larger overall inferred species average weight would also correlate with larger home range sizes (McHenry 1994) which would be consistent with both the committed terrestrial adaptation of these hominines and their inferred high quality diet.

The second correlate of increased female size has to do with reproductive strategy. In many primates small female size in sexually dimorphic species is associated with early first reproduction (Willner 1989). Everything else being equal, sexual dimorphism can be viewed in this context as a reproductive strategy which would maximize female reproductive success by lengthening the period of fertility. However such a strategy also has distinct costs. Sexually dimorphic females have been shown to incur relatively greater prenatal reproductive costs than monomorphic females (Willner 1989) and it is also expected that lactation would pose greater costs because of the smaller size of the mother in relation to the offspring.

The reproductive costs for smaller females may have been even greater if Walker & Ruff (1993) are correct in proposing that early *Homo erectus* would have already been giving birth to secondarily altricial offspring. In humans the large brain size is achieved by extending the rapid period of fetal brain growth into the first postnatal year (Martin 1983; 1990). This is also associated with the relative helplessness of the human infant in its early years of life. Ruff and Walker have argued that the pelvis in early *Homo erectus* would have been too small to produce infants with large enough brain sizes to achieve the known adult brain size without this human pattern of brain growth and increased infant dependency. If this is the case (and it is based on many levels of assumption), the reproductive cost for relatively small and highly sexually dimorphic females would have been even greater than suggested above. Female reproductive cost, in connection with the increased thermoregulatory stress, may have been an important correlate of the increase in female body size and the concomitant reduction in the degree of sexual dimorphism. There need not have been any significant change in social organization or mating strategy at this stage in human evolution to explain these changes.

This line of argument would suggest that *Homo erectus*, a relatively large-brained, bipedal, omnivorous, terrestrially adapted hominine with reduced sexual dimorphism, would be a direct result of adaptation to a hot, dry, open-country environment of a hominine that had preadaptions stemming from an arboreal climbing niche. This hominine might also be expected to have had at least one specific anatomical preadaption to speech in the form of a valvular larynx that allows the airway to be closed off (Whitcombe this volume). In speaking this permits pressure to build up below the larynx to be released during phonation. Air pressure below a closed larynx also stabilizes the chest to provide a fixed basis for the arm muscles in locomotion. All mammals with prehensile forelimbs have a valvular larynx which serves this locomotor function (Negus 1929). In this case bipedalism could also have been a preadaption to speech. In a committed terrestrial bipedal the selective pressure on the locomotor function of the valvular larynx would have been relaxed. This then would have permitted the development of less cartilaginous, and thereby more membraneous, vocal folds (cords) which would be associated with less harsh, more melodious sound.

Dietary changes correlated with the adaptive complex that has been discussed for *Homo erectus* may also have preadapted these hominines for speech by shortening and broadening the mandible. Movement of the tongue is dependent on the orientation of the muscles that control it and this orientation is ultimately dependent on the geometry of the mandible. Duchlin (1990) has demonstrated that the mandible of not only the Neanderthals but also *Homo erectus* would be within the human range of variation in relation to muscle orientation. Very literally, these hominines would have been able to get their tongues around the sounds that are fundamental to human speech.

It should be noted that these preadaptions are independent of the lowering of the larynx that has been argued by others to be essential to the production of full human phonemic speech (Lieberman *et al.* 1992). If a low larynx is dependent on the flexion of the base of the skull as has been claimed by both Laitman (Laitman *et al.* 1979) and Lieberman (Lieberman *et al.* 1972; Lieberman 1989), then the descent of the larynx would not have occurred until later in the human evolutionary sequence at the time of Steinheim or Kabwe, coincident with the second phase of hominine brain expansion. However the preadaptions described here would have been sufficient for the enhanced level of vocal communication that has been inferred for the *Homo erectus* stage of evolution (Aiello & Dunbar 1993).

One final point should be added to this discussion of the complex of adaptations and preadaptions associated with terrestrial open-country adaptation. Kien (1991) has pointed out that a large brain is not necessarily a more efficient brain unless there are also improvements in coping with the increasing amounts of data that the brain would have to handle. She suggests that a possible way of dealing with potential information 'over-load' would be to increase efficiency through the reduction or abstraction of the data into a form that would allow easy transfer among different parts of the brain. She suggests that these 'neural symbols' would not only free large amounts of the work space in the brain, transforming a larger brain into an even more efficient brain, but also form the basis for the human ability to symbolize.

It is impossible to know when this might have happened in human evolution. It is conceivable that it might be at the root of the apparent cognitive distinctions between Neanderthals and anatomically modern *Homo sapiens* that mark the inferred behavioural differences that have been noted across the

Middle to Upper Palaeolithic boundary (Davidson & Noble 1993). However some anthropologists consider this boundary to be more of an illusion than fact (Hayden 1993; Marshack 1989). Kien, herself, suggests that at least initial reorganization might have occurred during the early period of brain expansion leading to the appearance of *Homo erectus*. She is not alone in suggesting that the beautifully formed handaxes characteristic of the Acheulean tradition not only are highly sophisticated tools but also imply a certain amount of symbolic ability in their production (Toth & Schick 1993).

The question now becomes how capable the *Homo erectus* hominines were of human cognition. Traditional bases of inference which involve the anatomy of the endocranial casts or asymmetries of the brain (Holloway & de la Coste-Lareymondie 1982; Falk 1980; Tobias 1987; 1991) imply at least incipient symbolic and/or linguistic function. However current doubts in relation to the localization of cognitive function in the brain render these analyses difficult to interpret (Whitcombe 1995; this volume). It is instructive to consider whether or not there is anything in the anatomy of the early hominines that would suggest that these hominines had less than modern abilities. In this context there are two features that have recently been recognized in the anatomy of WT 15000, the almost complete skeleton of a *Homo erectus* (*Homo ergaster*) youth that dates to 1.6 MYA (Walker & Leakey 1993). These features suggest that vocally based symbolic communication may not have been a feature of hominine adaptation at this time. MacLarnon (1993) has demonstrated that modern humans have a unique expansion of their neural canals in the mid-thoracic region of their spinal column. She has also established that WT 15000 is similar to non-human primates in lacking this expansion. An expansion of neural tissue, and particularly grey matter, in the mid-thoracic region would be expected to be related to the enervation of local structures. On this basis, it has been suggested that this expansion in humans is likely to be a reflection of increased control of the thoracic and abdominal muscles which would be associated in finer control of respiration in the context of sustained vocalization associated with human speech. Because this expansion is lacking in WT 15000, it is likely that this hominine had yet to develop the muscular control that would be associated with human vocalization.

The second feature has to do with growth and development of this WT 15000. Smith (1993) has suggested that *Homo erectus* had not yet developed the adolescent growth spurt that is uniquely characteristic of human development. She bases this conclusion on the apparent contrast between the dental age of WT 15000 and its age inferred from epiphyseal closure. In its dental development, WT 15000 is more similar to the pattern characteristic of humans than of apes and has an inferred dental age of about 10 to 11 years. This contrasts with the human-based age of between 13 and 13.5 years that would be inferred from its stage of epiphyseal closure and the age of 15 years that would be suggested from its stature. The fact that the closure of the epiphyses (and stature) is in advance of dental development in this fossil suggests that the growth of the skeleton is in advance of dental development. This is the pattern that is seen in living chimpanzees which lack the adolescent growth spurt. Bogin (1988; 1990) has suggested that the function of the adolescent growth spurt is primarily to reduce the rate of growth of children, keeping them in a greater state of dependency for a longer period of time and facilitating the transference of symbolically based cultural knowledge. Smith (1993) speculates that the apparent absence of the growth spurt in WT 15000 would suggest that this had not yet become a feature of early *Homo erectus* life.

These speculations are relevant to the *Homo erectus* phase of evolution that lasted between approximately 1.7 and 0.25 million years ago (Fig. 8.1). This was a period where there was little or no significant increase in brain size. This long period of apparent stasis in brain size is also associated with a similar stasis in the material remains left by these hominines. In particular, the Acheulean tool tradition that appeared about 1.4 million years ago in Africa (Asfaw *et al.* 1992) and lasted until about 150,000 years ago showed little change or elaboration throughout this time (Gowlett 1992). This is all the more remarkable in the context of the relatively rapid change and diversification in stone tool traditions that followed this period. It was about 100,000 years before the disappearance of the Acheulean, at about 250,000 years ago, that improvements in manufacturing techniques and specialization of flake tools that foreshadowed the later Mousterian and Middle Stone Age traditions began to appear. This is about the same period that the second expansion in hominine brain size got underway.

Based on the arguments presented here, the *Homo erectus* phase of evolution can be interpreted as a direct response to the adaptation to an open country environment by a hominine that had postcranial preadaptions derived from an arboreal climbing locomotor heritage. The fact that there was little change in either the morphology of the *Homo erectus* hominines nor in their material culture over this long time period attests to the success of this level of adaptation in the variety of habitats they found themselves in as they spread throughout Africa and Eurasia. It also raises the question of why, after such a long period of apparently successful adaptation, brain size began to increase again in the later half of the Middle Pleistocene.

It is possible that this second phase of brain expansion was also a response to the necessity for increased group size (Aiello & Dunbar 1993). Only a very small annual increase in *Homo erectus* populations would have produced a significant increase in total population numbers over 1.6 million years of evolutionary history. The exponential increase in brain size matches quite closely the exponential increase in projected population numbers over this time, based on an increase of only 0.00000001 per annum (Foley in press). Living group size may have been forced to increase in response to overall population numbers to provide protection against other human groups in an increasingly densely populated environment (Alexander 1989; Aiello & Dunbar 1993). This could be likened to an evolutionary arms race for access to resources where the larger group would always win. An alternative, but not necessarily mutually exclusive possibility, would have to do with the dispersed nature of human populations and particularly with the advantages of a nomadic or migrating lifestyle. Such a wider-ranging lifestyle may have become increasing important as the hominines adapted to more severe habitats in the later Middle and early Upper Pleistocene. Whatever the cause, the brain size began to increase again at this time and hominines with smaller brain sizes, under about 1000 cc, disappeared completely from the record (Fig. 8.1).

Following on from the arguments presented in this discussion, the renewed increase in brain size would have had far-reaching implications for hominine lifestyle. If the need for larger population groups was driving the increase in brain size, the hominines would have been under renewed pressure to maintain group cohesion and this may have well involved the introduction of greater symbolic content in the vocal grooming behaviour. Such symbolic content would have enhanced the cohesive properties of vocal grooming. At the same time, the mechanics of vocal communication may have feed back on brain size. The sensory-motor integration and control involved in the production of more articulate speech would have demanded a larger and more complexly integrated nervous system (Whitcombe this volume). And again in the sense of Kien (1991) such a larger brain would have required improvements in information management that may well have taken the form of more refined neural symbols. Remembering the metabolic costs of the larger brain, this second phase of expansion would also have to be balanced against the metabolic costs of other organs of the body. The increasing appearance of fire in the sites of hominines at this stage of evolution may be more than a coincidence in that cooking would significantly improve the digestibility of both animal and plant food. It would be a way of externalizing part of the digestive process (Aiello & Wheeler 1995) thereby reducing the metabolic costs of digestion which could then be set against the larger brain size.

The renewed increase in brain size would also put increasing metabolic demand on the mother to meet the increased energetic demand of the larger-brained offspring (Foley & Lee 1991; Leonard & Robertson 1992). This may have had a profound effect on the social organization of these hominines by requiring an increased amount of paternal investment in the offspring to insure survival (Power & Aiello in press). One of the secure inferences that can be drawn from the later stages of the fossil record is that Neanderthals as well as anatomically modern humans were characterized by both absolutely and relatively large brain sizes in relation to the earlier hominines. It is not inconceivable that one of the advantages that the modern humans had over the Neanderthals was more monogamously based breeding units that would have involved significant economic cooperation among individuals thereby spreading the increased parental investment more equally between males and females (Soffer 1992; Power & Aiello in press).

Conclusion

The arguments that have been presented above are rooted in one of the best pieces of evidence that we have from the fossil record for human evolution, the pattern of brain size increase over time. It is clear from this evidence that there were two periods of relatively rapid brain expansion in our evolutionary history. The first of these marked the transition between the australopithecines and early members of the genus *Homo*, particularly *Homo erectus*, while the second is associated with the transition from *Homo erectus* to archaic *Homo sapiens*. By adopting a comparative framework and assessing the correlates of large brain sizes across primates and other mammals it is possible to speculate on the reasons behind, and the consequences of, the increase in brain size in the hominines. It emerges from this line of reasoning, that the first period of hominine brain size increase is most probably a direct consequence of the transition to a more open, hot and dry environment by a hominine that had skeletal preadaptions stemming from an earlier arboreal climbing niche. It is argued that both larger group sizes that are correlated with such environments and committed bipedal locomotion would demand an enlarged sensory-motor system with enhanced levels of fragmentation and integration. In other words it would demand an enlarged central nervous system including brain. It is also argued that a more omnivorous diet and reduced sexual dimorphism as well as certain preadaptions of the larynx and oral cavity for speech are direct consequences of this transition in habitat. The implication of this is that selection for symbolically based 'human' cognitive capabilities need not have been a factor at the *Homo erectus* stage of evolution and may not have been important until the later, second stage of brain evolution, that got underway in the second half of the Middle Pleistocene period. The long period of relative stability in *Homo erectus* brain size and material culture would suggested that cognitive capacities of this hominine were a correlate of, or emergent from, the increase in brain size that was responding to other selective pressures.

Acknowledgements

I would like to particularly thank Robin Dunbar, Peter Wheeler, Elizabeth Whitcombe and Camilla Power for collaboration in the development of the ideas presented in this discussion. I would also like to thank the following for discussion and criticism of some of the points presented here: John Fleagle, Kathy Homewood, Cathy Key, Ann MacLarnon, and Caroline Ross. I am grateful to Paul Mellars and Kathleen Gibson for organizing the workshop on 'Modelling the Early Human Mind' and inviting me to prepare this paper.

Leslie C. Aiello
Department of Anthropology
University College London
Gower Street
London
WC1E 6BT

References

Aiello, L.C., in press. Terrestriality, bipedalism and the origin of language, in *Evolution of Social Behaviour Patterns in Primates and Man*, ed. J. Maynard-Smith. London: British Academy.

Aiello, L.C. & M.C. Dean, 1990. *An Introduction to Human Evolutionary Anatomy*. London: Academic Press.

Aiello, L.C. & R.I.M. Dunbar, 1993. Neocortex size, group size and the evolution of language. *Current Anthropology* 34, 184–93.

Aiello, L.C. & P. Wheeler, 1995. The expensive tissue hypothesis: the brain and the digestive system in human and primate evolution. *Current Anthropology* 36, 199–221.

Alexander, R.D., 1989. The evolution of the human psyche, in Mellars & Stringer (eds.), 455–513.

Andrews, P. & L.C. Aiello, 1984. An evolutionary model for feeding and positional behaviour, in *Food Acquisition and Processing in Primates*, eds. D.J. Chivers, B.A. Wood & A. Bilsborough. New York (NY): Plenum Press, 429–66.

Asfaw, B., Y. Beyene, G. Suwa, R.C. Walter, T.D. White, G. WoldeGabriel & T. Yemane, 1992. The earliest Acheulean from Konso-Gardula. *Nature* 360, 732–5.

Behrensmeyer, A.K. & H.B.S. Cooke, 1984. Paleoenvironments, stratigraphy, and taphonomy in the African Pliocene and Early Pleistocene, in *Ancestors: the Hard Evidence*, ed. E. Delson. New York (NY): Alan R. Liss, 60–62.

Bogin, B., 1988. *Patterns of Human Growth*. Cambridge: Cambridge University Press.

Bogin, B., 1990. The evolution of human childhood. *BioScience* 40, 16–25.

Bromage, T.G. & F. Schrenk, 1995. Biogeographic and climatic basis for a narrative of early hominid evolution. *Journal of Human Evolution* 28, 109–14.

Calvin, W.H., 1983. A stone's throw and its launch window: timing, precision, and its implications for language and the hominid brain. *Journal of Theoretical Biology* 104, 121–35.

Calvin, W.H., 1993. The unitary hypothesis: a common neural circuitry for novel manipulations, language, plan-ahead, and throwing?, in Gibson & Ingold (eds.), 230–50.

Davidson, I. & W. Noble, 1993. Tools and language in human evolution, in Gibson & Ingold (eds.), 363–88.

Duchlin, L.E., 1990. The evolution of articulate speech: comparative anatomy of the oral cavity in *Pan* and *Homo*. *Journal of Human Evolution* 19, 687–98.

Dunbar, R.I.M., 1992. Neocortex size as a constraint on group size in primates. *Journal of Human Evolution* 22, 469–93.

Dunbar, R.I.M., 1993. Co-evolution of neocortex size, group size, and language in humans. *Behavior and Brain Sciences* 16, 681–735.

Falk, D., 1987. Hominid paleoneurology. *Annual Review of Anthropology* 16, 13–30.

Falk, D., 1980. Hominid brain evolution: the approach from paleoneurology. *Yearbook of Physical Anthropology* 23, 93–107.

Fleagle, J.G., J.T. Stern, W.L. Jungers, R.L. Susman, A.K. Vangor & J. Wells, 1981. Climbing: a biomechanical link with brachiation and with bipedalism, in *Vertebrate Locomotion*, ed. M.H. Day. (Symposium of the Zoological Society of London 48.) London: Zoological Society, 359–75.

Foley, R.A., 1987. *Another Unique Species*. Harlow: Longman Scientific and Technical.

Foley, R.A., in press. The evolutionary ecology of hominid dispersals, in *The Human Tide*, ed. O. Soffer.

Foley, R.A. & P.C. Lee, 1991. Ecology and energetics of encephalization in hominid evolution. *Philosophical Transactions of the Royal Society, London* B 334, 223–32.

Gibson, K.R., 1986. Cognition, brain size and the extraction of embedded food resources, in *Primate Ontogeny, Cognition and Social Behaviour*, eds. J.G. Else & P.C. Lee. Cambridge: Cambridge University Press, 93–105.

Gibson, K.R. & T. Ingold (eds.), 1993. *Tools, Language and Cognition in Human Evolution*. Cambridge: Cambridge University Press.

Gowlett, J.A.J., 1992. Tools — the Palaeolithic record, in *Cambridge Encyclopedia of Human Evolution*, eds. S. Jones, R. Martin & D. Pilbeam. Cambridge: Cambridge University Press, 350–64.

Grausz, H.M., R.E. Leakey, A.C. Walker, & C.V. Ward, 1988. Associated cranial and postcranial bones of *Australopithecus boisei*, in *Evolutionary History of the 'Robust' Australopithecines*, ed. F.E. Grine. New York (NY): Aldine de Gruyter, 127–32.

Hayden, B., 1993. The cultural capacities of neandertals: a review and re-evaluation. *Journal of Human Evolution* 24, 113–46.

Holloway, R. &. C. de la Coste-Lareymondie, 1982. Brain endocast asymmetry in pongids and hominids: some preliminary findings on the paleontology of cerebral dominance. *American Journal of Physical Anthropology* 58, 101–10.

Hunt, K., 1994. The evolution of human bipedality: ecology and functional morphology. *Journal of Human Evolution* 26(3), 183–202.

Ishida, H., H. Kumakura & S. Kondo, 1985. Kinesiological aspects of bipedal walking in gibbons, in Kondo (ed.), 59–80.

Kien, J., 1991. The need for data reduction may have paved the way for the evolution of language ability in hominids. *Journal of Human Evolution* 20, 157–65.

Kimura, T., 1985. Bipedal and quadrupedal walking of primates: comparative dynamics, in Kondo (ed.), 81–104.

Kimura, T., M. Okada & H. Ishida, 1979. Kinesiological characteristics of primate walking: its significance in human walking, in *Environment, Behavior and Morphology: Dynamic Interactions in Primates*, eds. M.E. Morbeck, H. Preuschoft & N. Gomberg. New York (NY): Gustav Fischer, 297–311.

Kondo, S. (ed.), 1985. *Primate Morphologphysiology: Locomotor Analysis and Human Pibedalism*. Tokyo: University of Tokyo Press.

Laitman, J.T., R.C. Heimbach & E.S. Crelin, 1979. The basicranium of fossil hominids as an indicator of their upper respiratory systems. *American Journal of Physical Anthropology* 51, 15–34.

Lee-Thorpe, J.A., N.J. van der Merwe, & C.K. Brain, 1994. Diet of *australopithecus robustus* at Swartkrans from stable carbon isotopic analysis. *Journal of Human Evolution* 27, 361–72.

Leigh, S.R., 1992. Cranial capacity evolution in *Homo erectus* and early *Homo sapiens*. *American Journal of Physical Anthropology* 87, 1–14.

Leonard, R.W. & M.L. Robertson, 1992. Nutritional requirements and human evolution: a bioenergetics model. *American Journal of Human Biology* 4, 179–95.

Lieberman, P., 1989. The origins of some aspects of human language and cognition, in Mellars & Stringer (eds.), 391–414.

Lieberman, P., E.S. Crelin & D.S. Klatt, 1972. Phonetic ability and related anatomy of the newborn, adult human, Neanderthal man, and the chimpanzee. *American Anthropologist* 74, 287–307.

Lieberman, P., J.T. Laitman, J.S. Reidenberg & P.J. Gannon, 1992. The anatomy, physiology, acoustics and perception of speech: essential elements in analysis of the evolution of human speech. *Journal of Human Evolution* 23, 447–68.

Lovejoy, C.O., 1974. The gait of australopithecines. *Yearbook of Physical Anthropology* 17, 147–61.

Lovejoy, C.O., 1988. Evolution of human walking. *Scientific American* 259, 82–9.

McHenry, H.M., 1994. Behavioral ecological implications of early hominid body size. *Journal of Human Evolution* 27, 77–87.

MacLarnon, A., 1993. The vertebral canal, in Walker & Leakey (eds.), 359–90.

Marshack, A., 1989. Evolution of the human capacity: the symbolic evidence. *Yearbook of Physical Anthropology* 32, 1–34.

Martin, R.D., 1983. *Human Brain Evolution in an Ecological Context*. (52nd James Arthur Lecture on the Evolution of the Human Brain.) New York (NY): American Museum of Natural History.

Martin, R.D., 1990. *Primate Origins and Evolution*. London: Chapman & Hall.

Mellars, P. & C. Stringer (eds.), 1989. *The Human Revolution: Behavioural and Biological Perspectives on the Origins of Modern Humans*. Edinburgh: Edinburgh University Press.

Menocal, P.B. de, 1995. Plio-Pleistocene African climate. *Science* 270, 53–9.

Negus, V.E., 1929. *The Mechanism of the Larynx*. London: William Heinemann.

Okada, M., 1985. Primate bipedal walking: comparative kinematics, in Kondo (ed.), 47–58.

Parker, S.T. & K. Gibson, 1977. Object manipulation, tool use and sensorimotor intelligence as feeding adaptations in great apes and cebus monkeys. *Journal of Human Evolution* 6, 623–41.

Parker, S.T. & K. Gibson, 1979. A developmental model for the evolution of language and intelligence in early hominids. *Behavioral and Brain Sciences* 2, 367–408.

Power, C. & L.C. Aiello, in press. Female proto-symbolic strategies, in *Women in Human Origins*, ed. L.D. Hager. London: Routledge.

Prost, J.H. 1980. Origin of bipedalism. *American Journal of Physical Anthropology* 52, 175–89.

Richman, B., 1976. Some vocal distinctive features used by gelada monkeys. *Journal of the Acoustic Society of America* 60, 718–24.

Richman, B., 1978. The synchronisation of voices by gelada monkeys. *Primates* 19, 569–81.

Richman, B., 1987. Rhythm and melody in gelada vocal exchanges. *Primates* 28, 199–223.

Rightmire, G.P., 1981. Patterns in the evolution of *Homo erectus*. *Paleobiology* 7, 241–6.

Rose, M.D., 1973. Quadrupedalism in primates. *Primates* 14, 337–58.

Sillen, A., 1992. Strontium-calcium ratios (Sr/Ca) of *australopithecus robustus* and associated fauna from Swartkrans. *Journal of Human Evolution* 23, 495–516.

Sillen, A., G. Hall & R. Armstrong, 1995. Strontium calcium ratios (Sr/Ca) and ^{87}Sr/^{86}Sr of *Australopithecus robustus* and *Homo* sp. from Swartkrans. *Journal of Human Evolution* 28, 277–85.

Smith, B.H., 1993. The physiological age of KNM-WT 15000, in Walker & Leakey (eds.), 196–220.

Soffer, O., 1992. Social transformations at the Middle to Upper Paleolithic transition: the implications of the European record, in *Continuity or Replacement? Controversies in* Homo sapiens *Evolution*, eds. G. Braüer & F.H. Smith. Rotterdam: Balkema, 247–60.

Spoor, F., 1993. The Comparative Morphology and Phylogeny of the Human Bony Labyrinth. Unpublished Ph.D. dissertation. University of Utrecht.

Spoor, F., B. Wood & F. Zonnefeld, 1994. Implications of early hominid labyrinthine morphology for evolution of human bipedal locomotion. *Nature* 369, 645–8.

Stern, J.T. & R.L. Susman, 1983. The locomotor anatomy of *Australopithecus afarensis*. *American Journal of Physical Anthropology* 60, 279–317.

Susman, R.L., J.T. Stern & W.L. Jungers, 1984. Arboreality and bipedality in Hadar hominids. *Folia Primatologia* 43, 113–56.

Tobias, P.V. 1987. The brain of *Homo habilis*: a new level of organization in cerebral evolution. *Journal of Human Evolution* 16, 741–62.

Tobias, P.V. 1991. *Olduvai Gorge*, vol. 4: *The Skulls, Endocasts, and Teeth of* Homo habilis. Cambridge: Cambridge University Press.

Toth, N. & K. Schick, 1993. Early stone industries and inferences regarding language and cognition, in Gibson & Ingold (eds.), 346–62.

Walker, A. & R.E. Leakey (eds.), 1993. *The Nariokotome* Homo erectus *Skeleton*. Cambridge (MA): Harvard University Press.

Walker, A. & C.B. Ruff, 1993. The reconstruction of the pelvis, in Walker & Leakey (eds.), 222–33.

Wheeler, P.E., 1984. The evolution of bipedality and loss of functional body hair in hominids. *Journal of Human Evolution* 13, 91–8.

Wheeler, P.E., 1985. The loss of functional body hair in man: the influence of thermal environment, body form and bipedality. *Journal of Human Evolution* 14, 23–8.

Wheeler, P.E., 1991a. The thermoregulatory advantages of hominid bipedalism in open equatorial environments: the contribution of increased convective heat loss and cutaneous evaporative cooling. *Journal of Human Evolution* 21, 107–16.

Wheeler, P.E., 1991b. The influence of bipedalism on the energy and water budgets of early hominids. *Journal of Human Evolution* 21, 117–36.

Wheeler, P.E., 1992a. The thermoregulatory advantages of large body size for hominids foraging in savannah environments. *Journal of Human Evolution* 23, 351–62.

Wheeler, P.E., 1992b. The influence of the loss of functional body hair on the water budgets of early hominids. *Journal of Human Evolution* 23, 379–88.

Wheeler, P.E., 1993. The influence of stature and body form on hominid energy and water budgets; a comparison of *Australopithecus* and early *Homo* physiques. *Journal of Human Evolution* 24, 13–28.

Wheeler, P.E., 1994. The thermoregulatory advantages of heat storage and shade-seeking behaviour to hominids foraging in equatorial savannah environments. *Journal of Human Evolution* 26, 336–50.

Whitcombe, E., 1995. *The Power of Speech: An Essay in Anatomy*. Oxford: Oxford University Press.

Willner, L.A., 1989. Sexual Dimorphism in Primates. Unpublished Ph.D. dissertation. University of London.

Wood, B., 1992. Origin and evolution of the genus *Homo*. *Nature* 355, 783–90.

Wood, B., 1993. Four legs good, two legs better. *Nature* 363, 587–8.

Part III

Comparative Studies of Behaviour

9

Systematics and Integrated Methods for the Modelling of the Pre-modern Human Mind

K.A. Robson Brown

Closer research links between archaeologists, hominid evolutionists, and psychologists are necessary if aspects of mind in pre-modern human groups are to be modelled, and one way forward towards this goal may be found in the application of phylogenetic systematic analysis. The role of evolution in the development of psychological functions in general should be evaluated carefully if hominid development in particular is to be understood. Integrated archaeological and psychological experimentation should attempt to provide empirical evidence for cognitive hypotheses. In addition, the reanalysis of Lower Palaeolithic material as evidence for behaviour allows the application of biological models, such as cladistic interpretations, which may prove revealing and which may make the complex relationships between behaviour, mind, and evolution more explicit.

One of the main issues at stake in the modelling of pre-modern human mind was summarized by Glynn Isaac.

> Beyond the dimly lighted upper rim is darkness extended away from the watcher . . . but in these depths is the gleam of light on water. If the well is not too deep, by straining our eyes we perhaps see a figure — a figure set in an unfamiliar context, but yet a familiar figure — familiar, because it is our own reflection. Archaeological enquiry into the very remote past has recently experienced comparable moments of truth. Awareness is dawning that in part we have been using archaeology and the early evidence as a mirror by which to obtain more or less familiar images of ourselves. (Isaac 1986, 237)

Glynn Isaac, whose pioneering work in this field has inspired the work of so many, expressed the hope that our scientific imagination would be able to rise to the challenge of enquiry presented by mind and behaviour in an evolutionary context. He believed that it is possible to conceive of patterns of behaviour and adaptation which could lead to the formation of familiar-looking archaeological evidence which was actually structured differently from the recent human material record. The aim of this paper is to present some suggestions for methods which could facilitate an understanding of such patterns. These suggestions rely on the premise that there need be nothing mystical about elements of mind. The term 'mind' is taken here to refer to psychological characters which may be at least partially understood through their expression as behavioural characters. The construction of models of mind, therefore, may involve the identification and investigation of observed component characters and the relationships between them. In addition, attempts to model the early human mind require consideration of evolutionary context. A second premise therefore, as Isaac (1986) has shown, concerns the possibility that pre-modern human hominids as individuals and populations should *not* be assumed to be modern, and indeed perhaps *not* human, in any familiar sense. Rather than becoming specifically concerned with drawing new boundaries for humanness, a useful way forward could be to look at how one might find a window on to behaviour in such extinct species. Just as palaeontologists or zoologists may attempt to assess the patterning of behavioural variables within an evolutionary context, even where incomplete or taxonomically uncertain, palaeoanthropology may be able to do the same. The hominid family may in some ways be disabled by its very association with modern 'humanness', and its behavioural record interpreted as the 'hopeful monster' of modern culture, but it remains a significant and unique data set within evolutionary biology. It is in many ways imperfect and incomplete, but it constitutes a remarkable record of complex behaviour within an evolutionary context. As such it is of extreme importance, and may be an invaluable hypothesis testing ground as well as a source of novel information, for the study of behaviour and mind.

In attempting to model the pre-modern human mind the evolutionary context may be seen as a confounding factor, in that it renders rules established for the study of mind in the present inadequate for the past. At the same time, however, the evolutionary context may be turned into a useful tool which permits the integration of information from many different sources, and acts as a regulating framework within which to assess variation and change. 'Systematics' is the broad term which describes several procedures employed to organize and make sense of such information. The procedures result in the ordering of groups of information into systems, where a system is a more inclusive entity whose existence depends on some natural process through which its elements are related. Many types of process may exist (de Queiroz 1988), but clearly within an evolutionary context the process assumption, or mechanism of ordering data, is 'evolution' itself. Systematics is an important feature of many disciplines, but it is in biology that the most intensive and comprehensive studies of systematics and its applications have been undertaken, because it is thought to underpin all areas of biological research. All comparative biology progresses through systematics, whatever the subdiscipline of interest, as the proponents of the 'Comparative Method' have shown (Harvey & Pagel 1991; Pagel & Harvey 1988; Ridley 1989).

During the history of evolutionary biology, several schools of systematic methodology have developed. Evolutionary taxonomy, also known as eclectic or synthetic systematics, was formulated by Mayr (1942; 1969), Dobzhansky (1935; 1937) and Simpson (1944; 1961). This school placed particular value on subjectivity within the procedures and considered systematics to be almost an art, with systems of biological taxa acting as a reflection of degrees of divergence, amount of diversification, and general patristic affinity. The methods used tended not to be standardized or made explicit, and largely for this reason more recent systematic methods have eclipsed them. Notable among the achievements of this school, however, was the redefinition of the species category to ensure that its members would be evolutionary units, a concept to which later systematists returned. The phenetic or numerical school headed by Sokal (1959; 1969) and Sneath (1957; 1968) developed as a direct result of attempts to quantify the processes and procedures used by taxonomists to classify and systematize organisms. The proponents of phenetics assume that organisms should be classified according to overall similarity assessed from a data matrix via a similarity matrix through the use of clustering algorithms without any *a priori* weighting. Inherent within this approach is the removal of an explicitly evolutionary axiom from the system, and it is this which is the main failing of phenetics in the eyes of most evolutionary theorists (Forey 1992).

The third major development in systematics this century has been the appearance of cladistics, or phylogenetic systematics, and it is the methodology of this school which may be most appropriate to the present context. This approach differs from the eclectic school in that it employs empirical methods to reconstruct phylogenies, not relying on intuition or authority, and it differs from phenetic approaches in that it seeks to find the genealogical relationships among taxa rather than overall similarity relationships. The core concept in this phylogenetic methodology is the use of derived or apomorphic characters to reconstruct monophyletic groups, common ancestry relationships, and the grouping of taxa based on evolutionary processes of common ancestry (Hennig 1950; 1966). Phylogenetic systematics makes explicit the assumption that the natural hierarchy is discoverable and effectively represented by a branching tree diagram within which characters may change their status at different levels. Importantly, the method also states that characters which are present in all members of the study group or have a wider distribution than the study group do not contribute to our understanding of the relationship between members within the group. Character congruence is thought to be the decisive criterion for identifying homology (similarity referring to common ancestry), and the principle of parsimony is held as maximizing character congruence.

One of the greatest strengths of the phylogenetic systematic approach is that the method and results are explicit, meaning that all decisions made during the analytical procedure are based on data which can be examined by any specialist interested in the results. Another is that because the reasoning delimits groups based on common ancestry an attempt can be made to reconstruct histories and from them develop a hierarchical ranking scheme. Phylogenetic groups are then a reflection of the order in the world. The third main advantage is that as a result of this, actual histories of specific characters may be described though the mapping of those characters on to phylogenies. This technique permits the reconstruction of ancestral states and shows where on the tree state changes have occurred.

In constructing a model of the pre-modern human mind, the adoption of the latter school of evolutionary systematic methodology may be extremely useful. Systematic procedure permits the incorporation of concepts surrounding process mechanisms into data analysis. If, for example, it is thought that the evolutionary process is an important context of archaeological assemblage formation, then this hypothesis can be explored through the estimation of archaeological 'systems' which depend on assumptions about underlying evolutionary models. In addition, systematic procedure permits explicit argumentation at every level of analysis. This is particularly useful for experimental research, allowing details referring to taxonomic units, characters, and process models to be addressed independently or as sets. Finally, systematic procedure produces trees of relationships between taxonomic units, but if the process model is thought to be accurate and if the taxonomic units are suitable, these trees may be interpreted as genealogies of ancestor–descendant phyla.

TAXA	Outgroup	A	B	C
CHARACTER STATE	0	1	1	1

The following example shows how these phylogenetic procedures may contribute to an understanding of character evolution. In comparing the performance (type 0 and 1) of three species (A, B, and C) on some behavioural task testing spatial cognition, for instance, an interesting question might arise as to why species C exhibits a performance type 1. To answer this question, one can map the trait 'performance type' on to a phylogeny derived from independent traits for those species. Figure 9.1 illustrates this procedure, so that Figure 9.1a shows a phylogenetic tree constructed from characters other than the 'performance type' trait, such as morphological characters, and Figure 9.1b shows the same phylogenetic tree with notation (-) indicating that the distribution of 'performance type 1' is correlated with the phylogenetic relationship of species A, B, ad C in such a way that it acts like a homologous trait of the clade. The evolutionary interpretation is that the trait originated in the common ancestor of the clade A+B+C and has persisted in all the descendent species. In other words, the presence of this performance type is a manifestation of phylogenetic constraints within the group. Additional explanations such as cost-benefit analyses involve character maintenance rather than character evolution for the trait. Analyses which attempt to explain the adaptive significance of the evolution of the trait must, therefore, refer to the ancestor.

This example shows how closer scrutiny of the internal structure of a topology may reveal more information about the transformation of character states. The procedure involves the assessment of patterns of homology and homoplasy, and therefore the adaptational context of such characters. This context might also be referred to as an adaptational hypothesis or evolutionary scenario. To a limited extent,

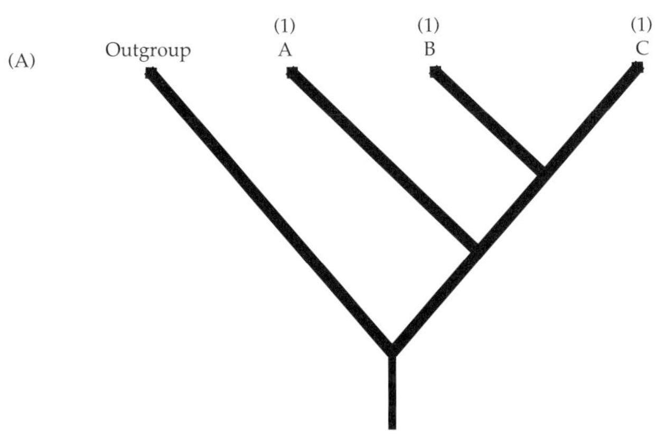

Figure 9.1a. *Hypothetical phylogenetic tree with taxa scored for performance type.*

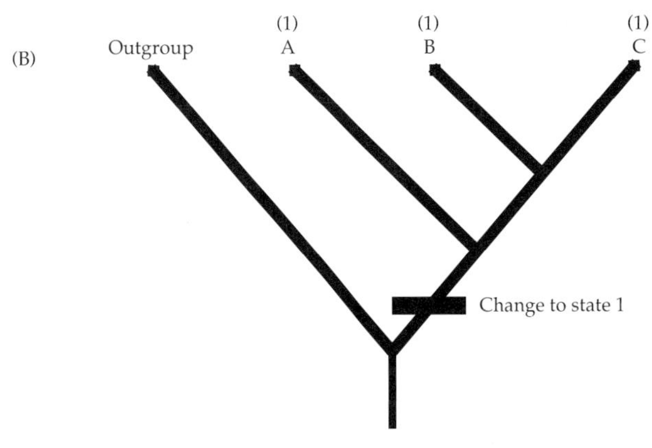

Figure 9.1b. *Hypothetical phylogenetic tree showing the appearance of performance type (1).*

therefore, adaptational hypotheses referring to inter-taxon evolutionary patterning may be tested. A cladistic framework provides an ideal context for such tests because it provides an historical definition of adaptation and the potential for an operational contextual test for evolutionary adaptations. If in cladistic terms an adaptation refers to an apomorphic function promoted by natural selection, as such it may be a conditional, hierarchical, comparative description of a state. In other words, to examine the hypothesized adaptations of any given group requires a methodology which permits testing of both evolutionary events, and stabilizing selection which maintains the trait arising within those events. Cladistic methodology provides just such a procedure.

While this school of systematics provides a useful evolutionary framework within which the development of the pre-modern human mind may be modelled, it is within specialist fields that information suitable for such analysis must be found. It has been suggested above that the term 'mind' may be taken to refer to psychological characters which may be at least partially understood through their expression as behavioural characters. Although there may be many means of identifying characters which describe constituent elements of mind, three important categories of information which may be particularly useful can be quite briefly summarized. These refer to comparative psychology, integrated archaeological and psychological experimentation, and reanalysis of archaeological data.

Comparative psychology

If one interpretation of the term 'mind' refers to the constellation of psychological functions possessed by a subject, and the relationships between those functions, then attempts to model aspects of mind in an extinct subject requires some kind of evolutionary comparative psychological framework within which variation in function patterning may be assessed. There is a pressing need for such investigation of the role of evolution in the development of psychological functions in general, if any model of pre-modern human functions is to be useful and meaningful.

Psychological disciplines have often not been primarily concerned with evolutionary issues, although some of the founding members of modern ethology stressed a need for a 'phylogenetic view' of animal psychology (Heinroth 1911; 1971; Lorenz 1935; 1937; 1950; Tinbergen 1951; 1959). In particular, Lorenz proposed the idea that an active behavioural phenotype may play a causal evolutionary role, and stressed the importance of investigating the purposefulness of behaviour and its relation to the whole pattern of the animal's activities and evolutionary history. The latter he achieved through his commitment to the 'comparative morphology of behaviour' (Lorenz 1935). This concern with comparative methods has arisen again in recent years within disciplines such as cognitive neuroscience and neuropsychology (Kesner & Olton 1990; Marchant & McGrew 1991; Mollon 1989), although the importance of phylogenetic context and the questions arising from it have still not been widely addressed (Hodos & Campbell 1990).

The fragility of the relationship between comparative psychology or ethology and phylogenetic systematics may be caused by concerns that behavioural characters are simply not suitable for systematic analysis of this kind. One major source of controversy is the identification of behavioural homology (Pribram 1958; Atz 1970; Hodos 1976; Lauder 1986; 1994; Greene 1994). Many criteria have been proposed, ranging from the structure of underlying neural or neuromuscular control systems (Baerends 1958; Hodos 1976), to the form of gross morphological features used in the behaviour (Haas & Simpson 1946; Atz 1970), and its functions (Mayr 1958). In practice comparative psychology proceeds very successfully using a combination of these methods depending on the details of the study. The suggestion that the importance of experiential effects on behavioural characters is necessarily so great that phylogenetically congruent similarities might not indicate genealogical relationships is generally based on an allied assumption that other types of characters, such as morphometric characters, are free of these effects, but these two views are equally unsubstantiated (Hollister 1918; Lauder 1986; 1994; Wenzel 1992; Greene 1994; Nelson 1994; Rieppel 1994). The primary test of homology is congruence with other characters and there is no reason to assume that one group of characters (such as behavioural characters) will not be useful within phylogenetic analysis. There is a need for research programmes which explicitly test the question of historical congruence among characters referring to behaviours, functions and morphological patterns in an attempt to identify general features of transformations in various aspects of the phenotype. The fields which have already begun this type of study give proponents of this approach cause for optimism. In many cases behaviour has been shown to be no more variable, nor more difficult to describe, measure, and compare across taxa than morphology (de Queiroz & Wimberger 1993; Greene 1994). Further research has provided evidence supporting the suggestion that because behaviours of many kinds may relate closely to development, physiology and morphology, they may be subject to the same evolutionary processes and should change in ways which reflect underlying phylogenetic relationships and evolutionary mechanisms (Brooks & McLennan 1991). Behavioural characters have been successfully combined with morphological characters to estimate phylogeny (Carothers 1984), they may be mapped onto independently derived phylogenies to analyze character evolution (McLennan *et al.* 1988), they may act as an arbiter in the selection of phylogenetic reconstruction (Dobson 1985), and they may be used as independent sources of information on which to construct phylogenetic relationships (Di Fiore & Rendall 1994; McLennan *et al.* 1988).

There are certainly many existing pools of comparative psychological information on which this kind of systematic analysis could draw. Two examples of questions based on comparative psychological data which may be addressed using this method are summarized below. The first concerns the phylogenetic context of mammalian sleep traits, and the second the relation between structural and behavioural variables of diurnality and colour vision in primates. In both cases the method assumes that organisms are the result of a dynamic interaction among genetic, developmental, physiological, morphological, and behavioural systems through time. It also, therefore, holds that there is no *a priori* reason to believe that one system contains more information about evolutionary relationships than any other. Also, in both cases the method translates into what is sometimes known as a 'top-down' phylogenetic analysis of behavioural and structural variables. Such analysis involves the examination of the patterning of characters of interest among closely related extant subjects, and then the assessment of possible evolutionary pathways for the development of that patterning. It does not, in other words, rely on fossil evidence for model building.

The first example, which considers the phylogenetic context of mammalian sleep traits, is a good place to start because it draws upon a comprehensive existing body of comparative data. Much of the current understanding of psychological and clinical aspects of human sleep is based on a knowledge of sleep characteristics in other species (Zepelin 1994). Many of the controversies surrounding the functions of sleep equally rely on evolutionary scenarios built upon comparative data (Allison & Ciccetti 1976; Berger 1995; Siegel 1995; Zepelin & Rechtschaffen 1974). Yet despite continued interest in the comparative context of this suite of characters, there has been little focus on their phylogenetic analysis. A simple case showing how such an approach may provide a new view of observed variation involves the relationship between the length of REM sleep and other sleep traits. One suggestion which has been made in the literature states that in mammals major necessary correlates of an increased REM sleep time include immaturity at birth and safe sleep conditions, and that REM sleep and non-REM sleep states have a linear evolutionary relationship (Allison *et al.* 1972). It is possible to test this kind of suggestion using phylogenetic systematics. In Figure 9.2 a simple three-taxon statement referring to mammalian phylogeny is reproduced. Mapped onto this phylogeny are a series of characters describing various aspects of sleep behaviour, all collected from the literature (Allison *et al.* 1972; 1976; Jouvet-Mounier *et al.* 1970; Siegel *et al.* 1994; Van Twyer 1969; Zepelin 1989; 1994; Zepelin & Rechtschaffen 1974). This analysis is interesting because it shows that although monotremes share safe sleeping conditions and immaturity at birth with other mammalian orders, they do not demonstrate REM sleep. According to phylogenetic systematic reasoning, there are three statements which can be made on this basis. First, the rule that in mammals as a whole immaturity at birth and safe sleep conditions correlate with increased REM sleep time is not universal. Secondly, within the context of the existing analysis, it is most parsimonious to suggest that an evolutionary event occurred in the common ancestor to the marsupial+eutherian clade on the phylogeny in which REM sleep appeared. Thirdly, that attempts should be made to assess the sister group to this three taxon statement to confirm the polarity of the REM sleep character. The relationship

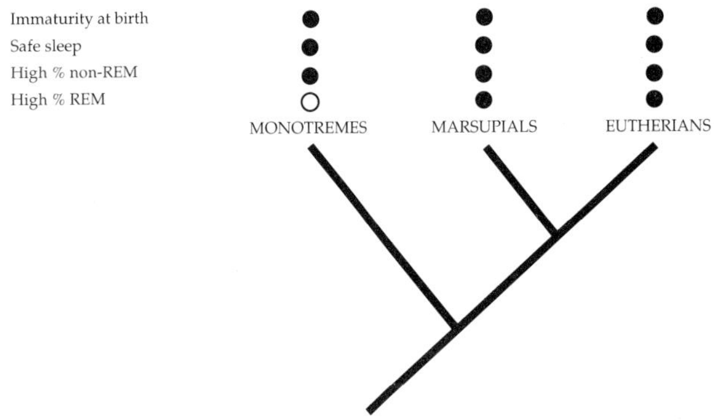

Figure 9.2. *A phylogenetic systematic analysis of mammalian sleep traits.*

between REM and non-REM sleep is clearly very complex, and it may be that the evolutionary relationship between them is not a simple linear one. It may be that both sleep states have evolved out of a primitive third state, as has been suggested elsewhere (Allison *et al.* 1972; Siegel 1995).

A second area of comparative psychology which could prove to be a valuable testing ground for this method involves the relation between structural and behavioural variables of diurnality and colour vision. This may be a fruitful area for research because these characters are well documented in the psychological literature, and because they have relatively straightforward anatomical and behavioural correlates. As a result any problems with the procedure, therefore, should become quite obvious. This research is ongoing, but the procedure which might be adopted is a simple one, mirroring that presented in Figure 9.1. Restated, it should begin with the estimation of a phylogeny of the relevant Old and New World primates, constructed with one of widely available phylogenetic systematics programs, such as PAUP 3.1.1 (Swofford 1993), using independent anatomical data. The second step is to code the characters chosen for analysis. Coding involves the assessment of the state of a character within each species, to allow the investigation of patterning of that character across all groups. This study includes two behavioural and two anatomical characters; these are the parvocellular layers in LGN, orbit size, diurnality, and colour psychophysics. These variables cluster as two coupled pairs which may serve as controls for each other, and therefore are particularly suited to such an analysis. The third step involves the mapping of character states onto the independently derived phylogeny, and comparisons between them. Potential evolutionary scenarios concerning the developmental histories of characters should then be possible. Statistical calculations such as consistency indices and estimates of homoplasy are available to indicate the robusticity of relevant cladograms, and these will be put to use.

Systematic analysis of this kind could serve to bring evolutionary issues back into psychology. There are, however, many more problems which need to be confronted. We need more empirical evidence to show how this kind of phylogenetic approach is useful in other contexts. We need to examine whether behaviours (and their psychological substrates) are evolving with structural features, or whether behaviours have their own independent patterning. In the context of modelling the pre-modern human mind such new comparative methods may in the future allow us to investigate the context of the appearance of new psychological functions, and may also permit the construction of hypotheses describing ancestral character states. In this way we might be able to assess the robusticity of the evolutionary scenarios concerning mind which have long been a focus of the human evolutionary literature.

Integrated archaeological and psychological experimentation

A second avenue for research involves the integration of experimental archaeology and psychology. As this workshop has shown, archaeology has seen the gradual development of a new interest in general cognitive issues. Within Palaeolithic archaeology, this has often been expressed in terms of psychological functions and general abilities, and systems specific to manufacturing procedure or 'planning depth' (see for example Gibson & Ingold 1993; Wynn 1989).

The development of these interests has led to two main fields of research within archaeology. On the one hand, an investigation of complex cognitive systems and their relationships with stone tool technology has encouraged a contextual view of human technological ability. Examples of such contextual systems include the role of language, sex differences, foraging and sociality (e.g. Alexander 1989; Falk 1993; Humphrey 1988; Ingold 1993; Kempler 1993; Pope & Keates 1994; Whiten & Byrne 1988; Wynn 1988). On the other hand, attention has also been drawn away from these high level cognitive systems and towards the examination of individual categories of cognitive ability, and the extent to which stone tool technology can provide evidence for these functions (Calvin 1993; Gowlett 1986; Greenfield 1991; Robson Brown 1993; Wynn 1989). It is in this second field of research that there may be much to gain in a closer relationship between experimental archaeology and psychology. Although shrouded in complexity, at heart the problem is a simple one. If the suggestion is put forward that a specific psychological function is demonstrated by stone tool manufacture, then perhaps this should be accompanied by attempts to test that suggestion experimentally.

One of the familiar characters which appears in many descriptions of the cognitive constellations related to lithic tool manufacture, is that of imagery, or 'mental templates'. Concepts of imagery, visualization, and spatial relations are extensively discussed in the literature of cognitive psychology, and the current understanding of these factors of visual perception is generally felt to be good (see Carroll 1993 for a review). Elsewhere, for example (Robson Brown 1993), it has been suggested that some types of knapping (where the tool face to be modified is held out of sight as the blow is made so that the waste flake intentionally falls from underneath, away from the knapper) may require such imagery on the part of that knapper. The action of the blow on the hidden face is not seen as it occurs, but it is *known*. It is precisely this kind of hypothesis which may be suitable for testing. There is an understanding among many psychologists that cognitive mechanisms, representing and operating on images overlap with those perceiving and manipulating real world objects (Pellegrino *et al*. 1984; Poltrok & Agnoli 1986). If this is accepted, then the performance of an image task may be susceptible to interference from concurrent spatial perceptuo-motor task. This leads to the intriguing inference that one might be able experimentally to establish the coding mechanisms employed in the performance of a specific perceptuo-motor task (Segal & Fusella 1970). Such an experiment should involve the assessment of repeated performance of subjects skilled in the given perceptuo-motor task under different interference conditions. These conditions would require the subject to undertake concurrently a propositional task such as description of known spatial information. The results of performance under these condition should then be judged against a value scale in some way, perhaps against a body of data provided during experimental control runs with not interference. As a further control it might be useful to ask subjects to perform another task which required no imagery, or no propositionally based mechanism. It would be possible to assess results and show that one class of interference task had a greater effect on performance than the other. This would lead to the conclusion that the type of coding involved in the interference class was also that involved in the performance of the perceptuo-motor task (Pellegrino *et al*. 1984).

This kind of relatively simple cognitive experiment would be suited to the investigation of lithic tool manufacture. Those who advocate the multiple code organizational mechanism for cognition have long believed the non-propositional visuo-spatial coding system to be very ancient (Carroll 1993). This multiple code paradigm challenges the assumption that thought is inherently symbolic or driven by quasi-linguistic entities. It argues that at minimum not all cognition may even by describable in terms of propositions, and therefore propositions cannot retain the structure of all replicators. Thus visual, acoustic, and motoric codes have also been inferred, and of these the most interesting and perhaps best supported is that of the visual code. Given these suggestion, an investigation such as is described here might provide a gauge of minimum necessary competence in modern humans with which to approach these archaeological problems more directly. If the results were to confirm that imagery and visualization mechanisms underlie the performance of this knapping task, even in our own language-oriented species, then given the broader character of the material record during this period a good case could be made for the importance of visuo-spatial image coding in the performing hominids. In addition, a research programme into the psychological functions acting within knapping situations which focused on specific, named, psychological characters could greatly facilitate comparative analyses such as those suggested above, and would shift attention away from the wholesale adoption of developmental Piagetian schemes which in archaeology at least, is coming under increased criticism (Plotkin 1988; Robson Brown 1993; 1995).

Reanalysis of archaeological data

In the introduction it was suggested that in this evolutionary context the term mind may be taken to refer to psychological characters which may be at least partially understood through their expression as behavioural characters. Consequently attempts to model the pre-modern human mind may to some extent proceed through the investigation of characters of behaviour. If we in Palaeolithic archaeology, which specifically refers to a set of tool behaviours, have an interest in aspects of mind, this should extend to an awareness of the complex relationship between the behavioural phenotype and specific psychological subsets of that phenotype, and the evolutionary implications of its fitness.

Two questions arise from this suggestion. The first asks how it is possible to consider Lower Palaeolithic archaeology as primary evidence of behaviour, and the second asks how this approach helps us to get at aspects of mind. The term 'behaviour' is perhaps too broad in meaning to have much value for research such as this without a more specific working definition. Most general meanings refer to 'having possession of' or 'being characterized by' features, but a more useful definition would pinpoint that particular element of the phenotype which is of interest. In recognition of this problem Plotkin (1988) has suggested that a general working definition should concern the 'interaction of the whole animal with the environment, such as when a single cell organism moves down some chemical gradient or when a bird migrates from one region to another' (Plotkin 1988, 7), and that a more constrained definition should describe a 'restricted form of interaction, for example the manual manipulation of a twig by a chimpanzee or a vocal warning signal of a vervet monkey' (Plotkin 1988, 7). Indeed, the suggestion that behavioural features require a dual level definition is now almost standard within the literature of behavioural biology. Piaget (1979) made this particularly clear when he proposed that behaviour could be distinguished from other phenotypic attributes because the former refers to 'all action directed by organisms toward the outside world in order to change conditions therein or to change their own situation in relation to these surroundings' (Piaget 1979). Piaget's distinction is a useful one. The unrestricted view of behaviour approaches and deals with behavioural data as a single subset of phenotypic attributes (Odling-Smee 1988). The restricted view assumes that such a subset should itself be subject to further subdivision. Waddington (1960) for example, has suggested that semi-autonomous 'choice' is the controlling structure at the heart of the 'exploitive system' which acts as a co-agent for evolutionary change. Thus, a restricted view of behaviour inherently involves the concept of a choosing intelligence, which in itself already brings us back to the concept of mind.

For the purposes of this paper the definitions of behaviour summarized above will be accepted, although it is clear that the theoretical considerations they involve require some more attention. Here, it is also important to note that there is in the literature a theoretical dichotomy. According to the equilibrium view of behaviour, variation is investigated in relation to the environment and explanations of similarities and differences are sought in terms of selection forces and environmental variables. According to the transformational view, patterns of similarity and difference among organisms, species and higher taxa may be used to construct evolutionary sequences. The latter view focuses on historical patterns of structure, function, and behaviour, and relies of phylogenetic hypotheses to provide a basis for interpreting historical sequences and testing evolutionary concepts. The suggestions presented here in effect attempt to bring together the two viewpoints, in light of the main points of interest raised by these definitions and models for archaeological research.

Any attempt to integrate the various factors that appear to be influencing artefact, fossil, and faunal assemblage variability during the Lower Palaeolithic is essentially an attempt to construct a model of behaviour, in the unrestricted sense. As summarized above, the unrestricted definition of behaviour concerns whole animal interactions with the environment, and tends to refer to a single subset of phenotypic attributes. Examples of hominid behavioural hypotheses which are particularly relevant to this model might include migration behaviour, clines of phenotypic variation, and the potential evidence for 'character release' in new environments. Clearly, however, this general definition is inadequate when a much more specific environmental interaction is demonstrated by the data, such as in the particular cases of lithic artefact production, use, and discard, and faunal assemblage accumulation. In these situations the scenario is one of complex behaviour. and perhaps a subtle feedback system of selection pressures, learning and environmental potential. It seems that in this case a restricted model of behaviour is required, with its concept of a choosing intelligence as the control feature of such a system. In other words, the model of behaviour is directly related to the model of mind.

Many major attempts to make behavioural subset scenarios explicit have been made. The main themes of this work involve the identification of cultural/artefactual 'progression' in time and space, the control of fire, subsistence patters, and more recently psychological functions. For instance since 1981 (Binford 1981) there has been heated debate, focusing largely on faunal data, as to the nature of the hominid subsistence behaviour, and this has become perhaps the most controversial issue in the Palaeolithic archaeology of the 1980s. While the debate has often been polarized around 'hunting versus scavenging' there are in fact three related issues; whether hominids were eating a substantial quantity of meat; whether that meat was transported to 'home bases' and shared; and whether the meat was acquired by hunting or scavenging. From the hunting 'Osteodontokeratic' scenario proposed by Dart (1953), to the scavenging food sharing 'home base' scenario which Isaac (1978) suggested, the first behavioural interpretations of the faunal and fossil breathed life into a discipline which was becoming increasingly dogged by taxonomy and typology. By modern criteria, however, they seem to involve claims for which support is weakening. The 1980s saw a change in this field from 'scenario led' to 'data led' research programmes which involved the investigation of site taphonomy, formation, structure, and cut marks as statistically significant evidence for subsistence patterns. Models which arose out of this shift in research methodology included marginal scavenging scenarios (Binford 1981; Blumenschine 1986), substantial carnivory (Bunn *et al.* 1980; Bunn & Kroll 1986), sporadic and reduced carnivory (Potts & Shipman 1981), and tendon and skin scavenging (Shipman 1981; 1986). Where site structure included the investigation of artefact clustering, such has FLK 22 Zinj, behavioural scenarios referred to windbreak construction (Leakey 1975), the delimination of food preparation and consumption (Davis 1978), 'home bases' (Isaac 1978), and binary-episodic inhabitation of sites (Binford 1987). Clearly the main conclusion to be drawn from this dialogue is that the evidence remains open alternative and competing interpretations, because although the hypothesis that hominids were transporting lithic artefacts and some animal body parts within the landscape appears to be robust, the details of this behaviour remain contentious.

In all these cases one drawback arises through the complexity of the context; a limited understanding of the dynamics of the system. In a recent paper Bower (1993) has suggested that the only way forward for palaeoanthropology to gain a greater understanding of hominid behaviour and the system in which it is set, is through the intensification of field work and the production of more data. This view may be overly pessimistic. More focused fieldwork is certainly to be attempted, but as part of a multifaceted programme of research. New data will not be useful if the theoretical and methodological frameworks we employ in its interpretation have not matured and developed.

It has been suggested that Lower Palaeolithic data may be interpreted as a feature of the behavioural subset of the phenotype. As such it has the status of restricted behaviour. Viewed in this way, an obvious avenue of research should be the investigation and assessment of the character and structure of that phenotypic subset, as other biological subsets have been investigated and assessed. Models referring to systematics, rates and trends of change, and ecological hypotheses, which have been constructed in the context of other biological features should be readily applied to the subset data. Only when the nature of the relation between biological models and the behavioural data is more completely known will the understanding of the behavioural-evolutionary system be within the grasp of biological anthropology and evolutionary biology. As suggested above, one example of how the application of a biological model to archaeological data may begin to increase our understanding of these systems is found in the adoption of cladistic methodology (Robson Brown 1995; in prep.). To illustrate this procedure, the following description summarizes some main methodological steps. In this case, because one concern is to show how systematic methodology permits the integration of information from different sources, these steps refer to the selection of archaeological taxonomic units, the selection of characters with which to describe them, and then the comparison of this information to that known for the fossil hominid context itself.

For this example the archaeological data were selected to represent large-scale variations between regions and time periods during the Lower Palaeolithic. The following Old World biogeographic areas (localities) were defined: Africa (East Africa, South Africa) is represented by the notation AO (Oldowan), ADO (Developed Oldowan), A1, A2, A3 (three stages of Acheulean); East Asia (China) is represented by EA1 and EA2; South Asia (India and Pakistan) is represented by SA1 and SA2; and Europe is represented by EU1 and EU2. Data were drawn from the primary literature on assemblage composition (date, character means, sample sizes, estimates of character variability) from a sample of sites. Data from a total of 133 sites were included. The broad chronological position of each was recorded, and taxon boundaries were initially drawn to take account of locality/horizon clusters of archaeological material and traditional units. The assemblages from these sites have all been described by such categories as the Oldowan, Developed Oldowan and Acheulean. Lithic assemblages recorded for *Pan troglodytes* were used as an outgroup.

Twenty diagnostic features were selected as characters for the data matrix. These features were drawn from classification systems that relied on typology and general descriptions of assemblage quantitative structure. Classical artefact typology is better suited to cladistic analysis than more recent systems because the latter have not been widely applied, emphasize mode of production rather than end product, and function rather than large-scale geographic patterning. Five of the twenty characters selected for analysis refer to discrete states, and fifteen to continuous variables. The characters do not represent an exhaustive list of all possible characters which have been associated with Lower Palaeolithic assemblages. Some important characters, such as blank form, were omitted not because they are thought to be irrelevant, but because their documentation in the published literature was found to be uneven.

The final list included characters describing raw material, proportions of 'finished' tools, percussion technique, retouch, the number of flake tool categories, the proportion of bifaces, picks, discoids, choppers, heavy duty scrapers, flake scrapers and other standard types. Where the archaeological taxa demonstrated the strong presence (especially binary) of synchronic and sympatric variants, the characters in question were ascribed 'polymorphic' status. To assess the relationships between the taxa which the chosen characters implied, a statistical algorithm was then applied to the matrix using the program PAUP 3.1.1 (Swofford 1993), and the most parsimonious solution was accepted as representing the relationships between the taxa. This solution is shown in Figure 9.3a. Figure 9.3b illustrates how individual character states may be traced onto the topology, permitting further investigation of character evolution.

There are many interesting features of this result. Strikingly, the group of early Asian and European sites clearly show many primitive traits, despite being more recent than early African industries, while the main body of African groups fall together, and the most recent Asian and European taxa appear to express the greatest number of derived characters. The consistency index of this topology is 0.783. A bootstrap analysis of this data set also provides evidence that this topology is robust; the clade (ADO A1 A2 A3 SA2 EA2 EU2) is supported in 93 per cent of replicates, the clade (A1 A2 A3 SA2 EA2 EU2) is supported in 85 per cent of replicates, the clade (A2 A3 SA2 EA2 EU2) is supported in 65 per cent of replicates and the clade (A2 A3) is supported in 83 per cent of the replicates. When the distribution of all possible tree lengths for this data matrix is examined, it is found that the distribution is strongly skewed. This may be interpreted as demonstrating a strong phylogenetic signal for the data (Hillis 1991; Hillis & Huelsenbeck 1992).

As has been demonstrated in many other systematic analyses (Brooks & McLennan 1991), the integration of information from different sources is often an important part of this methodology. To illustrate this point independent phylogeny for the genus *Homo*, and the data matrices from which it was constructed, was taken from the literature to provide a comparative context (e.g. Andrews 1984; Chamberlain & Wood 1987; Groves 1989; Rightmire 1984; 1990; Stringer 1984; 1987; Tobias 1991; Wood 1984; 1991; 1992). Figure 9.4 illustrates one method of comparing such data. The scoring of hominid taxa for archaeological characters was undertaken simply on chronological and biogeographic criteria. According to these results, increase in cranial capacity does not predict increased behavioural complexity as measured by these traits. Several suggestions can be made on the basis of this cladistic analysis of archaeological data. For example, the results may contribute to the current debate concerning hominid migration patterns, their tempo and mode (Lahr & Foley 1994; Roebroeks & van Kolfschoten 1994; Dean & Delson 1995). In particular, controversy surrounds the timing of the first migrations of *Homo* into Europe (Gabunia & Vekua 1995) and Asia (Swisher *et al.* 1994). With an earliest data of approximately 1.8–1.9 Myr for early African *Homo erectus* (Wood 1994), until recently conservative

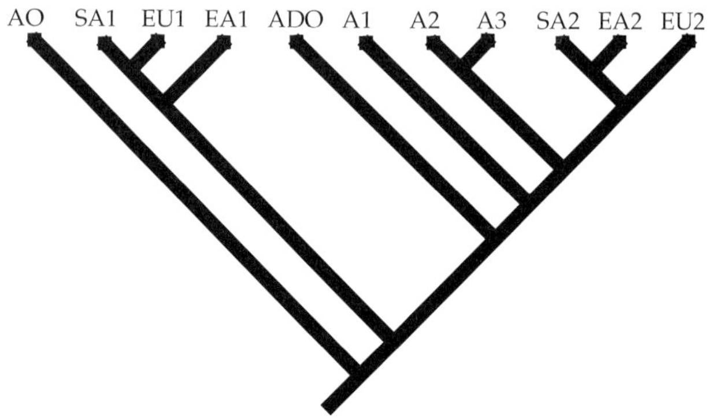

Figure 9.3a. *Topology for the complete archaeology data matrix calculated using the exhaustive search algorithm under the conditions of parsimony. (After Robson Brown 1995.)*

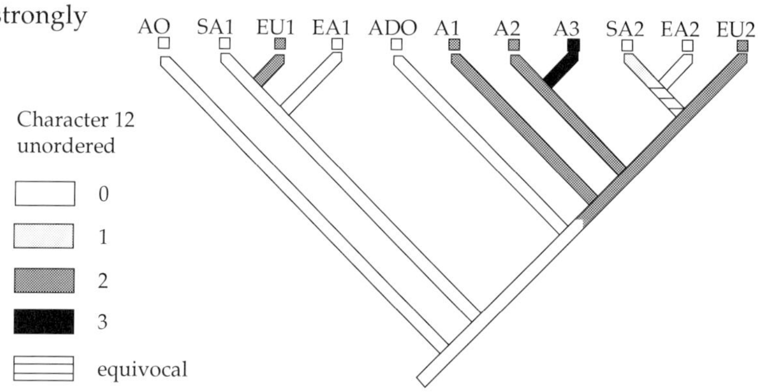

Figure 9.3b. *Percentage score of cleavers in assemblages traced onto the topology for the complete archaeology data matrix. State 0 (0%), state 1 (up to 5%), state 2 (6–10%), state 3 (11–25%). (After Robson Brown 1995.)*

estimates for Eurasian occupation based on the dating of fossils hovered at around 0.5 Myr (Rightmire 1990). In contrast, estimates based on the archaeological record have consistently suggested a much earlier migration out of Africa, but these estimates have generally been regarded with some scepticism. The earliest Chinese sites from Lantian, the Nihewan Valley and Kehe are documented as lying within palaeomagnetically reversed sediments, clustering around an approximate date of one million years (Wu & Olsen 1985; Jia & Wei 1987; You *et al.* 1978; Wei 1985). Similarly early dates for several European sites have been documented (Bonifay & Vandermeersch 1991). These earlier dates have been given greater credibility, however, by the publication of two reports announcing new dates for fossil remains of Eurasian *Homo erectus*. The first of these reports (Swisher *et al.* 1994) describes dates of 1.6–1.8 Myr for two sites in Indonesia, and the second (Gabunia & Vekua 1995) an approximate date of 1.8–1.6 Myr for the Dmanisi mandible. Taken together, this new information does appear to suggest an early migration of *Homo erectus* out of Africa, which was followed by a hiatus until the next documented occurrence of Eurasian hominids at around 1.0 Myr. Possible explanations for this hiatus have included the competition presented by large terrestrial carnivores or a harsh periglacial climate which could not be overcome until 'cultural adaptation, involving toolkit technology, and perhaps broader social evolution' (Dean & Delson 1995, 473) paved the way.

The archaeological systematic analysis presented here appears to show that there is more similarity between non-African late Acheulean groups than the familiar concept of a Movius line might suggest. In addition, the fact that this analysis highlights greater similarities between early east Asian and African industries, than between early east Asian and late east Asian industries may provide evidence for a flow of information concerning lithic tool using across the Old World from Africa at some time between about 1.7–1.0 Myr. It is most likely that this movement of information corresponded to a physical migration of hominids. While this scenario cannot cast any light on the issue of the hiatus suggested by the fossil record of this period, it does present further support for an early (pre-0.5 Myr) occurrence of hominids in Eurasia. An interesting feature of this scenario is that it implies some kind of continuity of information between these early African and early Eurasian populations. In other words, early Eurasian populations displayed a 'chopper-core' behaviour because the migration our of Africa of their ancestral population was pre-Acheulean. If the suggestion of a second wave of hominid migration at about 1.0 Myr is correct, then a clustering of late Pleistocene archaeological taxa from Africa and Eurasia is actually to be expected. However, the extent to which this second stage of information flow across the Old World reflected physical migration of hominids is difficult to assess on the basis of archaeological evidence alone. Where the systematic analysis may contribute more is in the suggestion that direct continuity between early and late Pleistocene Eurasian archaeological taxa within closed, independent areas of behavioural evolution, is unlikely. It is clear, therefore, that this systematic approach may contribute to the debates surrounding hominid behaviour, and this is also true in the specific context of the modelling of mind.

One of the most important and influential schemes for assessing cognitive information in an evolutionary context presented in the recent literature is the neo-Piagetian model described by Wynn (e.g. 1989). Although this is perhaps not the place for a discussion of Piaget's contribution to the debate, it is interesting to note that while his work as a psychologists was remarkable and innovative, his quite extensive writings on the nature of behaviour and evolution indicate a strong Lamarckian position. It has been suggested that the Piagetian developmental scheme may well not be as directly applicable to an evolutionary scheme as wither Piaget himself, or more recent proponents, have suggested; it may be that the complexity of the behavioural/cognitive

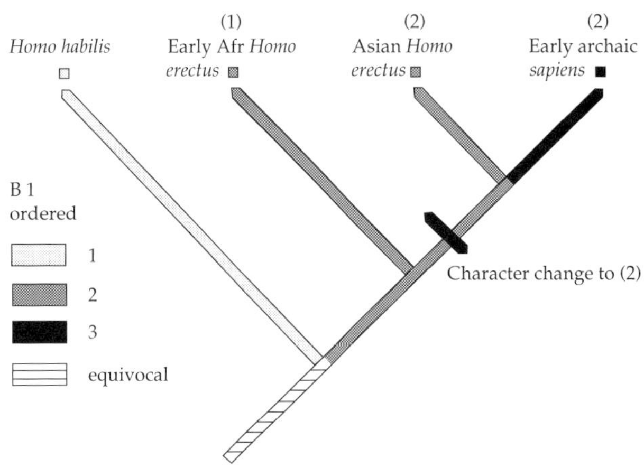

Figure 9.4. *Character tracing of endocranial capacity and the proportion of multiple-edged tools in each industry. Endocranial capacity (B1) state 1 (c. 500 ml), state 2 (c. 750 ml), state 3 (c. 1000 ml). The proportion of multiple edged tools in each industry state 1 (25–75%), state 2 (75–25%).*

system can be conceived as a hierarchical or modular network (Plotkin 1988; Robson Brown 1993). By focusing both on morphological and behavioural trends and changes in this light, it should be possible to perceive of cognitive modes within a scheme of mosaic evolution. For example, one of Wynn's most important contributions to the debate is the suggestion that at about 0.3 Myr there is something of a radical change, in cognitive terms, to be seen in the material record. This is the apparent appearance of operational thought, described in terms of qualitative displacement, spatio-temporal substitution, and congruency. Yet one of the drawbacks of the Piagetian stance is that the scheme has limited flexibility to describe patterns not familiar to the modern human developmental or experimental psychologist. Piaget's scheme assumes a fundamental co-development of cognitive abilities which may simply not be applicable to a pre-modern human cognitive complex.

Despite the criticisms which have been directed towards the application of a Piagetian scheme to an evolutionary context, however, the systematic analysis summarized here does provide some support for Wynn's (1989) hypothesis of a mid Pleistocene 'watershed'. Mapping individual behavioural characters on the independently derived phylogeny suggests that while early African archaic sapiens may be derived in their morphology, they share some behavioural conditions with Asian *Homo erectus*. Increase in cranial capacity does not predict increased behavioural complexity as measured by these traits. Robusticity as measured by the definition of cranial prominences, and complexity of tool production technique are also compared. Thus while the Asian *Homo erectus* taxon is derived in the morphological character, the behavioural trait is a synapomorphy concordant with that in the early archaic group. This kind of discrepancy could be seen as further evidence for a shift in the pattern of behavioural change in the mid-Pleistocene such as Wynn (1989) and others have suggested. Similarly, it might be suggested that a Movius Line referring to the lack of handaxes in east Asia reflected a significant cognitive difference between the populations of Africa and Asia which was persistent through much of the Pleistocene. A cladistic interpretation does not support this view, however, since late Asian industries appear within the same clade as late African industries, suggesting that any cognitive changes are more parsimoniously explained as occurring in the 'ancestor' of both sets.

These examples go some way to demonstrating that it is possible to consider archaeological evidence as specific behavioural information, and that as a result we may begin to investigate the articulation of behavioural and evolutionary processes within hominid evolution. By modelling aspects of the behavioural phenotype in this way the broad relationship between behaviour and mind may be suggested. This may provide a more secure context for the modelling of the psychological functions, as previously suggested. In addition we may be able to bring into a model of mind vital information concerning evolutionary mechanisms which render the model meaningful in terms of hominid development as a whole.

Conclusion

It seems, therefore, that the complex relationship between mind, behaviour, and evolution is not impenetrable. One way forward is to adopt an integrated approach to modern human and palaeoanthropological evidence, which may begin to reveal some features of the pre-modern human mind, and may permit research is this field to contribute to the general evolutionary debate concerning behaviour and evolution.

The role of systematic methods in this research could be very significant, and the three main areas of research discussed here should not be considered independent. Each will provide new data, and new frameworks with which to approach them, which will filter across disciplinary boundaries. Closer research links between archaeologists, hominid evolutionists, and psychologists are absolutely necessary if aspects of mind in pre-modern human groups are to be understood. The role of evolution in the development of psychological functions in general should be evaluated carefully if hominid development in particular is to understood, and integrated archaeological/psychological experimentation should attempt to provide empirical evidence for cognitive hypotheses. To increase our understanding of the context of this work, the reanalysis of Lower Palaeolithic material data as evidence for behaviour is also essential. This approach will permit the application of biological models which may make the complex relationship between behaviour and mind more explicit, without necessarily mirroring 'images of ourselves' (Isaac 1986).

Acknowledgements

Some of work presented here was conducted as part of my doctoral thesis at the University of Cambridge, and of course I would like to thank my supervisor

Rob Foley for his help and criticism. In addition I thank Peter Grossenbacher for leading me into the world of psychology, and the participants of the Workshop on Modelling the Early Human Mind for their comments on this paper. The research was funded by an N.E.R.C. grant, Newnham College, and Downing College, Cambridge.

Kate Robson Brown
Department of Biological Anthropology
University of Cambridge
Cambridge
CB2 1DQ

References

Alexander, R., 1989. Evolution of the human psyche, in *The Human Revolution: Behavioural and Biological Perspectives on the Origins of Modern Humans*, eds. P. Mellars & C. Stringer. Edinburgh: Edinburgh University Press, 455–513.

Allison, T. & D.V. Ciccetti, 1976. Sleep in mammals: ecological and constitutionao correlates. *Science* 194, 732–4.

Allison, T., H. Van Twyer & W.R. Goff, 1972. Electrophysiological studies of the echidna, *Tachyglossus aculeatus* I: Waking and sleep. *Archives Italiennes de Biologie* 110, 145–84.

Andrews, P., 1984. An alternative interpretation of the characters used to define *Homo erectus*. *Courier Forschungsinstitut Senckenburg* 69, 167–75.

Atz, J., 1970. The application of the idea of homology to animal behavior, in *Development and Evolution of Behavior*, eds. L. Aronson, E. Tobach, D. Lehrman, & J. Rosenblatt. San Francisco (CA): W.H. Freeman, 53–74.

Baerends, G., 1958. Comparative methods and the concept of homology in the study of behavior. *Archives Neenderlandaises de Zoologie Supplement* 13, 401–17.

Bailey, G. & P. Callow (eds.), 1986. *Stone Age Prehistory*. Cambridge: Cambridge University Press.

Binford, L.R., 1981. *Bones: Ancient Men and Modern Myths*. London: Academic Press.

Binford, L.R., 1987. Searching for camps and missing the evidence?, in *The Pleistocene Old World: Regional Perspectives*, ed. O. Soffer. New York (NY) & London: Plenum Press, 17–31.

Blumenschine, K., 1986. *Early Hominid Scavenging Opportunities*. (BAR 283.) Oxford: BAR.

Bonifay, E. & B. Vandermeersch, 1991. *Les Premiers Europeens*. Paris: Comitie Travaux Historiques Scientifiques.

Bower, J., 1993. Keynote Address. Pithecanthropus Centennial Conference, Leiden.

Brooks, D. & D. McLennan, 1991. *Phylogeny, Ecology and Behavior: a Research Program in Comparative Biology*. Chicago (IL): University of Chicago Press.

Bunn, H., J.W.K. Harris, Z. Kaufulu, E. Kroll, K. Schick & A.K. Behrensmeyer, 1980. FxJj 50. An early Pleistocene site in northern Kenya. *World Archaeology* 12, 109–36.

Bunn, H. & E. Kroll, 1986. Systematic butchery by Plio-Pleistocene hominids at Olduvai Gorge, Tanzania. *Current Anthropology* 27, 431–52.

Byrne, R. & A. Whiten, 1988. *Machiavellian Intelligence: Social Expertise and the Evolution of Intellect in Monkeys, Apes and Humans*. (Oxford Science Publications.) Oxford: Clarendon Press.

Calvin, W.H., 1993. The unitary hypothesis: a common neural circuitry for novel manipulations, language, plan-ahead, and throwing?, in Gibson & Ingold (eds.), 216–29.

Carothers, J.H., 1984. Sexual selection and sexual dimorphism in some herbivorous ancient dispersals. *American Naturalist* 124, 244–54.

Carroll, J.B., 1993. *Human Cognitive Abilities: a Survey of Factor-analytic Studies*. Cambridge: Cambridge University Press.

Chamberlain, A.T. & B.A. Wood, 1987. Early hominid phylogeny. *Journal of Human Evolution* 16, 119–33.

Dart, R., 1953. The predatory transition from ape to man. *International Anthropological and Linguistic Review* 1, 201–19.

Davis, D.D., 1978. Lithic assemblage variability in relation to early hominid subsistence strategies at Olduvai Gorge, in *Lithics and Subsistences: the Analysis of Stone Tools in Prehistoric Economies*, ed. D.D. Davis. (Vanderbilt University Publications in Anthropology 20.) Nashville (TN): Vanderbilt University, 35–86.

Dean, D. & E. Delson, 1995. *Homo* at the gates of Europe. *Nature* 322, 496–7.

de Queiroz, K., 1988. Systematics and the Darwinian revolution. *Philosophy of Science* 55, 238–59.

de Queiroz, K. & P. Wimberger, 1993. The usefulness of behavior for phylogeny estimation: levels of homoplasy in behavioral and morphological characters. *Evolution* 47(1), 46–60.

Di Fiore, A. & D. Rendall, 1994. Evolution of social organisation: a reappraisal for primates by using phylogenetic methods. *Proceedings of the National Academy of Science, USA* 91, 9941–5.

Dobson, F., 1985. The use of phylogeny in behavior and ecology. *Evolution* 39, 1384–8.

Dobzhansky, T., 1935. A critique of the species concept in biology. *Philosophy of Science* 2, 344–55.

Dobzhansky, T., 1937. *Genetics and the Origin of Species*. New York (NY): Columbia University Press.

Falk, D., 1993. Sex differences in visuospatial skills: implications for hominid evolution, in Gibson & Ingold (eds.), 194–216.

Forey, P., 1992. Fossils and cladistic analysis, in *Cladistics: A Preactical Course in Systematics*, eds. P. Forey, C. Humphries, I. Kitching, R. Scotland, D. Siebert & D. Williams. Oxford: Clarendon Press, 124–36.

Gabunia, L. & A. Vekua, 1995. A Plio-Pleistocene hominid from Dmanisi, East Georgia, Caucasus. *Nature* 373, 509–12.

Gibson, K.R. & T. Ingold, 1993. *Tools, Language and Cognition in Human Evolution*. Cambridge: Cambridge University Press.

Gowlett, J., 1986. Culture and conceptualisation: the Oldowan–Acheulean gradient, in Bailey & Callow (eds.), 243–60.

Greene, H., 1994. Homology and behavioural repertoires, in Hall (ed.), 369–91.

Greenfield, P., 1991. Language, tools and brain: the ontogeny and phylogeny of hierarchically organised sequential behavior. *Behavioral and Brain Sciences* 14, 531–95.

Groves, C.P., 1989. *A Theory of Human and Primate Evolution*. Oxford: Clarendon Press.

Haas, O. & G. Simpson, 1946. Analysis of some phylogenetic terms, with attempts at redefinition. *Proceedings of the American Philosophical Society* 90, 319–49.

Hall, B. (ed.), 1994. *Homology: the Hierarchical Basis of Comparative Biology*. San Diego (CA): Academic Press.

Harvey, P. & M. Pagel, 1991. *The Comparative Method in Evolutionary Biology*. Oxford: Oxford University Press.

Heinroth, O., 1911. Beitrage zur Biologie, namenlich Ethologie und Psychologie der Anatiden. *Verhandlungen des V, Internationalen Ornithologischen-Kongresses Berlin*, 589–702.

Heinroth, O., 1971. *Oskar Heinroth, Vater der Verhsltenforschung, 1871–1945*. Stuttgart: Wissenschaftliche Verlagsgesellschaft.

Hennig, W., 1950. *Grundzuge einer Theorie der phylogenetischen Systematik*. Berlin: Deutscher Zentralverlag.

Hennig, W., 1966. Phylogenetic systematics. *Annual Review of Entomology* 10, 97–116.

Hillis, D., 1991. Discrimination between phylogenetic signal and random noise in DNA sequences, in *Phylogenetic Analysis of DNA Sequences*, eds. M. Miyamoto & J. Cracraft. Oxford: Oxford University Press, 278–94.

Hillis, D. & J.P. Huelsenbeck, 1992. Signal, noise and reliability in molecular phylogenetic analyses. *Journal of Heredity* 83, 183–95.

Hodos, W., 1976. The concept of homology and the evolution of behavior, in *Evolution, Brain and Behavior: Persistent Problems*, eds. R. Masterton, W. Hodos & H. Jerison. Hillsdale (NJ): Lawrence Erlbaum Associates, 153–68.

Hodos, W. & C. Campbell, 1990. Evolutionary scales and comparative studies on animals cognition, in Kesner & Olton (eds.), 1–20.

Hollister, N., 1918. East African mammals in the United States National Museum, Part I: Insectivora, Chiroptera and Carnivora. *Bulletin of the United States National Museum* 99, 1–194.

Humphrey, N.K., 1988. The social function of intellect, in Byrne & Whiten (eds.), 13–26.

Ingold, T., 1993. Tool use, sociality and intelligence, in Gibson & Ingold (eds.), 429–46.

Isaac, G., 1986. Foundation stones: early artefacts as indicators of activites and abilities, in Bailey & Callow (eds.), 221–41.

Jia, L. & Q. Wei, 1987. Stone artefacts from the lower Pleistocene Donguto site near Nihewan (Nihowan), Hebei Province, China. *L'Anthropologie* 91, 277–93.

Jouvet-Mounier, D., L. Astic & D. Lacote, 1970. Ontogenesis of the the states of sleep in rat, cat and guineapig during the first postnatal month. *Developmental Psychology and Biology* 2, 216–39.

Kempler, D., 1993. Disorders of language and tool use. Neurological and cognitive links, in Gibson & Ingold (eds.), 193–215.

Kesner, R. & D. Olton (eds.), 1990. *Neurobiology of Comparative Cognition*. Hillsdale (NJ): Lawrence Erlbaum Associates.

Lahr, M.M. & R.A. Foley, 1994. Multiple dispersals and modern human origins. *Evolutionary Anthropology* 3(2), 48–60.

Lauder, G., 1986. Homology, analogy and the evolution of behaviour, in *Evolution of Animal Behaviour*, eds. M. Nitecki & J. Kitchell. Oxford: Oxford University Press, 9–40.

Lauder, G., 1994. Homology, form and function, in Hall. (ed.), 152–97.

Leakey, M.D., 1975. Cultural patterns in the Olduvai sequence, in *After the Australopithecines*, eds. K. Butzer & G. Isaac. The Hague: Mouton, 477–93.

Lorenz, K., 1935. Der Kumpan in der Umwelt des Vogels. *Journal fur Ornithologie* 83, 137–213; 289–413.

Lorenz, K., 1937. Biologische Fragestellung in der Tierpsychologie. *Zeitschrift fur Tierpsychologie* 1, 24–32.

Lorenz, K., 1950. The comparative method in studying innate behavior patterns. *Symposium of the Society of Experimental Biology* 4, 221–68.

McLennan, D.A., D.R. Brooks & J.D. McPhail, 1988. The benefits of communication between comparative ethology and phylogenetic systematics: a case study using gasterosteid fishes. *Canadian Journal of Zoology* 66, 2177–90.

Marchant, L.F. & W.C. McGrew, 1991. Laterality of function in apes: a meta-analysis of methods. *Journal of Human Evolution* 21, 425–38.

Mayr, E., 1942. *Systematics and the Origin of Species*. New York (NY): Columbia University Press.

Mayr, E., 1958. Behavior and systematics, in Roe & Simpson (eds.), 341–62.

Mayr, E., 1969. *Principles of Systematic Zoology*. New York (NY): McGraw-Hill.

Mollon, J.D., 1989. 'Tho' she kneeled in that place where they grew': the uses and origins of primate colour vision. *Journal of Experimental Biology* 146, 21–38.

Nelson, G., 1994. Homology and systematics, in Hall (ed.), 102–51.

Odling-Smee, F.J., 1988. Niche-constructing phenotypes, in Plotkin (ed.), 73–132.

Pagel, M. & P. Harvey, 1988. Recent developments in the analysis of comparative data. *Quarterly Review of Biology* 62, 413–40.

Pellegrino, J.W., D.L. Alderton & V.J. Shute, 1984. Understanding spatial ability. *Educational Psychologist* 19, 239–53.

Piaget, J., 1979. *Behavior and Evolution*. London: Routledge & Kegan Paul.

Plotkin, H.C., 1988. *The Role of Behaviour in Evolution*. Cambridge (MA): MIT Press.

Poltrok, S.E. & F. Agnoli, 1986. Are spatial visualization ability and visual imagery ability equivalent?, in *Advances in the Psychology of Human Intelligence*, vol. I, ed. R.J. Sternberg. Hillsdale (NJ): Lawrence Erlbaum Associates, 367–411.

Pope, G. & S. Keates, 1994. The evolution of human cognition and cultural capacity: a view from the Far East, in *Integrative Paths to the Past: Paleoanthropological Advances in Honor of F. Clark Howell*, eds. R. Corrucini & R. Ciochon. Englewood Cliffs (NJ): Prentice Hall, 531–67.

Potts, R. & P. Shipman, 1981. Cutmarks made by stone tools on bones from Olduvai Gorge, Tanzania. *Nature* 291, 577–80.

Pribram, K., 1958. Comparative neurology and the evolution of behavior, in Roe & Simpson (eds.), 140–64.

Ridley, M., 1989. Why not to use species in comparative tests. *Journal of Theoretical Biology* 136, 361–4.

Rieppel, O., 1994. Homology, topology and typology: the history of modern debates, in Hall (ed.), 64–101.

Rightmire, G.P., 1984. Comparisons of *Homo erectus* from Africa and southeast Asia. *Courier Forschungsinstitut Senckenberg* 69, 83–98.

Rightmire, G.P., 1990. *The Evolution of* Homo erectus. Cambridge: Cambridge University Press.

Robson Brown, K.A., 1993. An alternative approach to cognition in the Lower Palaeolithic: the modular view. *Cambridge Archaeological Journal* 3(2), 231–45.

Robson Brown, K.A., 1995. A Phylogenetic Systematic Analysis of Hominid Behaviour. Unpublished Ph.D. Thesis, University of Cambridge.

Robson Brown, K.A., in prep.. The application of phylogenetic systematic methods in the analysis of archaeological data. *Journal of Archaeological Science*.

Roe, A. & G. Simpson (eds.), 1958. *Behavior and Evolution*. New Haven (CT): Yale University Press.

Roebroeks, W. & T. van Kolfschoten, 1994. The earliest occupation of Europe — a short chronology. *Antiquity* 68, 489–503.

Segal, S. & V. Fusella, 1970. Influence of imaged pictures and sounds on detection of visual and auditory signals. *Journal of Experimental Psychology* 83(3), 458–64.

Shipman, P., 1981. *The Life History of a Fossil*. Cambridge (MA): Harvard University Press.

Shipman, P., 1986. Scavenging or hunting in early hominids: theoretical framework and tests. *American Anthropologist* 88, 27–43.

Siegel, J.M., 1995. Phylogeny and the function of REM sleep. *Behavioral Brain Research* 69, 29–34.

Siegel, J.M., P. Manger, R. Nienhuis, H.M. Fahringer & J. Petigrew, 1994. Novel sleep state organisation in the echidna: implications for the evolution of REM and non-REM sleep. *Sleep Research* 23, 37.

Simpson, G., 1944. *Tempo and Mode in Evolution*. New York (NY): Columbia University Press.

Simpson, G., 1961. *Principles of Animal Taxonomy*. New York (NY): Columbia University Press.

Sneath, P., 1957. The application of computers to taxonomy. *Journal of General Microbiology* 17, 201–26.

Sneath, P., 1968. International conference on numerical taxonomy. *Systematic Zoology* 13, 420–23.

Sokal, R., 1959. Comments of quantitative systematics. *Evolution* 13, 420.

Sokal, R., 1969. Review of Mayr's principles of systematic zoology. *Quarterly Review of Biology* 44, 209–11.

Stringer, C.B., 1984. The definition of *Homo erectus* and the existence of species in Africa and Europe. *Courier Forschungsginstitut Senckenburg* 69, 131–44.

Stringer, C.B., 1987. A numerical analysis for the genus *Homo*. *Journal of Human Evolution* 16, 135–46.

Swisher, C.C., G.H. Curtis, T. Jacob, A.G. Getty & A. Suprijo, 1994. Age of the earliest known hominids in Java, Indonesia. *Science* 263, 1118–21.

Swofford, D., 1993. PAUP: Phylogenetic Analysis Using Parsimony Version 3.1.1. Computer program distributed by the Illinois Natural History Survey. Champaign (IL): Illinois Natural History Survey.

Tinbergen, N., 1951. *The Study of Instinct*. Oxford: Clarendon Press.

Tinbergen, N., 1959. Behaviour, systematics, and natural selection. *Ibis* 101, 318–30.

Tobias, P.V., 1991. *Olduvai Gorge IV: the Skulls, Endocasts and Teeth of* Homo habilis. Cambridge: Cambridge University Press.

Van Twyer, H., 1969. Sleep patterns of five rodent species. *Physiology and Behavior* 4, 901–5.

Waddington, C., 1960. Evolutionary adaptation, in *The Evolution of Life*, ed. S. Tax. Chicago (IL): University of Chicago Press.

Wei, Q., 1985. Paleoliths from the lower Pleistocene of the Nihewan beds in the Donggutuo site. *Acta Anthropologica Sinica* 4(4), 289–300.

Wenzel, J., 1992. Behavioural homology and phylogeny. *Annual Review of Ecology and Systematics* 23, 361–81.

Wood, B.A., 1984. The origin of *Homo erectus*. *Courier Forschungsginstitut Senckenburg* 69, 99–111.

Wood, B.A., 1991. Who is the real *Homo habilis*? *Nature* 327, 187–8.

Wood, B.A., 1992. Early hominid species and speciation. *Journal of Human Evolution* 22, 351–65.

Wu, R. & J. Olsen, 1985. *Palaeoanthropology and Palaeolithic Archaeology in the People's Republic of China*. Orlando (FL) & New York (NY): Academic Press.

Wynn, T., 1988. Tools and the evolution of human intelligence, in Byrne & Whiten (eds.), 217–84.

Wynn, T., 1989. *The Evolution of Spatial Competence*. Urbana (IL): University of Illinois Press.

Zepelin, H., 1989. Mammalian sleep, in *Principles and Practice of Sleep Medicine*, eds. M. Kryer, T. Roth & W.C. Dement. 1st edition. Philadelphia (PA): W.B Saunders, 30–49.

Zepelin, H., 1994. Mammalian sleep, in *Principles and Practice of Sleep Medicine*, M. Kryer, T. Roth & W.C. Dement. 2nd edition. Philadelphia (PA): W.B. Saunders, 69–80.

Zepelin, H. & A. Rechtschaffen, 1974. Mammalian sleep, longevity and energy metabolism. *Brain Behaviour and Evolution* 10, 425–70.

10

Consciousness and Evolution in Neocortex

Peter G. Grossenbacher

What can brain evolution tell us about the human mind? This paper explores neuroanatomical-evolutionary constraints on cognition and perception by considering the evolutionary origin and psychological function of neural systems located in neocortex. A brief review of cortical neuroanatomy reveals intriguing correlation between mental and evolutionary aspects of brain organization. Two hypothetical links between evolution of cortical systems and consciousness are considered. Correlation between phylogenetic recency and phenomenological salience suggests that representations computed in recently evolved cortical systems are prioritized in consciousness. This hypothesis provides a plausible account for disparity in levels of conscious mediation among cortical systems. In addition, a complementary hypothesis suggests that conscious mediation of cortical function could facilitate evolutionary development of relevant neural networks. That is, neural mechanisms underlying consciousness may support new, tentative mental capacities which might otherwise fail to function sufficiently well to provide adaptive advantage. If correct, this idea could help explain the positively accelerated rate of cognitive development evident in the recent evolution of our own species. The speculations presented here are motivated by recent advances in comparative neuroanatomy and cognitive neuroscience.

In biology, evolution constitutes a fundamental dimension for explaining how things have come to be as they are today. Ultimately, evolution could provide an important axis for understanding almost every part of biology, including psychology. Many existing psychological mechanisms depend on ancestral adaptations (Cosmides & Tooby 1987; Parker & Gibson 1979). Therefore the modern human brain must contain neural systems which evolved to support our current repertoire of perceptual and cognitive abilities. By relating psychological functions to neuroanatomy, knowledge of brain evolution can help to illuminate the murky origin of human and pre-human minds.

Generally speaking, mammals share a number of sense modalities, including vision, hearing, smell and touch. Though the peripheral sensory organs (e.g. retina, cochlea) differ substantially across species, at least as much variation is found in the networks of brain cells (neurones) which process the sensory information. Many sensory brain systems are highly content-specific and can be meaningfully compared across genera. For example, comparison of modality-specific systems in primate brains reveals that during the evolution of pre-human ancestors, visual brain systems became greatly elaborated, whereas *olfactory* systems, which process smells, became smaller and less important (Stephan & Andy 1970). Both conscious experience and unconscious psychological mechanisms must be constrained by functional limits of modality-specific brain systems.

This analysis can be extended to compare isolable perceptual dimensions within a sense modality. For example, human brain systems appear to be biased so that colour occupies conscious awareness more than do other perceptible attributes such as location of objects in visual space. Several other authors in this volume mention that colour was exceedingly important for early humans. This salience may generalize across many fruit-eating primates, and indeed fructivorous diet offers a plausible adaptive impetus for colour salience in primates (Mollon

1989). In Galagos, a primitive primate genus, the cortical colour processing area V4 seems to be lacking, indicating that this brain area possibly developed specifically within the primate radiation (Preuss *et al.* 1993). Might the priority for colour in conscious perception relate to the recent evolutionary development of cortical colour processing?

This chapter presents a framework for explaining human mental function with respect to an evolved brain. The following questions are addressed: What is the evolutionary history of perceptual brain mechanisms? Does the evolutionary development of brain mechanisms constrain their psychological functions? How does evolutionary recency relate to consciousness? The following section reviews relevant findings from comparative neuroanatomy, highlighting the importance of cytoarchitecture for assessing evolutionary development of cortical systems. The subsequent section describes the functional organization by which neural tissue computes mental operations to perform psychological tasks. After this brief and selective neurological survey, the rest of the chapter offers speculation on neural underpinnings of consciousness and relations between mental and neuroanatomical evolution.

Comparative neuroanatomy

Because brains are composed of soft tissue which does not preserve in the fossil record, brain evolution is best illuminated by comparative neuroanatomy of existing species considered within a phylogenetic framework. We will first consider inter-species variation among mammalian brains. Comparison of brain organization in rat and monkey will serve to illustrate an evolutionary trend toward sense modality segregation. Of course, we must be careful not to equate differences among extant species with evolutionary history: monkeys did not evolve from rats. According to the theory of evolution, rats and monkeys share a common ancestral species from which both lines diverged. However, phylogenetic relations between existing species can be estimated via cladistic analysis, a method for reconstructing parts of the ancestral tree (see Chapter 9 by Robson Brown this volume, for explanation of this method). The marked divergence of primates from general mammalian trends does show a clear pattern of uniquely primate evolutionary development. After pointing out important aspects of primate neuroanatomy, we will proceed to comparison among primates, with emphasis on the genus *Homo*. The final part of this section describes evolutionary aspects of cortical architecture.

This chapter exclusively considers the neocortex without describing what is known about the evolutionary history and psychological relevance of other brain structures. Though many subcortical brain systems are clearly important to psychology, a 'cortical chauvinism' narrows the scope of discussion in this paper in order to facilitate efficient comparison of a few neural systems which clearly play a role in conscious perception. During the last decade, several mental functions have been physiologically localized, at least in part, to particular areas of cortex. For example, visual area V4 contains neural networks involved in processing colour in both humans and monkeys (Schein & Desimone 1990; Zeki *et al.* 1991). This functional-anatomical mapping provides a method for relating mind to brain in a specific, though piece-meal and probably incomplete fashion (Heywood *et al.* 1992). Combined with knowledge of brain evolution, recent results from neurocognitive studies offer a new, evolutionary-neuroanatomical framework for explaining how the human mind has come to be as we find it today.

Sensory specialization in primate brains

How do primate brains differ from other mammalian brains? Rat brain contains a relatively undifferentiated, highly interconnected cortex with incomplete segregation of sense modalities (Miller & Vogt 1984; Paperna & Malach 1991; Reep *et al.* 1984). Cortical areas specialized for the sense modality of vision nonetheless receive signals from other modalities as well (Haenny *et al.* 1988). Unlike the rat, primate cortex boasts fully independent primary receiving areas which do not interconnect. These areas appear to process elemental features of stimulus attributes, and outputs are sent via axonal projection fibres to modality-specific association areas (Hubel & Livingstone 1990; Hubel & Wiesel 1972). As a group, primates have evolved increasingly specialized input (and output) cortical areas for interfacing with the world. Primary sensory areas, which receive incoming sensory signals and send signals on to secondary and higher-order cortical areas, are comprised of the most recently evolved types of cortical tissue (Pandya *et al.* 1988). Subsequent evolutionary changes in cortex have not produced new forms of cortical tissue. Instead, new circuits have developed using previously existing neuronal arrangements.

Consider the psychological ramifications of increased segregation among sense modalities. As substantial cortical processing is increasingly segregated by sense modality, the means by which sensory information enters the nervous system might be more readily apparent. Sensory information could be more readily tagged as heard, seen, etc.. Concomitant with increased modality segregation, the number of modality-specific association areas in cortex appears to increase. The proliferation of areas devoted to particular components of sensory processing presumably enriches the sensory dimensions available for perception and imagery (Zatorre & Halpern 1993). In such a system, combination of information across modalities requires integration among highly differentiated sensory channels. The set of possible sensory channels requiring integration is larger and differentiated with greater specificity than in systems having less modality segregation. In this way, increasingly abstract representation could derive from the neuroanatomical consequences of increasing modality segregation.

The information processing accomplished in human primary sensory areas may not differ much from early cortical processing in other primates. However, in addition to the modality-specificity typical of primate brains, humans have evolved highly complicated central processing capacities. Some of these specializations are shared with anthropoids, such as some of the association areas in temporal lobe (Preuss & Goldman-Rakic 1991). Yet in *Homo sapiens* we see greater expansion of phylogenetically old parts of cortex compared to other primates. Uniquely human frontal, temporal and parietal circuits handle sensory information which has been pre-processed by modality-specific primary areas. What does this tell us about the early human mind? In primates, sense modalities are more highly segregated than in any other family. And humans have the largest mass of neural circuitry positioned to make sense of these parallel, independent input streams. No doubt our abstract reasoning skills depend on these neural circuits. The puzzle is that these uniquely human brain networks have developed in cortical tissue of an ancient form.

A brain with highly segregated input pathways must somehow make connections between modalities, so that, for example, the smell and sight of a fruit can be associated. In most mammalian brains (e.g. the rat) this association can happen relatively early in sensory processing via direct anatomical connectivity. It is not known in detail how primate brains, which lack direct connectivity among primary sensory areas in cortex, achieve multimodal integration. Multimodal association areas in cortex contain cells which probably play an important role. It seems very telling that it is precisely these areas which are greatly expanded in humans. For example, Wernicke's area is situated in a multimodal region receiving input from unimodal association areas, and is considered necessary for meaningful language (Galaburda & Pandya 1982; 1983; Geschwind 1965). Frontal association areas, as discussed by Robbins in this volume, are strongly involved in plans and other representations concerned with guiding brain output. Multimodal areas in parietal and temporal lobes, on the other hand, are involved in perception and imagery of space and objects (Roland 1982; Roland *et al.* 1987; Roland & Friberg 1985; Roland & Skinøj 1981). Later we will examine the claim that some multimodal brain areas play a major role in mediating conscious awareness.

Relating neuroanatomy to evolution: cytoarchitecture and cortical recency

Different kinds of cortical organization can be distinguished within a single brain, according to the distribution of large and small neurones among different layers lying tangential to the cortical surface. *Cytoarchitecture* is crudely specified by the number and thickness of cortical layers found in an area (Zilles *et al.* 1982). For example, primary visual, auditory and somatosensory cortical areas (V1, A1 and S1) have highly similar cytoarchitecture, all containing six clearly distinguishable layers, including a thick fourth layer. A more complete specification of cytoarchitecture includes the number and types of neurones found in each cortical layer. For example, primary sensory areas in primates are characterized by a dense distribution of small neurones in the fourth cortical layer, resembling a thick layer of dust when viewed through a microscope, hence the name konio- (dust) cortex. Primary areas V1, A1 and S1 also share a similar anatomical connectivity, as it is the small cells in the fourth layer to which afferent axonal fibres (from thalamus) project, bringing sensory signals to cortex. Although the primary sensory areas in these three modalities are located far from each other in distinct brain lobules, they resemble each other in physiological and anatomical aspects.

Following the lead of Sanides, Pandya and others have inferred the evolutionary history of cortical development in order to explain cortical structure (Pandya & Yeterian 1985; Sanides 1969; 1972). In primates, primary sensory cortical areas share a recently evolved cytoarchitecture which is unique to these areas. In contrast, multimodal areas of cortex which receive inputs from multiple sense modalities have a different cytoarchitecture, one resembling that found in mammals with small, primitive-looking cortices. Koniocortex, the most recently evolved type of cortical tissue, comprises the modality-specific input interface of cortical processing (primary sensory areas). Archicortex, the older type of cortical tissue, is anatomically connected so as to receive inputs from multiple modalities, and handles some of the most evolutionarily advanced cognitive functions (e.g. language) (Braak 1978; Pandya & Yeterian 1985). Over the course of evolution, newly specialized forms of cortical tissue developed to pre-process sensory signals before delivery to other sensory cortical areas. The change in signals provided by this new sensory interface may have forced the neural systems built in more ancient cytoarchitecture to become modified. Yet because the older cytoarchitecture was preserved, any evolutionary changes in the multimodal association systems did not require modification of the basic tissue structure. Indeed, the evolutionarily ancient forms of cortical tissue may be ideally suited for neural circuits which must handle complex or evolutionarily new patterns of information across a wide range of content domains.

Functional organization of an evolved brain

The anatomical detail delivered by cytoarchitectonic description does not explain how brains go about their business. Something more is needed in order to understand how neural networks carry out mental operations.

This section describes functionally important levels of brain organization which emphasize dynamic aspects of information processing and neural communication. By 'circuit' let us mean a set of synaptically connected neurones which, through a particular pattern of inhibitory and excitatory connections with each other, perform computations by transforming input signals into output signals which can be taken as input by other circuits. Note that a circuit is not merely an anatomical grouping based solely on connectivity. Rather, a set of neurones comprise a circuit by virtue of their being *functionally related* to each other. Though the number of neurones constituting different circuits might range over a few orders of magnitude, it may be useful to distinguish 'networks' as higher-order assemblages of functionally related circuits.

Functional brain areas and neural pathways

Human brains have anatomically localized neural networks which support particular mental operations. For example, a cortical area in left occipital lobe is metabolically active during perception of visually presented words, but this localized network is so content-specific that it does not appear active during perception of character strings which do not closely resemble words (Posner *et al.* 1989). Localization of visual word representation in occipital cortex reveals several important aspects of brain organization. First, it demonstrates the degree to which a cortical network can be dedicated to handling highly specific contents. Perhaps this visual area originated for some ancestral visual task such as identification of plants, but today can be specialized for word recognition. Second, given that children do not learn to recognize words until they are several years of age, we see that learning is necessary for the visual word area to develop its functional specialization. Thus experience can strongly affect functional brain organization. Finally, it would be a mistake to think that neural circuitry in this visual word area operates independently from circuits located in other brain areas. For example, visual feature analysis of letters or similar stimuli involves an area in right parietal cortex (Posner *et al.* 1989). When a literate person views a word, as visual information becomes available to these two cortical areas, the featural output produced by the parietal visual area may serve as input to the occipital area involved in word recognition. Yet performance in letter discrimination tasks can be influenced by whether the letter appears in a word or not (Reicher 1969), suggesting that the visual processing required for letter discrimination involving, say, the parietal visual area, must somehow be influenced by the psychological computations conducted by the occipital word area. How does the brain achieve two-way communication between anatomically separate areas which compute different levels of representation (i.e. letters and words)?

The brain continuously receives neural messages originating from a variety of sensory receptors which signal physical events inside and outside the body. How does this information flow through the brain? Some parts of the brain deal exclusively with a single sense modality, such as vision or touch, whereas other parts receive inputs from multiple modalities. Within a single sense modality, incoming afferent signals proceed through a sequence of distinct cortical areas which together comprise a modality-specific pathway. Typically, visual, auditory, and tactile pathways lead to multimodal areas of cortex which receive information from all three exteroceptive senses (Pandya & Seltzer 1982). Within a neural pathway, signalling occurs in both 'bottom-up' (afferent) and 'top-down' (efferent) directions (Cynader *et al.* 1988). Because cortical areas within a pathway interconnect with both feed-forward and reciprocal feed-backward connections, incoming sensory signals can be modulated by top-down control signals. This two-way signalling within a pathway provides a mechanism by which modality-specific processing can be influenced by higher-ordered associations computed further along the pathway. This neural mechanism might well mediate the psychological interaction between visual word and letter representations discussed above.

Let us now consider two neural pathways which convert sensory information into object representations. Faces and other purely visual aspects of objects are represented in a connected series of cortical areas in the temporal lobe (Baylis *et al.* 1987). The functional nature of this visual pathway is clearly indicated by neurophysiology: it contains cells which are selectively responsive to particular views of hands, heads and other objects (Perret *et al.* 1987). Damage to this area in humans results in visual agnosia, a syndrome in which objects cannot be recognised by sight, despite unhindered perception of elementary visual dimensions such as colour (Farah 1990). A similarly structured neural pathway runs parallel to the visual object recognition pathway along the top of the temporal lobe. The superior temporal sulcus contains a series of interconnected areas, each with cells which respond to multiple sense modalities (Bruce *et al.* 1981; Seltzer & Pandya 1989). This strip of adjacent, connected areas constitutes a neural pathway probably involved in representation of multimodal object attributes. Here the neural signals arising from the sound of a dog's bark and the sight of a swishing tail may come together.

In temporal cortex there appears to be two object recognition pathways, one multimodal and one purely visual. Did one of these systems evolve first, providing a prototype which was then anatomically replicated to form the other system? Generally speaking, cortical sensory areas have evolved such that new cortex is unimodal and old cortex is multimodal (Pandya & Seltzer 1982). Perhaps the visual object recognition system evolved as a specialized copy of the multimodal pathway. Alternatively, this replication may have occurred very early in the development of these pathways, after which the two could have shared a parallel developmental course. This latter account is favoured by the observation that, in both object recognition pathways, feed-forward signals project through a sequence of cortical areas which gradually shift from relatively new to relatively old cytoarchitectures (Pandya & Seltzer 1982). It seems that new cortical areas evolve to compute output representations adaptively suited as input to the recently evolved cortical area to which they project in a neural pathway.

Brain evolution and system development

Recently evolved brain mechanisms provide an organism with mental capacities which other organisms (lacking those brain mechanisms) may not share, thus providing an adaptive advantage. Consider two alternative models of brain evolution. Suppose that, as new networks appear through the course of evolutionary change, they are added as a new component to existing systems. Let us call this the 'independent addition model'. This model emphasizes modular aspects of brain organization, and posits aggregation of new structures which interact with evolutionarily prior neuroanatomy. As a first approximation, the independent addition model has some pedagogical value, but it cannot be correct in detail (Derryberry & Tucker 1991). More realistically, we can distinguish an alternative view which asserts that, as new networks evolve, previously existing networks are preserved but can be modified. This form of evolutionary change results in new circuitry which resembles but may diverge from the old system, combined with new components. This 'interdependent innovation model' emphasizes pathway and inter-network communication aspects of brain organization. Using the framework of interdependent innovation, it becomes possible to envisage older forms of cortical tissue adapting to recently evolved neural representation which provide input to the older cytoarchitecture. This line of inquiry should shed light on the means by which distinct mental modules communicate, a vital aspect of human intelligence (Mithen 1994; Rozin & Schull 1988).

Content-specific vs *content-general brain systems*

The brain in general, including the cortex in particular, contains many dedicated systems of interconnected neurones which operate only on particular kinds of inputs, perform particular kinds of computations, and deliver particular outputs. Systems specialized for particular sense modalities provide clear examples of such *content-specific* dedicated brain systems. For example, somatosensory cortex processes sensory information from touch and bodily senses, but never handles information regarding smell or colour. The functional specificity of the visual word area mentioned previously hints at the large number of content-specific networks which comprise much of the human brain's repertoire of mental skills.

Co-ordination among a large number of content-specific systems may require some mechanisms which have a more general nature. Take awareness of objects as an example. The same neural mechanisms may mediate awareness of an apple regardless of whether it was seen, heard, smelled or felt by hand. This type of awareness may depend on a brain network involving the anterior cingulate gyrus (Posner & Petersen 1990), a phylogenetically ancient form of cortical tissue which receives inputs from most every sense modality and also sends axonal projections into a large number of brain areas (Pandya *et al.* 1981). Anterior cingulate is metabolically active during many different kinds of perception and cognition (Pardo *et al.* 1990). Its level of metabolic activity correlates with the number of noteworthy stimuli perceived during a time period of about one minute (Posner *et al.* 1988). Thus consciousness, of sensory information at least, may depend on neural circuitry in anterior cingulate cortex (Posner & Rothbart 1991). In addition, lesions in this part of the brain result in complete lack of self-initiated movement (Damasio & Van Hoesen 1983), indicating a possible role in volition as well as perception. Though future studies may reveal that other brain parts are important for mediating conscious awareness, our purpose is served by the anterior cingulate gyrus providing an example of a content-general neural system empirically linked to conscious awareness.

Though there may be other content-general systems (e.g. brain structures involved in memory storage), this chapter restricts discussion to content-general networks important for conscious awareness and control. The distinction between content-specific and content-general neural systems suggests a constraint on access of perceptual information to conscious awareness: awareness may be biased toward those sensory dimensions which depend on content-specific networks having relatively profuse connections with content-general circuitry required for conscious awareness. What factors determine whether a particular sensory dimension is well connected (anatomically and psychologically)? Heritable brain organization traits must provide fundamental constraints, although these constraints leave room for great variability among individuals. Life experiences which modify functional organization within an individual brain play a powerful role. For example, a musician and a painter can have brains that differ as to the ease with which auditory or visual sensory dimensions impact awareness.

Awareness, control and brain evolution

Conscious awareness

By *consciousness* let us mean awareness of phenomena (e.g. perceived objects during perception, memories during remembrance, or thoughts during cogitation). The contents of this consciousness can be observed subjectively, from the first person perspective. For simplicity, we will focus on consciousness in perception. What determines the kinds of perceptual information which impact conscious awareness? Traditional psychological answers to this question have correctly emphasized the roles of arousal, motivation, attention, and stimulus intensity. But perhaps there is an additional class of theoretical constraints, based on brain evolution, which bear on this question.

In consciousness, certain kinds of perceptual information generally *dominate* over other kinds. That is, typically there is a preponderance of some sorts of sensory information, and not other sorts, in the contents of consciousness. This sort of dominance is discernible in subjective self-observation. As an easy thought experiment, ask yourself now: When you perceive something, what aspects occupy your conscious awareness? You could conduct this phenomenological exercise under a few different conditions, changing where you look, changing posture (sitting *vs* standing), etc..

Now see if your experience matches the following description. In comparing the relative salience of *where* something is and *what* it is, location in space usually serves as background information which is dealt with automatically, whereas aspects such as color and shape occupy conscious awareness more. In general, certain visual aspects (colour) often seem more salient than other visual information (location) or non-visual aspects (smell). Admittedly, exceptions to this rule are numerous and frequent: conscious awareness can certainly be absorbed by music or the fragrance of a flower. But in these cases, under what conditions do non-visual aspects most *fully* absorb conscious attention? Many people *close their eyes*, thereby blocking visual inputs, in order to best appreciate an orchestral melody or the scent of a rose. Even without such obvious control of sensory inputs, perceptual pathways with feed-backward neural connections allow for more subtle modulation via top-down control. Thus, we are able to temporarily focus awareness on (selectively attend to) particular stimuli, such as the feel of clothing contacting the skin, without making it obvious to others what we are attending to.

Controlled vs *automatic behaviour*

In addition to awareness of information delivered by the senses, the control of overt behaviour (e.g. via skeletal muscles) can also involve consciousness. Consider what happens when learning a complex skill. We will take automobile driving as a familiar example, but the logic applies equally well to other motor tasks such as making stone tools or climbing a tree. The novice is all too aware of each component action involved in shifting gears, steering, braking, etc.. With many hours of practice, the disturbingly detailed conscious control gradually gives way to less conscious involvement in performance. Extended practice can naturally lead to *learned automatization* of the behaviour. Automatic behaviours have been described as efficient yet inflexible compared to similar behaviours guided by conscious control (Schneider & Shiffrin 1977).

What happens in the brain of an individual during acquisition of learned automatization? Presumably, activity in circuits mediating conscious control mechanisms gradually becomes less important for performance. If neural signalling in these circuits decrease as automaticity is learned, then the remaining circuits (i.e. communication between content-specific brain systems) must change to embody the learned skill. Such learned changes can happen in brain circuits simply by modification of the synaptic links which mediate communication between neurones. In terms of neural network modelling, it is the relative weights of connections (between neurones) which change.

Once a skilled behaviour has been practised many times, through learned automaticity it can place reduced demands on conscious control mechanisms. Perhaps there is a similar transition in evolution where previously controlled processes can proceed with less conscious control. Of course, there must be important differences between the neural changes underlying *evolved automaticity* and learned automaticity.

Consciousness and communication among brain systems

Perception depends on patterns of activation in a complex hierarchy of sensory brain areas (Amir *et al.* 1993) which somehow supply contents of consciousness. Brain mechanisms of consciousness are not well understood, so the following account must be regarded as speculative, though it has been informed by current findings in neural science. One primary function of neural mechanisms underlying consciousness could be facilitation of inter-network communication. Perhaps it is by virtue of selected processes sharing their outputs in a functional foreground delineated by content-general brain systems that aspects of these shared outputs can be experienced in conscious awareness.

In brains which are increasing in complexity through evolutionary development, as the number of content-specific systems increases over time, how might interaction between content-general mechanisms and content-specific circuits be affected? On one hand, there may be an increase in generality of general mechanisms, as they are required to co-ordinate among a larger and more diverse assemblage of specialized systems. Thus, brain mechanisms involved in conscious awareness and control could accumulate increasing capacity for co-ordination between content-specific processes drawn from an increasingly large repertoire. In this way, the richness of conscious experience might increase during evolution as a function of the variety of potentially conscious content-specific processes. An intriguing ramification of increased generality for evolution of mind is discussed later in support of the Conscious Mediation Hypothesis.

On the other hand, under some conditions a content-specific system may become *less* involved with one or more general mechanisms during evolutionary development, either by replacement or automatisation. Suppose that the contribution of a dedicated system is gradually replaced by a functionally similar, but more adaptive, system. In this case, the original system might decreasingly interact with most systems with which it communicates, including conscious mechanisms. For example, as vision increased in neuroanatomical complexity and smell decreased during human evolution, the interaction of these sense modalities with brain mechanisms of consciousness probably shifted so that smell became less involved with consciousness.

But replacement may not be the only mechanism for decreasing interaction between a dedicated system and more general brain systems. As already mentioned, there may be an evolutionary analog to learned automaticity. A content-specific system which can run independently from consciously co-ordinated brain systems could become increasingly automated during evolutionary development. Rather than modification of synaptic strength (the probable mechanism for learned automaticity), evolved automaticity could be accomplished by an increase in the number of axonal connections linking a content-specific system to other, functionally related content-specific brain systems. Suppose, for example, that somatosensory brain networks in humans now have relatively less direct connectivity with content-general circuits involved in conscious awareness (in proportion to the total set of dedicated systems subject to conscious mediation) compared to our ancestors. This arrangement could reflect evolved automaticity wherein conscious awareness and control of bodily states decreased during our evolutionary development. Why might this happen? Evolved automaticity would be expected to develop for processes having inputs and computational demands which change little over large numbers of generations. This stability constraint does seem more likely to be met in perception of internal environments (interoception) than for perception of stimuli outside the organism (exteroception).

Cortical recency hypothesis: explaining the contents of consciousness

Though there is much mapping between mind and brain for cognitive neuroscience yet to complete, at this point we can take some of the early findings and sketch the beginning of a neuroanatomical evolutionary account of mental function. No doubt later versions of this account will prove more detailed and accurate. But even a preliminary enquiry using an anatomical-evolutionary framework might be useful. In particular, it offers a new set of constraints for modelling the mind. In order to detail one way in which evolutionary brain development can impact psychological theory, this section presents and discusses the Cortical Recency Hypothesis, intended as an example of how psychological organization could be explained in terms of brain evolution.

Contents of consciousness range widely over perceptual, motor and semantic processes. The richness of conscious experience must somehow depend on the complex web of interconnected cortical systems involved in perceptual, semantic and motor control. As already mentioned, some phylogenetically older cortical functions may become automated during evolutionary development, requiring less conscious guidance. Thus, in the course of evolution, conscious control capacities originally devoted to these older functions may become available for coordination with new functions. Perhaps for this reason, *more recently evolved cortical functions have greater access to consciousness than do older brain systems*. This Cortical Recency Hypothesis describes a speculative link between brain evolution and conscious awareness which provides a possible basis for explaining tendencies in human phenomenology.

Cortical recency and colour salience

Assuming that more recently evolved cortical systems contribute more to conscious awareness, the phylogenetic recency of the visual object recognition pathway can explain why human consciousness is so often dominated by visual perception. Moreover, to the extent that imagery relies on the same brain areas involved in perception (Boussaoud *et al.* 1991; Kosslyn 1988), the kinds of information most saliently represented during imagery should correspond with those which dominate consciousness during perception. This prediction is supported by phenomenological salience of imagined objects and object attributes such as color and form. Information relevant to visual object recognition may dominate conscious awareness in perception because the brain systems involved in processing this information are recently evolved, perhaps more recent than the neural mechanisms required for other perceptual functions such as spatial localization (Komatsu *et al.* 1992; Preuss & Goldman-Rakic 1991). As an aside, it may be worth noting that synesthesia, an unusual condition manifest in many individuals for whom information picked up in one sense modality is experienced in another, most often involves colour as the synthesised dimension (Baron-Cohen *et al.* 1987; Cytowic & Wood 1982; Karwoski *et al.* 1942; Marks 1975; Riggs & Karwoski 1934).

Language and cortical recency

To the extent that auditory perception and imagery can dominate consciousness, these phenomena should depend on recently evolved cortical brain systems. Preliminary evidence accords with this prediction (Roland *et al.* 1981). Modern language makes use of elaborated sensory representations and ability to image, as well as the ability to plan and sequence complex motor activity. Probably these elaborated input and output representations were necessary for language to develop. To the extent that 'mental speech' dominates consciousness, conscious verbalisation should depend on recently evolved cortical functions. Human language does depend on cortical systems located in multimodal association areas (Galaburda & Pandya 1982). It remains to be determined whether or not specifically *conscious* aspects of language (e.g. semantics) depend on recently evolved cortical systems more than do *unconscious* aspects (e.g. syntax).

Exceptions?

It may be important that several potential counter-examples to the recency hypothesis share a common aspect: the capacity to reliably usurp cognitive control on a short-term basis, apparently for purposes of survival of self or offspring. For example, pain, nausea, orgasm, rage, and fear all demonstrate that top-down volitional control does not have a sure grip on behaviour. Thus, urges toward fight or flight, mediated by brain structures outside neocortex, can take over control when need arises. Nonetheless, within cortex, it appears that the customary contents of conscious awareness are supplied by recently evolved circuitry.

Discussion

Methodological limitations

Because ancestral organisms are dead and their brains are not available for inspection, data acquisition in evolutionary neurobiology is indirect, and derives from comparative study of existing organisms. Psychological functions can be at least partially localized to particular brain areas, and many of these functionally distinct cortical areas can be distinguished from each other by the particular patterns in which neurones are arranged within the cortical tissue. These distinct physical arrangements of neurones within cortex can be analyzed according to evolutionary origin (Pandya & Seltzer 1982; Sanides 1969; 1972). Together, these functional, cytoarchitectural and evolutionary analyses make it possible to relate psychological function to brain evolution. The validity of this method depends on the accuracy of the phylogenetic trees constructed to represent evolutionary differentiation among related species. (See Chapter 9 by Robson Brown for elucidation of this approach applied to the early human.) Moreover, the line of thought developed in this chapter also relies on the neuroanatomist's ability to specify cortical cell organization in a phylogenetic framework.

Conscious mediation hypothesis: consciousness as enabling mechanism

The Cortical Recency Hypothesis claims that contents of consciousness are constrained by brain evolution. Could causation link anatomy and consciousness in both directions? Might access to consciousness affect the evolutionary development of content-specific cortical systems?

Neural mechanisms involved in conscious awareness might support recently evolved (content-specific) capacities which otherwise might flounder without such support. Suppose that a recently evolved content-specific circuit interconnects with a circuit which mediates consciousness (e.g. a content-general system involved in co-ordinating arbitrary sets of dedicated systems). Genetic variation must produce slightly different configurations of brain circuits within a population. Innovative functions enabled by small genetic perturbations in dedicated circuits could be facilitated via conscious mediation, and subsequently become consolidated in subsequent generations through natural selection. Thus, the functional limits of recently evolved neural circuits might be extended within the controlled framework of consciousness.

In order to understand that this line of thought does not take a Lamarkian approach, one need only accept the possibility that fitness can be affected by behaviours (Bateson 1988), and that some of these behaviours depend on content-specific brain systems sharing their outputs. The claim is not that mental skills acquired during an individual's life are passed on to its offspring. Rather, the hypothesis is based on two assumptions. First, capacity for inter-network communication can be inherited. Second, this capacity can affect the viability of content-specific systems, especially those having only marginal connectivity with other brain systems. The idea is that a larger capacity for inter-network facilitation increases the likelihood that a nascent neural system will contribute to fitness.

In this way, *consciousness might guide the leading edge of evolutionary development by facilitating communication between newly evolved mental functions and other mental functions.* The net effect could be an increase in the speed with which new mental capacities evolve. This Conscious Mediation Hypothesis provides a possible neurobiological basis for the very rapid increase in cognitive abilities realized by our hominid ancestors.

Summary

To the extent that mind and brain are two descriptive levels regarding the same organic system, it can be useful to study mental and neural mechanisms in a joint theoretical framework. *Cognitive neuroscience* is a recently adopted label for the integrated study of mind and brain. In practice this approach has shown little concern for evolution. Extending cognitive neuroscience along an evolutionary axis should yield two payoffs. First and foremost, a fundamental biological theory (evolution) becomes available for explaining data already in hand. Evolution offers a unique and important level of explanation of psychological and neural mechanisms which complements that provided by ontological development of the individual. Thus, the hypothesis that cortical recency facilitates salience in conscious awareness offers an explanation of contents of consciousness in terms of evolved anatomy. Second, new questions may present themselves when psychological issues are approached from an understanding of brain evolution. We already know that conscious experience can strongly influence cortical circuitry during an individual life span. Can conscious mediation also facilitate evolutionary development of cortical systems?

Peter G. Grossenbacher
Laboratory of Psychology and Psychopathology
National Institute of Mental Health
Building 10, Room 4C110
10 Center Drive MSC 1366
Bethesda, MD 20892-1366
USA

References

Amir, Y., M. Harel & R. Malach, 1993. Cortical hierarchy reflected in the organization of intrinsic connections in Macaque monkey visual cortex. *Journal of Comparative Neurology* 334, 19–46.

Armstrong, E. & D. Falk (eds.), 1982. *Primate Brain Evolution*. New York (NY): Plenum Press.

Baron-Cohen, S., M.A. Wyke & C. Binnie, 1987. Hearing words and seeing colours: an experimental investigation of a case of synaesthesia. *Perception* 16, 761–7.

Bateson, P., 1988. The active role of behaviour in evolution, in *Evolutionary Processes and Metaphors*, eds. M.-W. Ho & S.W. Fox. Chichester: John Wiley & Sons, 191–207.

Baylis, G.C., E.T. Rolls & C.M. Leonard, 1987. Functional subdivisions of the temporal neocortex. *Journal of Neuroscience* 7, 330–42.

Boussaoud, D., R. Desimone & L.G. Ungerleider, 1991. Visual topography of area TEO in the macaque. *Journal of Comparative Neurology* 306(4), 554–75.

Braak, H., 1978. On magnopyramidal temporal fields in the human brain — probable morphological counterparts of Wernicke's sensory speech region. *Anatomy and Embryology* 152, 141–69.

Bruce, C., R. Desimone & C.G. Gross, 1981. Visual properties of neurons in a polysensory area in superior temporal sulcus of the macaque. *Journal of Neurophysiology* 46, 369–84.

Cosmides, L. & J. Tooby, 1987. From evolution to behaviour: evolutionary psychology as the missing link, in *The Latest on the Best: Essays on Evolution and Optimality*, ed. J. Dupré. Cambridge (MA): MIT Press, 277–306.

Cynader, M.S., R.A. Andersen, C.J. Bruce, D.R. Humphrey, V.B. Mountcastle, H. Niki, G. Palm, G. Rizzolatti, P. Strick, N. Suga, W. von Seelen & S. Zeki, 1988. General principles of cortical operation, in *Neurobiology of Neocortex*, eds. P. Rakic & W. Singer. Chichester: John Wiley & Sons, 353–71.

Cytowic, R.E. & F.B. Wood, 1982. Synesthesia 1. A review of major theories and their brain basis. *Brain and Cognition* 1, 23–35.

Damasio, A.R. & G.W. Van Hoesen, 1983. Emotional disturbances associated with focal lesions of the limbic frontal lobe, in *Neuropsychology of Human Emotion*, eds. K. Heilman & P. Satz. New York (NY): Guilford, 85–99.

Derryberry, D. & D.M. Tucker, 1991. The adaptive base of the neural hierarchy: elementary motivational controls on network function, in *Nebraska Symposium on Motivation*, vol. 38: *Perspectives on Motivation*, ed. R. Dienstbier. Lincoln (NA): University of Nebraska Press, 289–342.

Farah, M.J., 1990. *Visual Agnosia: Disorders of Object Recognition and What They Tell Us About Normal Vision*. Cambridge (MA): MIT Press.

Galaburda, A.M. & D.N. Pandya, 1982. Role of architectonics and connections in the study of primate brain evolution, in Armstrong & Falk (eds.), 203–16.

Galaburda, A.M. & D.N. Pandya, 1983. The intrinsic architectonic and connectional organization of the superior temporal region in the rhesus monkey. *Journal of Comparative Neurology* 221, 169–84.

Geschwind, N., 1965. Disconnexion syndromes in animals and man. *Brain* 88, 237–94, 585–644.

Haenny, P.E., J.H.R. Maunsell & P.H. Schiller, 1988. State dependent activity in monkey visual cortex, II: Retinal and extraretinal factors in V4. *Experimental Brain Research* 69(2), 245–59.

Heywood, C.A., A. Gadotti & A. Cowey, 1992. Cortical area V4 and its role in the perception of color. *Journal of Neuroscience* 12, 4056–65.

Hubel, D.H. & M.S. Livingstone, 1990. Color and contrast sensitivity in the lateral geniculate body and primary visual cortex of the macaque monkey. *Journal of Neuroscience* 10(7), 2223–37.

Hubel, D.H. & T.N. Wiesel, 1972. Laminar and columnar distribution of geniculocortical fibers in the macaque monkey. *Journal of Comparative Neurology* 146, 421–50.

Karwoski, T.F., H.S. Odbert & C.E. Osgood, 1942. Studies in synaesthetic thinking, II: The role of form in visual responses to music. *Journal of General Psychology* 26, 199–222.

Komatsu, H., Y. Ideura, S. Kaji & S. Yamane, 1992. Color selectivity of neurons in the inferior temporal cortex of the awake Macaque monkey. *Journal of Neuroscience* 12, 408–24.

Kosslyn, S.M., 1988. Aspects of a cognitive neuroscience of mental imagery. *Science* 240, 1621–6.

Marks, L.E., 1975. On colored-hearing synesthesia: cross-modal translations of sensory dimensions. *Psychological Bulletin* 82, 303–31.

Miller, M.W. & B.A. Vogt, 1984. Direct connections of rat visual cortex with sensory, motor, and association cortices. *Journal of Comparative Neurology* 226, 184–202.

Mithen, S., 1994. From domain specific to generalized intelligence: a cognitive interpretation of the Middle/Upper Palaeolithic transition, in *The Ancient Mind*, eds. C. Renfrew & E. Zubrow. Cambridge: Cambridge University Press, 29–39.

Mollon, J.D., 1989. 'Tho' she kneel'd in that place where they grew. . .': the uses and origins of primate colour vision. *Journal of Experimental Biology* 146, 21–38.

Pandya, D.N. & B. Seltzer, 1982. Association areas of the cerebral cortex. *Trends in Neurosciences* 5, 386–90.

Pandya, D.N. & E.H. Yeterian, 1985. Architecture and connections of cortical association areas, in *Cerebral Cortex*, vol. 4: *Association and Auditory Cortices*, eds. A. Peters & E.G. Jones. New York (NY): Plenum Press, 3–61.

Pandya, D.N., G.W. Van Hoesen & M. Mesulam, 1981. Efferent connections of the cingulate gyrus in the rhesus monkey. *Experimental Brain Research* 42, 319–30.

Pandya, D.N., B. Seltzer & H. Barbas, 1988. Input–output organization of the primate cerebral cortex, in *Comparative Primate Biology*, eds. H.D. Steklis & J. Erwin. New York (NY): Alan R. Liss, 39–80.

Paperna, T. & R. Malach, 1991. Patterns of sensory intermodality relationships in the cerebral cortex of the rat. *Journal of Comparative Neurology* 308, 432–56.

Pardo, J.V., P.J. Pardo, K.W. Janer & M.E. Raichle, 1990. The anterior cingulate cortex mediates processing selection in the stroop attentional conflict paradigm. *Proceedings of the National Academy of Sciences of the United States of America* 87, 256–9.

Parker, S.T. & K. Gibson, 1979. A developmental model for the evolution of language and intelligence in early hominids. *Behavioral and Brain Sciences* 2, 367–407.

Perret, D.I., A.J. Mistlin & A.J. Chitty, 1987. Visual neurones responsive to faces. *Trends in Neuroscience* 10, 358–64.

Posner, M.I. & S.E. Petersen, 1990. The attention system of the human brain. *Annual Review of Neuroscience* 13, 25–42.

Posner, M.I. & M.K. Rothbart, 1991. Attentional mechanisms and conscious experience, in *The Neuropsychology of Consciousness*, eds. D. Milner & M. Rugg. New York (NY): Academic Press, 91–111.

Posner, M.I., S.E. Petersen, P.T. Fox & M.E. Raichle, 1988. Localization of cognitive operations in the human brain. *Science* 240, 1627–31.

Posner, M.I., J. Sandson, M. Dhawan & G.L. Shulman, 1989. Is word recognition automatic? A cognitive-anatomical approach. *Journal of Cognitive Neuroscience* 1, 50–60.

Preuss, T.M. & P.S. Goldman-Rakic, 1991. Architectonics of the parietal and temporal association cortex in the strepsirhine primate Galago compared to the anthropoid primate Macaca. *Journal of Comparative Neurology* 310, 475–506.

Preuss, T.M., P.D. Beck & J.H. Kaas, 1993. Areal, modular, and connectional organization of visual cortex in a prosimian primate, the slow loris (*Nycticebus coucang*). *Brain, Behaviour and Evolution* 42(6), 321–35.

Reep, R.L., J.V. Corwin, A. Hashimoto & R.T. Watson, 1984. Afferent connections of medial precentral cortex in the rat. *Neuroscience Letters* 44, 247–52.

Reicher, G.M., 1969. Perceptual recognition as a function of meaningfulness of stimulus material. *Journal of Experimental Psychology* 81, 275–80.

Riggs, L.A. & T. Karwoski, 1934. Synaesthesia. *British Journal of Psychology* 25, 29–41.

Roland, P.E., 1982. Cortical regulation of selective attention in man. A regional cerebral blood flow study. *Journal of Neurophysiology* 48, 1059–78.

Roland, P.E. & L. Friberg, 1985. Localization of cortical areas activated by thinking. *Journal of Neurophysiology* 53, 1219–43.

Roland, P.E. & E. Skinøj, 1981. Focal activation of the cerebral cortex during visual discrimination in man. *Brain Research* 222, 166–71.

Roland, P.E., E. Skinøj & N.A. Lassen, 1981. Focal activations of human cerebral cortex during auditory discrimination. *Journal of Neurophysiology* 45, 1139–51.

Roland, P.E., L. Eriksson, S. Stone-Elander & L. Widen, 1987. Does mental activity change the oxidative metabolism of the brain? *Journal of Neuroscience* 7, 2373–89.

Rozin, P. & J. Schull, 1988. The adaptive-evolutionary point of view in experimental psychology, in *Stevens' Handbook of Experimental Psychology*, vol. 1: *Perception and Motivation*, eds. R.C. Atkinson, R.J. Herrnstien, G. Lindzey & R.D. Luce. New York (NY): John Wiley & Sons, 503–6.

Sanides, F., 1969. Comparative architectonics of the neocortex of mammals and their evolutionary interpretation. *Annals of the New York Academy of Sciences*, 167, 404–23.

Sanides, F., 1972. Representation in the cerebral cortex and its areal lamination patterns, in *Structure and Function of Nervous Tissue*, ed. G.F. Bourne. New York (NY): Academic Press, 329–453.

Schein, S.J. & R. Desimone, 1990. Spectral properties of V4 neurons in the macaque. *Journal of Neuroscience* 10(10), 3369–89.

Schneider, W. & R.M. Shiffrin, 1977. Controlled and automatic human information processing, 1: Detection, search and attention. *Psychological Review* 84, 1–66.

Seltzer, B. & D.N. Pandya, 1989. Intrinsic connections and architectonics of the superior temporal sulcus in the rhesus monkey. *Journal of Comparative Neurology* 290, 451–71.

Stephan, H. & O.J. Andy, 1970. The allocortex in primates, in *The Primate Brain*, eds. C.R. Noback & W. Montagna. New York (NY): Appleton-Century-Crofts, 109–35.

Zatorre, R.J. & A.R. Halpern, 1993. Effect of unilateral temporal-lobe excision on perception and imagery of songs. *Neuropsychologia* 31, 221–32.

Zeki, S., J.D.G. Watson, C.J. Lueck, K.J. Friston, C. Kennard & R.S.J. Frackowiak, 1991. A direct demonstratrion of functional specialization in human visual cortex. *Journal of Neuroscience* 11(3), 641–9.

Zilles, K., H. Stephan & A. Schleicher, 1982. Quantitative cytoarchitectonics of the cerebral cortices of several prosimian species, in Armstrong & Falk (eds.), 177–201.

11
Inferring Cognition from Social Behaviour in Non-Humans

P.C. Lee

The recent formulation of the social hypothesis for the evolution of intellect has provoked considerable interest in the ways in which social species of non-human mammals manipulate others within their social group, with cognitive intent and individual recognition as the basis for these manipulations. It becomes important to assess the nature of sociality among different species, in order to highlight the potential commonalty underlying social sophistication, and to unravel the effect of life histories on cognitive abilities. This paper aims to relate cognitive skills such as assessment and social maintenance to patterns of sociality. The diverse modes of sociality existing among mammals, and the life history solutions to problems of reproduction and mortality are discussed in relation to these social skills. Apparently sophisticated cognitive capacities can arise from the conjunction of simple social skills and complex social contexts; thus contexts as well as skills need to be assessed. The existing 'social templates' may act as unique solutions to specific ecological problems, and the precursors to the radical reorganization in cognitive ability in hominid evolution are to be found within these social solutions.

What are the behavioural continuities between animal minds and human minds? This question has been posed many times from Darwin through to recent studies (e.g. Gibson 1993), and numerous detailed investigations of animal cognition have been undertaken. Comparative psychologists have examined the learning and thinking capacities of apes (e.g. Kohler 1927) and other animals (Griffen 1992) over the past five decades. The importance of this question lies in the attempt to determine the nature and extent of differences between animals and humans in the ways in which they respond cognitively to their world. Studies of animal behaviour can potentially shed insight into the processes of learning and memory, of integration and perception, and of the capacity to solve problems (Roitblat *et al.* 1984). Since the human mind can only have arisen from the basic capacities of its ancestors, the animal, and specifically, primate mind at the root of human abilities should provide a useful starting point in such investigations. A brief review of recent findings in non-human primate and other mammals' cognitive skills follows.

With the now well developed 'social hypothesis' for the evolution of intellect in primates (see Jolly 1966; Humphrey 1976; Byrne & Whiten 1988) comes a focus on behaviour in the social context among mammals. Brain size and structure have been related to simple social parameters, such as group size (Dunbar 1992), with an underlying assumption that social complexity is causal in determining cognitive complexity. Debates as to the role of food types (Milton 1984; 1988), food extraction (Gibson 1986), learning of foraging techniques (Byrne 1995), temporal-spatial memory of resources (Sigg 1986) and complexity of procuring food resources (McGrew 1992), in promoting certain levels of cognitive skills, which can then be applied socially, persist and are unlikely to be easily resolved. However, the focus of this brief review is on the other unresolved problems underlying behaviour observed within the social context of non-human mammals.

While elements of cognition can be distinguished, and analyzed separately, it seems more productive to start with the holistic view, espoused by Washburn (1960), that thoughts, brains, behaviour and communication — the elements that we call 'human' — are all part of a single package, with implications for growth, parenting, reproduction, survival and death; all these components are subsumed under the description of a life history. No single element can be used to distinguish the human cognitive package from the primate ancestral package; rather we need to focus on the connections and inter-relations between the diverse elements making up the abilities of each species (Lee 1991). Furthermore, it can be expected that if one part of the package changes, there will be repercussions on other components, which are forced into concomitant changes. Thus rather than generate further debate over the continuum or discontinuity between animal and human minds, I hope to define some starting points within an evolutionary perspective.

The problem of sociality

As a first step, an understanding of the causes of sociality among mammals is central to determining the capacities and expressions of social skills. Sociality among mammals exists for a variety of different reasons, which have been well discussed (see Krebs & Davies 1993). The factors promoting sociality are presented in Table 11.1. It should be noted that the causal factors promoting sociality can be very different from the factors which maintain sociality (Dunbar 1988), and it is this maintenance of sociality which is essential for relating to cognitive complexity and the expression of cognitive skills. In this paper, the term sociality is applied to species where individuals co-reside and interact with each other over time, and develop relationships as a result of these interactions. Thus, sociality is taken here to mean those species who maintain relationships, rather than all species which aggregate on a temporal basis in the face of specific ecological pressures (Lee 1994). The number of such species is relatively small, and specifically excludes the obligate social insects, while including a diverse array of species from swans to ground squirrels to elephants, and the vast majority of the non-human primates.

As a further important point, not all sociality is the same. To say that a species is social, and that the sociality has a specific form overlooks the ways in which the sociality has arisen, how it is maintained, and the consequences for individual behaviour within a social context. What is apparently a similar form of social structure in comparisons between species masks the differences in underlying causation, which can have marked implications for our expectations of cognitive skills. It is worth emphasizing that the possible expression of social life is limited in numbers of options (Foley & Lee 1989; Lee 1994; Fig. 11.1). As a result of these limitations, there is equifinality in social structure; the same mode of sociality can have arisen through a number of different routes.

Differences in causation, despite any resulting similarity in social structure, necessitate differences in the cognitive ability of the animal to cope with its specific evolutionary and extant social domain. Thus we expect, and indeed find, differences in brain sizes and structures of the more social species. A simple interspecific comparison of brain size in relation to body size (Encephalization Quotient or EQ) has commonly been used to explore these differences among mammals (Jerison 1973; Martin 1989; Dunbar 1992). One such approach was taken by Eisenberg (1981) in his calculations of relative brain size for a large number of different species. Although his predictions of brain size for body weight use a slope that may be too low (e.g. Martin 1981), his data none the less reveal some interesting trends (Table 11.2). Within a taxonomic group, the mammalian species that are social have brains that are relatively large for their body size! This simple relationship has been observed in a number of studies (see Clutton-Brock & Harvey 1980; Milton 1981; Parker 1990), suggesting that sociality does indeed necessitate some form of brain expansion for its existence and perpetuation. These studies also highlight the relative rarity of maintained sociality; both the large brains and the social system may be energetically and reproductively expensive to run. It is interesting that the one group where no such finding is apparent from the Eisenberg EQ measures is that of the non-human primates. In fact, all higher primates have enormous brains for their body size, and are conspicuously and permanently social. Dunbar's (1992) exploration of relative primate neocortex complexity in relation to group size suggests that even across the different species of primates, larger brains are associated with more

Table 11.1. *Proposed causal variables determining sociality for mammalian species. (From Lee 1994.)*

Benefits	Costs
• Protection from predation (dilution, defence, swamping)	• Conspicuousness to predators
• Increased foraging efficiency (food location, share time costs)	• Competition; inter & intragroup
• Information exchange (transmission, learning)	• Competition; intragroup
• Access to mates (choice, availability)	• Competition; intragroup
• Thermoregulation (lower metabolic costs)	• Disease transmission through proximity
• Access to infant caretakers (infant survival, lower energy costs to mother)	• Infanticide; competition for reproduction

Figure 11.1. *Options for sociality — based on the association and distribution of individuals of each sex relative to the same sex and to the opposite sex (from Foley & Lee 1989). Only three categories are possible, in that individuals can be distributed either singly (alone), with one or more member of their same sex who is related to them, or with one or more individual who is unrelated. Each cell represents the potential combined distributions of both sexes, and is divided into two halves which represent either a stable relationship between the opposite sexes or a temporary or transitory relationship between males and females. It should be noted that some cells are unlikely to be occupied by species, and exist as hypothetical social space only.*

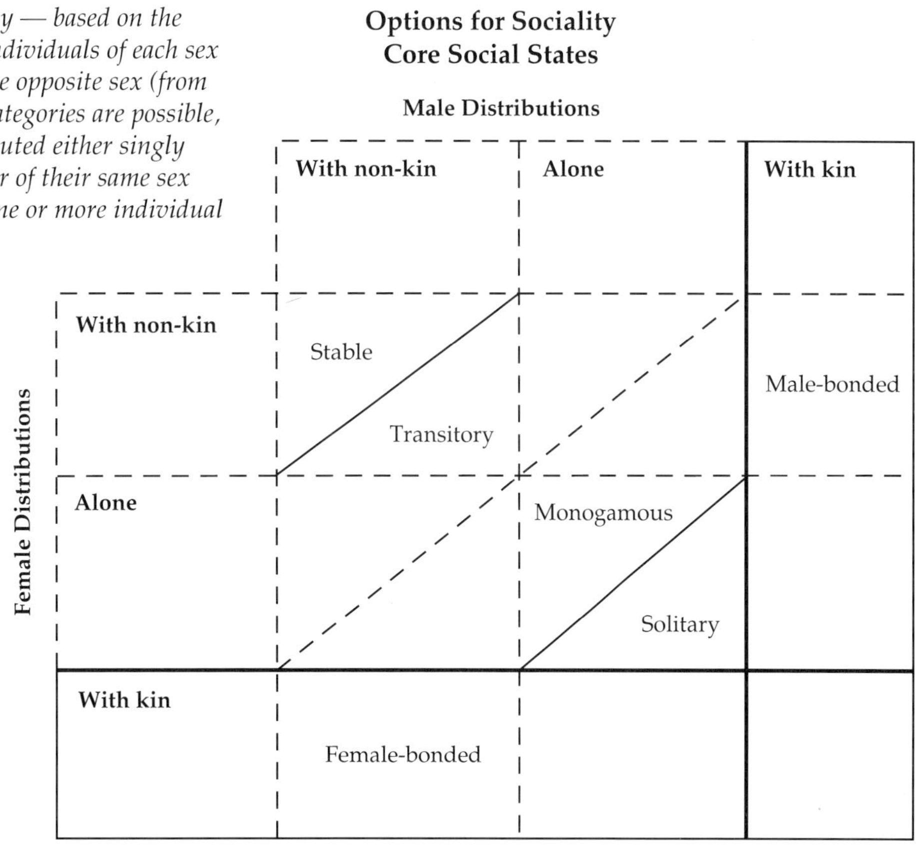

extensive social problems. Sociality and large brains *jointly* characterize this group of mammals, suggesting that the roots of primate cognitive ability lie deep in their evolutionary past.

These observations, while not trivial, do not specifically relate to any cognitive abilities. Eisenberg (1981) has, however, proposed a number of specific and shared behaviours associated with a high EQ among mammals. These are: 1) Active (rather than cryptic) anti-predator avoidance strategies; 2) Locomotion in three dimensions; 3) Complex foraging tasks; 4) Requirements for rapid and variable information processing (both visual and auditory); 5) Requirements for information transmission between individuals (auditory and behavioural). Only the last element can be linked directly to sociality, while the others are in fact associated with the causal factors promoting sociality (see Table 11.1). Information exchange between individuals thus appears to be one critical social component in promoting large and complex brains.

Table 11.2. *Eisenberg's Encephalization Quotient (EQ) and sociality. (Data from Eisenberg 1981.)*

Social Species	EQ	Taxonomic Group	Median EQ
Ground Squirrel	1.5–1.7	Rodents	0.81
Vampire Bats	1.23	Bats	0.94
Ringtail Lemur	1.5	Strepsirhine Primates	0.94
Macaque	2.34	Haplorhine Primates	2.34
Wolf	1.3	Carnivores	1.36
Elephant (female)	2.05	Ungulates	0.94

Life history and large brains

The life history of an organism is made up of the reproductive and mortality events which characterize its lifespan (reviewed in Stearns 1992). These are typically defined for mammals in the form of a number of parameters such as gestation length, duration of lactation or infant dependence, age at first reproduction, interval between reproductive events, litter size, and lifespan. Further variables, such as the rate of mortality at each stage of the lifespan are also included. Early work (e.g. Western 1979; Clutton-Brock & Harvey 1980; Eisenberg 1981) has highlighted the importance of the life history in describing differences between small- and large-brained species. In classical terms, the latter are species who have traded off the energy allocated to reproduction for time allocated to reproduction over the lifespan, while the former are species who allocate greater energy to reproduction either as a unique event or over a short lifespan. The differences between the small-bodied, quick-living species (the so-called 'r-strategists') and the large-bodied, long-lived species (K-strategists; see Pianka 1970) can be characterized in terms of the rates of reproduction and death, and mortality appears to be a major influence in the evolution of the different strategies (Promislow & Harvey 1990).

One reason that large brains are so rare (e.g. Parker 1990) is due to the costs of growing a large brain (Martin 1983; Foley & Lee 1991) and the effects that these costs have on life histories. Large-brained mammals share certain life history traits; they have long gestations, long periods of lactation and infant dependence, late ages at first reproduction, slow rates of reproduction and long lifespans. These extended periods and events within the lifespan are in excess of those predicted for the animal's body size, and indeed most of these species are large mammals to begin with. The life history costs of relative brain expansion over and above that expected for an animal's body size are large, especially among the primates (Harvey *et al.* 1987). The importance of the role of life history variables in understanding brain expansion lies in their interdependence; when one variable changes in the evolutionary history of a species, other changes are produced as a result. It becomes exceedingly difficult to tease out any causal relations among the mass of inter-correlated changes. Thus, as many of the known variables as possible within the 'package' need to be defined. Furthermore, the role of the ecology of the species is crucial, both to sustain the high costs of brain metabolism and growth, and to sustain reproduction and lactation (Martin 1981; 1984). Originally, it was proposed that stable or predicable environments would produce species who could make the trade-off of energy for time, while living in uncertain or risky environments would constrain the temporal patterning of reproduction for these species and increase energy costs (see Pianka 1970; Stearns 1992). Among the primates, similar relationships can be observed (Ross 1988), even though the primates as a group have already specialized in the large-brained, slow living mode. Even between individuals or populations, subtle differences in life histories can be causally related to local ecological differences (e.g. Cheney *et al.* 1988). The species-specific patterns have resulted from their evolutionary history in specific environments, further modified by the ecological variation in the range of environments inhabited by populations of the species. This understanding makes the application of a life history approach potentially useful in understanding the ecological origin of brain expansion in *Homo*.

Life history and ecology are also linked to patterns of sociality, with further implications for brain size and cognitive skills (see Table 11.3). The long period of development required for brain and body growth takes place in a social context for learning, where the mode of sociality is itself a reflection of ecological problems and the needs of the dependent young for nutrition, transportation and protection from sources of mortality. Causality and consequence are again difficult to separate, promoting a more holistic view of the complex links between ecological, social and life history variables in the cognitively and communicatively sophisticated species. What we need to determine are the ways in which the variables interact, in order to pinpoint how changes in one variable will produce dynamic changes in other variables.

Social behaviour among non-human mammals

While the causation of sociality varies from species to species, some elements of behaviour common to social species can in fact be determined. It should be emphasized that the distinction between behaviour and its outcomes must be made (Griffen 1992), and that behaviour alone is not cognition. Simple stimulus-response mechanisms can produce apparently sophisticated outcomes, and do not necessitate highly complex modes of thought. However, researchers working with animal cognition (e.g. Roitblat *et al.* 1984) can define further levels of behaviour where cognitive abilities do come into play. Some level of cognitive skill beyond simple stimulus-response is required for animals where behaviour is governed by rules. Such rules can be simple rules of order, of matching or of category, and relatively little cognitive complexity is involved in their learning or performance. These are the so-called 'rules of thumb' (Krebs & Davies 1993) which produce what on the surface is a highly complex outcome specific to a single context. Such rules are cheap to learn, quick to operate, and effective in producing a result without necessitating laboured or time-consuming thought!

Table 11.3. *Life history traits associated with sociality.*

Trait	Associated Sociality	Physical/Ecological Variables
1. Large body size	Competitive mating	Sexual dimorphism
2. Long lifespan	Kinship	Resource defence
3. Slow rate of infant development	Maternal Care	Slow growth
	Paternal Care/Alloparents	Rich/Abundant resources

They may in fact be rather more common in human behaviour than intentional thought. There is a further category of rule, that of assessment, which does require more complex cognition. Assessment rules appear to be common in aggressive interactions, such as those between males in contests over access to females (red deer: Clutton-Brock & Albon 1979; elephants: Poole 1989). Both matching (e.g. of size) and category (reproductive or non-reproductive; level of motivation etc.) are involved, but the further dimension is that of the assessment of individual A's size and condition *relative* to that of individual B.

This type of skill can be elaborated into the potential for *thinking*, by adding a dimension of contingency. If A performs one act, then B will follow with another, which is contingent on the first individual's behaviour, on the second interactant's previous experiences, and capable of modification throughout the interaction. Such contingent behaviour underlies alternative tactics (e.g. Dunbar 1984), transactions (e.g. Cheney & Seyfarth 1990) and accommodation (de Waal 1993), and ultimately provides the possibility for intentional manipulation and deception, which have been proposed as central to social intellect (Byrne & Whiten 1988; 1992).

The common elements to assessment, contingent thinking, and the potential for intentionality, are found in the nature of sociality. For the majority of social species of mammals, some degree of structuring through kinship is apparent (see Fig. 11.1). Kin structures are especially marked, in that infant dependence on the mother is extended through lactation, promoting a mother–offspring relationship that is dynamic, changing with the development of the young and the condition and reproductive status of the mother. Much of the core of social thinking could be generated in this special and persisting relationship between mother and young, which is forced by physical development to remain labile. Furthermore, it contains the components necessary for learning of rules of matching, category, and assessment with the changes in state and status of the young relative to that of the mother (see for example, Hauser 1988; 1993). This fundamental mammalian social relationship probably lies at the heart of some aspects of the evolution of the elaborate cognitive abilities seen among the hominids.

Kin structures in most social species extend beyond the mother–offspring pair, in that typically one sex of young will co-reside for extended periods with one or both of its parents. Among the higher primates, and indeed many social mammals, a common residence pattern is that of stable groups of female kin, with dispersing males. Among such species, interactions are most frequent between close kin and such interactions tend to be cooperative,

Table 11.4. *Social cognition; attributes and examples.*

Trait		Species	Reference
Individual recognition		Birds, most social mammals (e.g. ground squirrels, wolves, elephants, primates, dolphins)	Griffen 1992
Attribute recognition	*Classification*		
a) Environmental	Predators as categories	Vervet monkeys	Cheney & Seyfarth 1990
	Food patch size & quality	Toque macaques, spider monkeys, chimpanzees	Snowdon 1993
b) Social	Strangers and numbers	Lions, gibbons	McComb *et al.* in press; Raemaekers & Raemaekers 1985
	Relative rank and sexual state	Elephants, red deer	Poole 1989; Clutton-Brock & Albon 1979
	Kinship	Many primates	Cheney & Seyfarth 1990
Relationship recognition	*Between Others*		
	Mother–infant	Macaques	Dasser 1988
	Status relations	Vervets, macaques, baboons	Cheney & Seyfarth 1990
	Consortships	Hamadryas baboons	Bachmann & Kummer 1980
Transactions	*Behavioural Outcomes*		
	Kin affiliation and avoidance	Macaques, baboons	Harcourt & de Waal 1992
	Meshing	Vervet mothers & infants	Hauser 1993
	Alliance formation and maintenance	Primates	Harcourt & de Waal 1992
	Deceit	Primates	Byrne & Whiten 1992
	Reconciliation	Macaques, chimpanzees	de Waal 1993
	Empathy	Chimpanzees	Povinelli *et al.* 1992

while those with less distantly related individuals tend to be more competitive (see Wrangham 1982).

Individual recognition is a fundamental ability of animals living within a stable or maintained social context. The identity of another can be reliably recognized when those individuals interact frequently, when the interactions are patterned over time, and when interactions have a specific context (Hinde 1976). Furthermore, the ability to discriminate between classes of individuals (kin, non-kin, stranger, familiar, male, female, infant, etc.) and the specific attributes of those individuals (such as size, status, age etc.), are all necessary to social individuals (see Table 11.4).

The next stage in this cognitive world is that of transaction. A transaction between individuals, each of whom have specific interests in the outcome of the interaction, produces a result which can generally fall into the categories of competitive or cooperative (e.g. Alexander 1974). Inherent in stable social groups is the iteration of interaction; repeated contacts between the same individuals, in different categories of interaction over time, allow for the development of relationships (e.g. Hinde 1976). Relationships are dynamic, and require further knowledge over and above that of the immediate interaction. The implication of relationships is that their maintenance requires the ability to recognize and retain attributes of others in relation to attributes and experiences of self. The next step is the recognition of the relationships existing between others, which has been well documented for the primates (see Table 11.4).

Further to the transaction within the context of an existing relationship, which is itself recognized as being embedded in a network of relationships, is the element of accommodation within the transactions. The maintenance of stable forms of sociality frequently requires that the transaction is not purely a selfish one, accomplishing individual goals. Rather, some form of interactive compromise is reached between the conflicting individual goals, which may be expressed through reconciliation after aggression (de Waal 1993). Individuals who demonstrate reconciliation, defined as affiliative contact seeking or gestures after an aggressive conflict has occurred, can be seen as compromising their own immediate goals with the need to coexist with another in the future.

The ability to compromise, to reconcile, and to manipulate others future behaviour through an interaction may also lead to deceit. The ability to produce an exaggerated or incorrect assessment on the part of others is central to current theories as to animal mind (Byrne & Whiten 1988; 1992). But this still leaves the question of intentionality to be dealt with. It could be argued that deception can be accomplished through a simple rule, such as 'dishonest' (exaggerated or costly) advertisement of motivation, status or size, which then alters the simple assessment, producing the apparent manipulated outcome without changing the underlying rules or their application in cognitive terms. In some social species where reconciliation is common, this, rather than deceit, may be the social cement resolving individual conflicts of interest.

Some recent work has suggested that primates, at least, form a concept of their social world, and that this is learned during development through 'punishment' of inappropriate behaviour (Bernstein & Ehardt 1986). Such socialization goes beyond that required by simple active sociability, and appears to require complex rules about behavioural expectations, roles, and indeed the nature of primate social life.

Do primates thus have a 'theory of mind' (Premack & Woodruff 1978), with consistency across behavioural domains and contexts. Are they sensitive to others' emotions and intentions, and able to recognize appropriate responses in relation to others' mental as well as physical states. Research on humans, and especially learning-impaired children (see Archer 1992), suggests that there is a potential for the development of a non-verbal theory of mind, so some form of this capacity may lie within the primate domain. As yet, the evidence for this is limited among monkeys (Cheney & Seyfarth 1990), but suggestive for the apes (Povinelli *et al.* 1992). The richness of social responses among primates, both in cooperative and conflict situations, suggests that their cognitive skills are greatest in the interactive and relationship areas which could ultimately lead to a theory of mind, but such mentalization of the world is probably unnecessary in most social contexts, and would impose high energy costs of developing and maintaining a large brain (Foley & Lee 1991).

In summary, sociality at least provides the opportunities for complex forms of cognition; whether these options are taken may depend further on the life history of the animal, on the stability of sociality and on the reasons why that sociality arose during the species' evolutionary past. While the expression of sociality is limited, the causes for the evolution of sociality are complex and specific to each species. Thus, similarity of forms of maintained sociality masks the underlying diversity of evolutionary pressures and phylogenetic histories among the social species. Furthermore, the complexity and richness of the social context among these species promotes the appearance of very sophisticated behaviour, which nonetheless may have relatively simple underlying rules for its operation. The cognitive abilities, and limitations, of the non-human mammals need to be recognized in constructing the framework for understanding how and when human abilities arose, the constraints on those abilities, and why the constraints exist. By starting to define these constraints and opportunities in relation to the evolution of complex behaviour in animals, we should have a greater insight into the nature of cognition in the evolution of humanity.

P.C. Lee
Department of Biological Anthropology
University of Cambridge
Downing Street
Cambridge CB2 3DZ
UK

References

Alexander, R.D., 1974. The evolution of social behaviour. *Annual Review of Ecology and Systematics* 5, 325–83.

Archer, J., 1992. *Ethology and Human Development*. Hemel Hempstead: Harvester Wheatsheaf.

Bachmann, C. & H. Kummer, 1980. Male assessment of female choice in hamadryas baboons. *Behavioural Ecology and Sociobiology* 6, 315–21.

Bernstein, I.S. & C. Ehardt, 1986. The influence of kinship and socialization on aggressive behaviour in rhesus monkeys (*Macaca mulatta*). *Animal Behaviour* 34(3), 739–47.

Byrne, R., 1995. *The Thinking Ape*. Oxford: Oxford University Press.

Byrne, R. & A. Whiten (eds.), 1988. *Machiavellian Intelligence: Social Expertise and the Evolution of Intellect in Monkeys, Apes and Humans*. (Oxford Science Publications.) Oxford: Clarendon Press.

Byrne, R. & A. Whiten, 1992. Cognitive evolution in primates: evidence from tactical deception. *Man* 27, 609–27.

Cheney, D.L. & R.M. Seyfarth, 1990. *How Monkeys See the World*. Chicago (IL): University of Chicago Press.

Cheney, D.L., R.M. Seyfarth, S.J. Andelman & P.C. Lee, 1988. Reproductive success in vervet monkeys, in *Reproductive Success*, ed. T.H. Clutton-Brock. Chicago (IL): University of Chicago Press, 384–402.

Clutton-Brock, T.H. & S.D. Albon, 1979. The roaring of red deer and the evolution of honest advertisement. *Behaviour* 68, 145–70.

Clutton-Brock, T.H. & P.H. Harvey, 1980. Primates, brains and ecology. *Journal of Zoology (London)* 190, 309–23.

Dasser, V., 1988. Mapping social concepts in monkeys, in Byrne & Whiten (eds.), 85–93.

de Waal, F.M.B., 1993. Reconciliation among primates: a review of empirical evidence and theoretical issues, in *Primate Social Conflict*, eds. W.A. Mason & S.P. Mendoza. New York (NY): State University of New York Press, 111–44.

Dunbar, R.I.M., 1984. *Reproductive Decisions*. Princeton (NJ): Princeton University Press.

Dunbar, R.I.M., 1988. *Primate Social Systems*. London: Croom-Helm.

Dunbar, R.I.M., 1992. Neocortex size as a constraint on group size in primates. *Journal of Human Evolution* 20, 469–93.

Eisenberg, J.F., 1981. *The Mammalian Radiations*. Chicago (IL): University of Chicago Press.

Else, J.G. & P.C. Lee (eds.), 1986. *Primate Ontogeny, Cognition and Social Behaviour*. Cambridge: Cambridge University Press.

Foley, R.A. & P.C. Lee, 1989. Finite social space, evolutionary pathways and reconstructing hominid behaviour. *Science* 243, 901–6.

Foley, R.A. & P.C. Lee, 1991. Ecology and energetics of encephalization in hominid evolution. *Philosophical Transactions of the Royal Society* Series B 334, 223–32.

Gibson, K.R., 1986. Cognition, brain size and the extraction of embedded food resources, in Else & Lee (eds.), 93–105.

Gibson, K.R., 1993. Animal minds, human minds, in Gibson & Ingold (eds.), 3–15.

Gibson, K.R. & T. Ingold (eds.), 1993. *Tools, Language and Cognition in Human Evolution*. Cambridge: Cambridge University Press.

Griffen, D.R., 1992. *Animal Minds*. Chicago (IL): University of Chicago Press.

Harcourt, A.H. & F.M.B. de Waal (eds.), 1992. *Coalitions and Alliances in Humans and Other Animals*. Oxford: Oxford University Press.

Harvey, P.H., R.D. Martin & T.H. Clutton-Brock, 1987. Life histories in comparative perspective, in *Primate Societies*, eds. B.B. Smuts, D.L.C. Cheney, R.M. Seyforth, R.W. Wrangham & T.T. Struhsaker. Chicago (IL): University of Chicago Press, 181–96.

Hauser, M.D., 1988. Variation in maternal responsiveness in free-ranging vervet monkeys: a response to infant mortality risk? *American Naturalist* 131, 573–87.

Hauser, M.D., 1993. Do vervet monkey infants cry wolf? *Animal Behaviour* 45(6), 1242–4.

Hinde, R.A., 1976. Interactions, relationships and social structure. *Man* 11, 1–17.

Humphrey, N., 1976. The social function of intellect, in *Growing Points in Ethology*, eds. P. Bateson & R.A. Hinde. Cambridge: Cambridge University Press, 303–17.

Jerison, H.J., 1973. *Evolution of the Brain and Intelligence*. New York (NY): Academic Press.

Jolly, A., 1966. Lemur social behavior and primate intelligence. *Science* 153, 501–6.

Kohler, W., 1927. *The Mentality of Apes*. London: Routledge & Kegan Paul.

Krebs, J.R. & N.B. Davies, 1993. *An Introduction to Behavioural Ecology*. Oxford: Blackwell Scientific.

Lee, P.C., 1991. Biology and behaviour in human evolution. *Cambridge Archaeological Journal* 1(2), 207–26.

Lee., P.C., 1994. Social structure and evolution, in *Behaviour and Evolution*, eds. P. Slater & T. Halliday. Cambridge: Cambridge University Press, 266–303.

McComb, K., C. Packer & A. Pusey, in press. Roaring and numerical assessment in contests between groups of female lions. *Animal Behaviour*.

McGrew, W.C., 1992. *Chimpanzee Material Culture*. Cambridge: Cambridge University Press.

Martin, R.D., 1981. Relative brain size and basal metabolic rate in terrestrial vertebrates. *Nature* 293, 57–60.

Martin, R.D., 1983. *Human Brain Evolution in an Ecological Context. 52nd James Arthur Lecture on the Evolution of the Brain*. New York (NY): American Museum of Natural History.

Martin, R.D., 1984. Scaling effects and adaptive strategies in mammalian lactation. *Symposium of the Zoological Society of London* 51, 87–117.

Martin, R.D., 1989. *Primate Origins and Evolution*. London: Chapman & Hall.

Milton, K., 1981. Food choice and digestive strategies of two sympatric primate species. *American Naturalist* 117, 476–95.

Milton, K., 1984. The role of food processing factors in primate food choice, in *Adaptations for Foraging in Non-Human Primates*, eds. P.S. Rodman & J.G.H. Cant. New York (NY): Columbia University Press, 249–79.

Milton, K., 1988. Foraging behaviour and the evolution of primate intelligence, in Byrne & Whiten (eds.), 285–305.

Parker, S.T., 1990. Why big brains are so rare: energy costs of intelligence and brain size in anthropoid primates, in *'Language' and Intelligence in Monkeys and Apes*, eds. S.T. Parker & K.R. Gibson. Cambridge: Cambridge University Press, 129–54.

Pianka, E.R., 1970. On 'r' and 'K' selection. *American Naturalist* 104, 592–7.

Poole, J.H., 1989. Announcing intent: the aggressive state of musth in African elephants. *Animal Behaviour* 37(1), 140–52.

Povinelli, D.J., K.E. Nelson & S.T. Boysen, 1992. Comprehension of role reversal in chimpanzees: evidence of empathy? *Animal Behaviour* 43, 633–40.

Premack, D. & G. Woodruff, 1978. Does the chimpanzee have a theory of mind? *Behavioural Brain Sciences* 1, 515–26.

Promislow, D.E.L. & P.H. Harvey, 1990. Living fast and dying young: a comparative analysis of life-history variation among mammals. *Journal of Zoology (London)* 220, 417–37.

Raemaekers, J. & P.M. Raemaekers, 1985. Field playback of loud calls to gibbons (*Hylobates lar*): territorial, sex-specific and species-specific responses. *Animal Behaviour* 33, 481–93.

Roitblat, H.L., T.G. Bever & H.S. Terrace (eds.), 1984. *Animal Cognition*. Hillsdale (NJ): Lawrence Erlbaum Associates.

Ross, C., 1988. The intrinsic rate of natural increase and reproductive effort in primates. *Journal of Zoology* 214, 199–207.

Sigg, H., 1986. Ranging patterns in hamadryas baboons: evidence for a mental map, in Else & Lee (eds.), 87–91.

Snowdon, C.T., 1993. A comparative approach to language parallels, in Gibson & Ingold (eds.), 109–28.

Stearns, S.C., 1992. *The Evolution of Life Histories*. Oxford: Oxford University Press.

Washburn, S.L., 1960. Tools and human evolution. *Scientific American* 203, 63–75.

Western, D., 1979. Size, life history and ecology in mammals. *African Journal of Ecology* 17, 185–204.

Wrangham, R.W., 1982. Mutualism, kinship and social evolution, in *Current Problems in Sociobiology*, eds. King's College Sociobiology Group. Cambridge: Cambridge University Press, 269–90.

12
Egalitarianism and Machiavellian Intelligence in Human Evolution

David Erdal & Andrew Whiten

Archaeological evidence for an ancestral human hunting niche means that living hunter-gatherers offer important behavioural data for modelling the early human mind. Following the principles of strategic evolutionary modelling, 24 detailed hunter-gatherer ethnographies from four continents were examined for universal behaviour patterns likely to reflect fundamental adaptation to this human niche. Egalitarian behaviour is one of the most clearly documented universals, consistent with Knauft's suggestion that human evolution can be described by a U-shaped curve in which non-egalitarian ape societies gave way to hunter-gatherer egalitarianism, to be replaced in turn by social inequalities as settlement and resource surpluses became more recently established. Over 100 particular observations further document the unique nature of human hunter-gatherer egalitarianism, demonstrating an absence of social hierarchy and sharing of resources which goes beyond the explanatory power of either kinship or reciprocation. Individuals do sometimes attempt to obtain a disproportionate share of resources or influence for themselves, but this is contained through vigilance and counter-dominant behaviour by their group members. We propose a model of human social evolution in which ancestral apes' Machiavellian tactics spiralled close to, but were intrinsically unable to settle at, a ceiling at which egalitarianism becomes the only viable strategy.

Introduction

Materials for modelling the early human mind

Modelling the minds of living individuals is hard enough: attempting the same for minds long past is daunting indeed, when all that remains of those individuals is a patchy assemblage of bones and artefacts. By comparison with what we can infer from those remains, the chance to board some magicial time machine and observe all the richness of the actual behaviour of our ancestors would elevate our understanding of their minds almost beyond measure.

Fortunately, direct analysis of living action is still possible for two major classes of individuals that for different reasons give pointers to what we might have observed from our time machine. The two classes are hunting-gathering peoples, whose way of life may indicate important features of the late evolutionary period of *Homo sapiens*, and non-human primates, whose conduct may reflect aspects of the behaviour of the common ancestor they share with us. In both cases we need to recognize that any parallels are not direct, but are distorted in systematic ways.

We are doubly fortunate in the substantial nature of the material available for study, for in recent times there have been many different hunter-gatherer societies extant, and close to two hundred species of non-human primate. In both cases these numbers appear destined to suffer drastic reduction even in our lifetimes, so we stand at a unique point in the study of human origins, and we bear a heavy responsibility in garnering the insights fleetingly within our grasp. Happily, the disappearance of many of these subjects was preceded — by the narrow squeak of a century — by the insights of Darwin and Wallace, rather than the reverse.

The value of these sources has been recognized for some time now, as exemplified in such classic works as Lee & DeVore (1968) and Washburn & DeVore (1961). But there has also been plenty of time for critical analysis to highlight the shortcomings and pitfalls to which the inferences drawn may be prone. The rationale for valuing observations on a hunting and gathering mentality has been, to quote just one of the many appeals to the same statistic, that 'basic human social forms, language, and human nature itself were forged during the 99 percent of human history when people lived in hunting and gathering camps' (Lee 1979, 1). But do we really know that? Certainly there is good evidence for the beginnings of settled agriculture from only about 10–12,000 years BP, and a contrasting dearth of such evidence for any periods earlier than this (Hole 1992). The 99 per cent figure thus rests on hunting and gathering being representative of the last million years. On this there is more dispute. Leakey & Lewin, having reviewed the evidence and sceptics' doubts about it, conclude that 'the earliest *Homo* were incipient hunters and gatherers, and that way of life would shape the human body and the human mind for more than two million years' (1992, 141–2). At the other extreme, it is possible to argue that what some take for archaelogical evidence of early *Homo* hunter-gatherer bases may instead reflect only scavenging and a variety of other natural processes (Binford 1981). Impressive evidence of hunting could be said to arrive only with the emergence of anatomically modern *Homo sapiens* as recently as 100,000 years ago. Direct archaeological evidence for the other side of the foraging pattern — gathering — is hard to find: but it is reasonable to assume that it took place.

Constraints on modelling the past from the present

The chronology of the history of hunting and gathering is not clear, but the logic of evolution implies that todays' hunter-gatherers offer models likely to be misleading the further into the past we attempt to cast our inferences. Even if Leakey & Lewin are correct in identifying a hunting and gathering way of life with the emergence of *Homo*, we should not expect an identity between the mentality of today's foraging peoples and that of *Homo habilis* or *Homo erectus*, with their significantly smaller brains.

There are additional constraints in the modelling process. Many modern hunter-gatherers have been relegated by pressure from agricultural and industrial societies to marginal areas, whereas before the advent of herding and agriculture foragers dominated the land occupied by hominids and could choose the most attractive areas. Further, most hunter-gatherers for whom there are good ethnographic records have some degree of symbiotic relationship with agricultural peoples, which may have influenced their behaviour (e.g. Leacock 1954; 1980; Hyde 1959, 20; Bailey *et al.* 1989).

Strategic modelling and hunter-gatherer core behaviour patterns

The constraints reviewed above clearly impose methodological limitations. We concur with Tooby & DeVore (1987) that referential modelling — in which one particular hunting-gathering society is selected as an optimal model for the past — is most likely to be misleading and should be avoided in favour of strategic modelling. The rationale of the latter is to use a much greater diversity of information to build general models of behavioural ecology, into which the known features of ancestral hominids may be fitted in order to produce inferences about yet unknown characteristics. A good example is the way in which our knowledge of the relationships between sexual dimorphism and mating patterns in primates and other mammals may be used to deduce the implications of dimorphism in fossil hominids (e.g. Foley 1988, 216; Potts 1992).

A basic methodological rule derived from this approach is to identify common patterns in the enormous diversity of cultures and ecologies associated with present-day foragers. If such patterns exist, then this suggests that they are inherent functional components of human hunting-gathering society, as likely to characterize the history we aim to reconstruct as the present we can observe. Such configurations do exist, and students of hunter-gatherers have discerned a cluster of features which appear to act as the functional core of the societies: egalitarianism, cooperation, and sharing, on a scale unprecedented in primate evolution, and sometimes dubbed 'primitive communism' (Lee 1979, 460).

Accordingly we focus on this core in the present paper. There are additional reasons for doing so. First, recent years have seen the accumulation of important new data on these behaviours in both human and non-human primates,which permit evolutionary modelling (e.g. Knauft 1991; Harcourt & de Waal 1992; Boesch & Boesch 1989). Second, there have been exciting developments in the theory of altruism, from Trivers (1971) to Frank (1988). And finally, insights about the evolution of this core have relevance for us today. Associated proximate psychological mechanisms are likely to be with us in modern times, and although they may not have the reproductive consequences which they did in the ancestral environments which moulded their evolution, they are likely to have material effects on psychological motivation and well-being. Structures of present-day living whose design is shaped by an understanding of these predispositions may prove much more productive than those which ignore them.

The egalitarian puzzle

The egalitarianism of modern hunter-gatherers is an apparent anomaly in evolutionary terms. The puzzle is that although dominance hierarchies are likely to have characterized the ancestor shared by chimpanzees and humans, and institutional hierarchies are characteristic of modern humans, the hunter-gatherers representing the intervening phase are almost entirely egalitarian in social structure and behaviour.

This anomaly was characterized by Knauft (1991) as a 'U' shaped curve in evolution: moving down the left arm from ape hierarchies to hunter-gatherer egalitarianism, and then up the right arm with agricultural and then modern hierarchical societies.

This curve is a useful way to conceptualize the evolution of social behaviour, but it must have been a heavily skewed 'U'. We do not know when the bottom of the left hand arm was reached — when egalitarian behaviour first appeared in evolution — but we do know that the left arm and the bottom together covered a period of several million years. We also know that the process at work involved biological evolution, with an enormous increase in brain size being a key result. If there are specific inherited predispositions for social behaviour, it must have been during this period that they were shaped by evolutionary processes.

On the other hand, the right hand arm turned upwards, into hierarchical social organizations, only about 10,000 years ago, with the advent of agricultural societies. There were possible precursors in the complex hunter-gatherer societies of the northwest coast of North America, perhaps five thousand years earlier (Kroeber 1939, 29; Suttles 1968, 105) and possibly earlier still in Europe (Mellars this volume). However, until the advent of agriculture this social complexity is not characteristic. The evolutionary pressures that led to modern levels of encephalization took effect in the long preceding period when humans and their precursors lived in simple foraging societies — associated, we therefore infer, with egalitarian social behaviour. The process that led to hierarchy was not an additional period of biologically driven evolution — there was not time for that, and no further significant biological changes are apparent. It was rather the result of the already evolved human psychology meeting new circumstances. As a result, the 'U' may look (perhaps appropriately), more like a question mark lying on its back (Fig. 12.1).

In order better to model the mentality underlying the different parts of the curve and to attempt to understand the processes which moulded it, we need a deeper appreciation of the nature of the egalitarianism in question. This is the task of the next section.

Hunter-gatherer egalitarianism

Hunter-gatherers are characterized by food-sharing and by a virtually complete absence of hierarchy or dominance. We will deal with these two subjects separately, and then draw out their common psychological elements.

Food-sharing

Meat is universally shared

Amongst hunter-gatherers food, and particularly meat, is shared. This has been observed by virtually all ethnographers and has been tested quite rigorously by modern anthropologists. There is not space here to quote widely from these studies, but the similarities are impressive, across all continents (Table 12.1 — column headed 'Meat-Sharing').

Of particular interest are the three more recent studies which have used techniques developed in the discipline of behavioural ecology to quantify their observations. These papers are marked with an

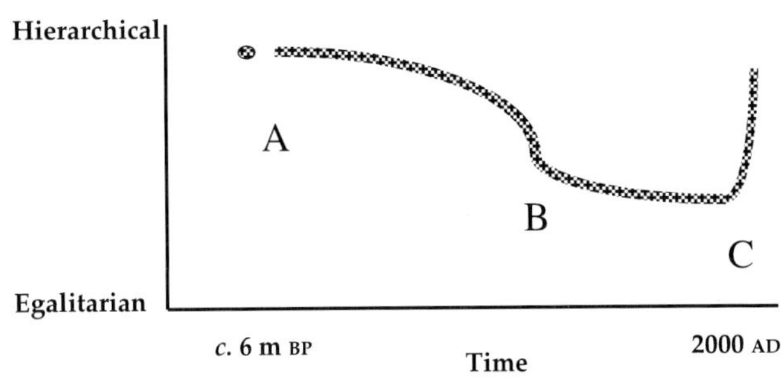

Figure 12.1. *Social structure in evolution. 'A' is the point at which the hominid line split from the chimpanzee line. 'C' is some 10–15,000 years* BP, *the beginning of complex societies. 'AC' covers several million years of biological evolution. It is not clear when 'B' — the beginning of egalitarian social behaviour, as shown by modern hunter-gatherers — occurred. 'BC' may have been millions of years or merely tens of thousands.*

asterisk in the 'Meat-Sharing' column in Table 12.1. Kaplan & Hill, in a rigorous study designed to test competing hypotheses about sharing, conclude that: 'all the data suggest that the Ache are pooling most of the food they acquire and are sharing the food among families according to the number of dependents present' (1985, 233). Bailey's work among the Efe focuses only on the sharing among hunters before they return to camp, where further sharing takes place, but his conclusion is similar (1991, 90). Finally, Hawkes *et al.* (1991, 87) find that the Hadza hunters choose a sharing strategy even though a selfish strategy apparently gives higher returns.

In all these cases, a psychological model which is based on rational self interest as the prime explanatory variable simply does not fit the data. This conclusion is thoroughly reinforced by the more traditional ethnographies, based on their observation of incidents, without quantification. Among the Walbiri, for example, in times of scarcity the members of the band disperse to join neighbouring groups. 'Even if the hosts' resources were also limited, they were shared until all the food was consumed. Then hosts and guests alike literally tightened their string-belts and scoured the countryside' (Meggitt 1962, 52).

In summary, modern hunter-gatherers, as long as they have not been drawn in to trading relationships or property systems based on other ways of life, all share food. They do not share only with kin; they do not share only with those who reciprocate; they share out what they have according to need, even when food is scarce.

The predispositions behind food-sharing

The cross-cultural evidence on food-sharing is so consistent that it invites the hypothesis that food-sharing is influenced by evolved, inherited predispositions.

The alternative explanation that sharing is produced by cultural processes would be better supported if there were greater cultural variation, with at least some pre-contact cultures generating an opposition to sharing. The evidence is that the specific ways in which food is shared are influenced in their details by the local culture: but the existence of generalised sharing is universal.

The argument that cultural processes explain the very existence of food-sharing implies that sharing is not an evolved predisposition: rather, that selfishness is the genetically effective 'strategy', which is only overcome by cultural training. Such a model implies that there is a tension between on the one hand inherited, adaptive, selfish tendencies and on the other cultural rules favouring sharing. In principle, the likelihood of such a tension being always and everywhere resolved in favour of sharing, with aberrations being minor and rare, is too low to give support to the cultural model.

Shepher's (1983) work on incest suggests a much more plausible model for the relationship between predispositions and culture. Shepher showed that there is a developmental process which leads to negative imprinting among young children brought up together, so that the varied and elaborate incest taboos found in different cultures are built on a basic, irreversible psychological process. Generalizing this model suggests that cultural elaborations are likely to be built on a foundation of evolved predispositions: they are not determined in detail by them, and the refinements may be very varied and complex indeed, but it is only very rarely that cultural elaborations systematically contradict the evolved predispositions on which they are founded. And when they do, as with the celibacy of priests, they are much honoured in the breach.

This model suggests that sharing is likely to be, or to be based on, an evolved, psychological predisposition. The cross-cultural consistency of sharing, in an environment which is in important respects likely to be similar to the environment in which we evolved, supports this picture of an evolved tendency. The various cultural elaborations of sharing may all be founded on an evolved predisposition to share, just as the elaborations of the incest taboo are founded on the evolved negative imprinting process among children.

Conflicts over food-sharing

If there are predispositions at work here, however, they are not straightforward. It is not simply that a human hunter-gatherer, carrying meat and faced with a hungry member of his or her band, is predisposed to give some of the meat away. Sometimes there are intense arguments and jealousies about the sharing of the meat. References are given in Table 12.1, in the column headed 'Vigilance/Argument'.

In Africa, Bailey speaks of the Efe's 'extremely heated vociferous arguments' (1991, 94). Turnbull says the Mbuti arguments 'are frequently long and loud' (1965, 158). Tanaka speaks similarly of the San (1980, 95). Among the !Kung, Marshall observes 'self interest and much jealous watchfulness' (1961, 350) and Thomas similarly describes tension in food-sharing (1959, 116–19).

Elsewhere, either hostility and suspicion are not reported, or food-sharing takes place with greater

Table 12.1. *Ethnographic references describing egalitarianism in hunter-gatherers, including vigilant food-sharing, informal leadership without authority, and counter-dominant behaviour. The numbers refer to pages in the texts. *These studies are based on numerical data.*

		Vigilant Sharing			Leadership		
Continent and Society	*Reference*	*Meat-Sharing*	*Vigilance/Argument*	*Cheating/Stealing*	*Informal Leadership*	*Absence of Authority*	*Counter-Dominance*
Africa							
Dorobo	Huntingford 1955	612					
Efe	Bailey 1991	90 *	83, 94	70		59	93
Hadza	Hawkes *et al.* 1991	√ *					
Mbuti	Turnbull 1965	158	158	198	126, 178, 181–2	180–81	178, 183
San - Naron	Bleek 1928	16			37	37	
San - !Kung	Lee 1979	118	372		343–8	xxii, 343	24, 244–6, 344
" "	Marshall 1961	357, 362	350		358		
San	Tanaka 1980	89, 95, 99	95	122	108, 123	93, 108	113
Australia							
Pitjandjara	Tindale 1972				261–2	261	
Tiwi	Goodale 1971	171–2			338		
Walbiri	Meggitt 1962	52	252	175	190, 245, 250	247, 249, 250	88
Asia							
Batek	Endicott 1988	115–18	124–5	125	112, 123	121–3	
Paliyan	Gardner 1972	415				415, 425–6	426
Pandaram	Morris 1982a [1982b]	180 [103]				[158–9]	
N. America							
Copper	Damas 1972	24–5			32	32	33
Cree	Rogers 1972	120			122	121	121
Dogrib	Helm 1972	69, 78, 80			78	79–80	77, 79
Naskapi	Leacock 1980 [1982]	[160]				27, 29–30	
Netsilik	Balikci 1970	117		178–9	72, 116, 149	xv	
"	Rasmussen 1931	142, 147, 162			13		
Northern Nomads	Riches 1982	70, 79, 83	71		19, 72, 77	45, 74	27
Utkuhikhalingmiut	Briggs 1970	88				112	
S. America							
Ache	Kaplan & Hill 1985	233 *					
Siriono	Holmberg 1950	31		33, 36, 61	25, 46, 60		

equanimity. Balikci speaks of 'laughter and merry-making' among the Netsilik (1970, 46). Damas' (1972) picture of the Copper Eskimo is similar in tone. The picture among the Asian and Australian groups is similarly lacking in reports of argument and conflict over food-sharing. However, the existence of conflict in the African groups is sufficient to demonstrate that there is no simple predisposition to share food.

Cheating and stealing in food-sharing

A second indication that there is no simple predisposition to share is that cheating and stealing sometimes occur. References are listed in Table 12.1, in the column headed 'Cheating/Stealing'.

Tanaka says of the San, for example, that 'the individual has a strong sense of possessiveness and usually tends to think egocentrically . . . A man may sneak a small catch of game into his hut to share only with his family, or otherwise fail to share food as he should' (1980, 122). Turnbull remarks that 'It would be a rare Mbuti woman who did not conceal a portion of the catch in case she was forced to share with others' (1965, 198). On the same page Turnbull tells the interesting story of how Pepei the thief was humiliated into changing his ways by massive gifts of meat, delivered with contempt (see also 1965, 120).

In the American groups, Balikci observes theft among the Netsilik, 'cheerfully practised in relation to strangers' (1970, 178), and the destruction of food belonging to someone out of favour (1970, 179). Holmberg (1950, 36) describes hoarding and unwillingness to share among the Siriono, although their history as agriculturalists who reverted to foraging makes them a poor candidate for inclusion in a strategic model (Stearman 1984).

In Australia, Meggitt mentions a case of a woman being thrashed for stealing food from another camp (1962, 175).

The function of food-sharing

These instances of conflict over sharing and of stealing or hoarding suggest that universal as it is among hunter-gatherers, food-sharing must be the outcome not of a simple predisposition to share, but of a complex set of behavioural predispositions, in interaction with key features of the hunting-gathering environment. The puzzle from an evolutionary view is not that there is conflict and stealing, which can easily be incorporated into a 'selfish gene, selfish person' model, but that these instances are rare, and that sharing itself is ubiquitous and normal.

At the functional level, the main argument is

that food-sharing reduces risk for all individuals, including the hunters (Damas 1972, 25; Lovejoy 1981; Wiessner 1982; Cashdan 1985; Smith 1988), with an additional suggestion that the successful hunters' largesse with meat helps them as individuals obtain the most capable women as wives (Riches 1984, 240) and more frequent adulterous mating opportunites (Kaplan & Hill 1985, 237; Hawkes 1993).

The risk-reduction model depends on two established facts. Firstly, there is a high variance in hunting returns, because game is distributed unpredictably and is hard to kill when it is found. And secondly, many game animals have much more meat than any single human can consume before it rots. So the marginal cost to a successful hunter of giving meat away is low, and the benefit to the hungry members of his band is high. Moreover, if he can count on receiving meat when he fails but others succeed, then the risk profile over time for each individual is significantly improved by sharing: each successful hunter is giving away meat which he cannot use, and receiving at a different time meat of great value to him. A similar balance of costs and benefits has led to blood sharing in vampire bats, purely through the evolution of behavioural dispositions, apparently without any cultural influence (Wilkinson 1984). The main difference is that the bats tend to reciprocate with particular individuals, whereas humans consistently share with all in the band.

The psychology of food-sharing

Moving from the functional to the proximate psychological level, it is still unclear what is going on. A 'rational self-interest' model would suggest that possibly each individual is making calculations, recognizing the improvement in his personal situation created by sharing, and so deciding to take part. A 'reciprocity' model would explain sharing as being in effect an exchange, which evens out returns over time (Trivers 1971). Neither of these models is supported, however, by the fact that consistently over time some hunters continue to contribute more meat than they and their families consume (Kaplan & Hill 1985, 234; Bailey 1991, 89). This is neither 'rationally self-interested' nor 'reciprocal'. It is perhaps possible that over the very long term there is rough reciprocity: a family with several productive adults makes significant net contributions for years, but then becomes a family of old and young, making significant net demands of the band. However, if this is the case, it does not occur through rational calculations, which take as their horizon timespans far shorter than generations.

The fact that food-sharing is sustained even when there is nothing like a full return of what is given away suggests that at the proximate, psychological level there is one or more evolved predisposition at work.

Leadership

The second area which demonstrates egalitarianism in hunter-gatherers, and so adds to the evolutionary anomaly, is the process of leadership.

Leaders with authority are not created by hunter-gatherers

There is no dominance hierarchy among hunter-gatherers. No individual has priority of access to food, which, as we have seen, is shared. In spite of the marginal female preference for the more successful hunters as lovers, access to sexual partners is not a right which correlates with rank — there is no *droit de seigneur*. In fact rank in that sense is simply not discernible among hunter-gatherers. This is a cross-cultural universal, which rings out unmistakably from the ethnographic literature, sometimes in the strongest terms: 'Individual authority is unthinkable' (Turnbull 1965, 181). References to some of the many studies which fill out this picture across all continents are given in Table 12.1, in the column headed 'Absence of Authority'.

Contact with agricultural or colonial societies frequently leads to the creation of leaders — or at least to the adoption of leadership titles. The titles remain empty in some cases, like the Hill Pandaram (Morris 1982b, 158–9), the Paliyan (Gardner 1972, 425–6), the Mbuti (Turnbull 1965, 27, 42, 45, 103) and the San (Lee 1979, 346, 348–9); in others real leadership positions develop in response to the pressures from invading peoples, as amongst some San (Lee 1979, 32) and the Plains Indians (Hyde 1959, 20; Provinse 1955, 344); and in others real authority is to some extent created by external government authorities, as among the Walbiri (Meggitt 1962, 250). In 1921 Bleek noted an example of the contact effect among the Naron: 'whenever (he) is speaking Dutch, which the others do not understand, he asserts that he is now chief, but when speaking Bushman he never ventures to claim this, but chimes in with the rest, that there are no chiefs . . . Certainly he had no authority now save among his own family' (1928, 36). And the early seventeenth-century Jesuit, Father LeJeune, reported to his superiors the reproaches of the Montagnais Naskapi at the uncivilized behaviour of the French, who 'fear our captains, while they laugh at and make sport of theirs . . . (and) . . . only obey their chief through goodwill towards him' (Leacock 1980, 29).

The consistent picture is that there is no permanent role of 'leader-with-authority' created by hunter-gatherers for themselves.

Leadership is others listening with respect

Without authoritative leaders, the process of group decision-making is universally consensual. Discussions develop naturally and informally. Everyone generally has the right to speak, and it is characteristic that people are always ready to make suggestions and to listen to others. So these conferences can be lengthy.

There is a pattern to the outcome of these discussions: effective individuals — the best hunters when hunting is the subject, the most knowledgeable person in a conference about ritual — are generally paid more attention than others who are seen to be less skilled. And frequently these experts gain agreement from the others to what they propose.

However, this is not always the case. Great hunters can be over-ruled even in hunting matters. And when a consensus has been reached, no-one has to follow it against their will — there is no enforcement mechanism.

In some cases there are named leadership roles — e.g. the Netsilik 'inhumataq' or 'thinker' — but even where there are named roles, the leaders are not 'obeyed', but rather 'listened to' (Riches 1982, 74) — and sometimes not listened to. References filling out this picture are also given in Table 12.1, in the column headed 'Informal Leadership'.

In general, expertise and personal characteristics lead to the granting of respect in particular circumstances. There is nothing permanent about that respect: in a different situation a different person will be given respect, and even in an identical situation the next day the same expert may not be heeded. Leadership is ultimately created by others' listening. Even combining expertise with a forceful personality is not enough to raise anyone into a permanently dominant position, however much individuals may try to achieve that.

Leaders are prevented from achieving dominance

It is characteristic of hunter-gatherers that they bring back to earth, often with a bump, anyone who tries to achieve dominance — no matter how well respected they may be for particular skills. Ambitions for status and domination are simply not tolerated. The various manoeuvres used to bring leaders back into line have been catalogued by Boehm (1993), across a broad range of societies. He calls this phenomenon a 'reverse dominance hierarchy'. However, since respect is still given to leaders in particular situations, the incipient hierarchy is not really reversed but rather prevented from developing beyond those particular situations where leadership is required. We feel the term 'counter-dominance' is more appropriate (see Erdal & Whiten 1994; Boehm 1994 for discussion). References to studies documenting counter-dominance are included in Table 12.1, in the column with that heading.

In Africa, among the Mbuti, Turnbull reported that 'Some men, because of exceptional hunting skill, may come to resent it when their views are disregarded, but if they try to force these views they are very promptly subjected to ridicule' (1965, 183, see also 178). This is a particularly important observation, making clear that the skilled habitually are heeded, and that this gives them pleasure. The taking of pleasure in gaining the respect of others in the group implies a psychological predisposition at work. The others seem to be aware of the pleasure, and beyond a certain point resist.

Lee summarizes: 'The !Kung are a fiercely egalitarian people . . . cutting down to size the arrogant and the boastful' (1979, 244. See also pp. 24 and 246 for surprising examples, and the beautiful !Kung explanation that if someone is too proud they want to 'cool his heart and make him gentle'). Bailey gives an interesting picture of how, when a powerful Efe hunter took a second wife, the group supported his first wife against his attempted self-aggrandizement, and reduced his share of meat dramatically. In response, he went through a transformation in his behaviour, changing from being highly assertive to being positively retiring, giving up any semblance of dominance (1991, 93).

In America, Balikci described how the recognized Netsilik 'thinkers', were respectful of others (1970, 116). This appears to be partly for fear of jealousy and hostility, which can be intense and dangerous (1970, 130, 137, 170, 173–5).

In Asia, Gardner describes among the gentle, anti-competitive Paliyans the resentful 'criticism of his assertiveness and . . . questioning of his stature' against polygynous 'headmen' (1972, 426).

Among the Australians, Boehm found only two categories of counter-dominance — ridicule and execution (1993, 232, table 1). Reports are hard to find. One reason may be that there are automatic social mechanisms at work. Meggitt describes how among the Walbiri, kin relationships mean that an individual with a considerable role to play one day might only have a minor role the next, and adds: 'this frequent variation in the extent of authority that an individual exercised from one situation to another militated against the emergence of a class of permanent leaders' (1962, 249). Because this is a feature of the kin system, egalitarianism does not depend on reactions by individuals, as it does elsewhere.

In summary, if individuals attempt to build on the respect that they are granted in situations where they have expertise, so as to become more permanent leaders, then in various ways the others around them put them back in their place — as respected equals, but nothing more.

The predisposition towards counter-dominance

Those who attempt dominance can be ignored, criticized, ridiculed, overruled, abandoned, ostracized or killed. Boehm (1993) explains this as a conscious, intentional, joint response by followers operating as a group, emphasizing the role of conscious decision-making in the process.

We have put the argument that food-sharing is better explained as a culturally influenced activity founded on an evolved disposition to share, rather than being a culturally created process in contradiction with purely selfish evolved predispositions. Exactly the same argument applies to counter-dominance. Counter-dominance and egalitarian behaviour depend on psychological dispositions which create and support them. If the dominance hierarchy had remained functionally effective in evolution, then the possibility of creating egalitarianism in hunter-gatherer bands simply would not have arisen. The psychological predispositions which underlie counter-dominance — for example, resentment at aggressive behaviour, discomfort or anger at being dominated, satisfaction when consensus is reached, respect for others' feelings — form part of our psychology because they were more effective functionally than the dominance/submission psychology out of which they must have evolved. There would be no ability to take conscious decisions to counteract dominant individuals — certainly not as a characteristic behaviour — if we did not have the underlying psychological disposition to do so.

The evolution of counter-dominance

How could the evolutionary change represented by the first part of the 'U' curve in Figure 12.1 have come about? We have argued that human egalitarianism represents a radical innovation when compared with the dominance structures of non-human primates.

However, anthropoid primates do exhibit social characteristics which make the evolutionary innovation less miraculous that it may seem at first sight. First, a substantial new literature shows that competition amongst monkeys and apes is mediated to a significant degree by brain as well as brawn. The social and cognitive expertise concerned has been called 'Machiavellian Intelligence' (Byrne & Whiten 1988). Skills such as tactical deception (Whiten & Byrne 1988; Byrne & Whiten 1992; Byrne this volume) permit the physical superiority of dominant individuals to be undermined, even if only to a limited extent.

Second, the formation of alliances, in which two or more individuals cooperate to displace a third who is capable of dominating each of them in a dyadic contest, has now been well documented in anthropoid primates (Harcourt 1988; Harcourt & de Waal 1992).

These characteristics appear to be special to anthropoids (Harcourt 1988; 1992). They are sufficiently widespread that it is uncontroversial to suggest that counter-dominance has been built on such foundations. That these foundations are limited in non-human primates, however, is emphasized by the fact that dominance relations continue to permit gross variations in access to resources.

In our closest relatives, chimpanzees, detailed studies of captive groups have revealed these political abilities to be particularly refined (de Waal 1982; 1992a). The operation of alliances incorporates a number of specific communicative abilities which function to recruit and reassure potential allies, in what are often protracted sequences of negotiation. Other political acts are deployed at times remote from the operation of alliances themselves. These include affiliative acts directed at potential allies before an attack on an adversary is launched; reward of supporters (through grooming, for example) in the aftermath of a conflict; interference in the formation of competing alliances; and revenge against those who switch allegiance. In addition, powers of deception and counter-deception reach new heights (Byrne & Whiten 1992; de Waal 1992b). Observational and experimental data converge in support of chimpanzees' abilities to read states of mind in others and exploit this ability in sophisticated social manipulation (Whiten 1991; 1993). As a result of this cognitive repertoire, chimpanzee dominance relations can be highly dynamic. De Waal (1982) charted the way in which each of three adult males succeeded in gaining alpha status, only to be ousted by a coalition of the other two: 'the strongest of the three competing parties almost automatically elicits cooperation against himself, because the weaker parties gain more by joining together and sharing the payoffs than by joining the strongest party, who will monopolize the payoffs' (de Waal cited in Leakey & Lewin 1992, 290).

Such instability and political skill suggest the sort of social preadaptations from which human counter-dominance is likely to have evolved. Chimpanzees are not our ancestors, of course, and to treat them as such would court the dangers of referential modelling. In the spirit of the more robust approach of strategic modelling referred to earlier, it is important to acknowledge those features of Machiavellian behaviour outlined above, especially alliance formation, which are widespread in anthropoid primates and therefore reliably inferred attributes of our last non-human ancestor. In conjunction with this picture, chimpanzees demonstrate how far an ape encephalized to an extent similar to *Australopithecus* can push the manipulation of dominants.

Clearly, such temporary and partial counter-dominance could have formed the evolutionary precursor to the more comprehensive counter-dominance we have described for hunter-gatherers. Answers to how and why the shift occurred must at present be speculative. Erdal & Whiten (1994) have suggested that a small increase in encephalization would permit a refinement in counter-dominance, which could set in train a spiralling selection pressure for further political skills. Robert Frank has suggested to us that among the new skills a greater emphasis on attractive as against coercive tactics would be predicted — a strategy of 'dominance through charm'. We envisage the evolutionary spiral eventually reaching a ceiling where the Machiavellian skills of the population at large were so refined that it would no longer have been strategically practicable to expend energy in an effort to dominate other individuals physically. This logic is parallel to that of Trivers (1971), who proposed that an arms race of subtle cheating in a population of reciprocal altruists could escalate to a ceiling where the only viable strategy is generalized altruism.

At this point different behaviour would have become functionally viable — more viable than attempted dominance. Instead of wasting time and energy in a futile effort to dominate others, individuals who devoted enough of their personal resources to counteracting others' dominance, but did not waste time and energy by themselves trying to achieve dominance, would be able to devote much more energy to productive foraging and social behaviour. Those who remained trapped in the old dominance/submission patterns would be wasting their time by comparison. In particular, the ability of a group of individuals arranged in a dominance hierarchy to exploit resources with the unpredictably scattered, large-package characteristics of the hunter-gatherer ecological niche is severely limited. The successful strategy in that niche includes the daily dispersion of the hunters and gatherers, the return to a home base and the sharing of food (Isaac 1978, 92). This implies a complex set of social skills involved in making trustworthy commitments. The underlying psychological predispositions are quite different from those of the old dominance/submission spectrum. In particular, a switch is required from an individual 'scare off the others and get what I can' strategy, to one of fair play, of identification with the others in the group, of pleasure in relationships of equality rather than dominance, of emotional commitment to share and to help others in need, of confidence that others will help. These positive feelings are seen, for example, in Rasmussen's observation among the Netsilik that 'People living together in a hunting camp feel closely attached to one another in many ways' (1931, 147), and in Tanaka's report that after a night of dancing 'Feelings of peace and security flood the San camp' (1980, 115).

An important element in this fair-play complex is the inclination in each individual to resist being dominated or taken advantage of, either in food-sharing or in leadership: hence the vigilance seen among hunter-gatherers in sharing, and in bringing leaders to heel when they step out of line.

The psychology of egalitarianism

If something like this evolutionary process occurred, then the resulting psychology would be expected to be very complex. Each individual must still retain the disposition to 'look out for number one', at least on an opportunistic basis. In the case of sharing, this will include the inclination to hoard and perhaps even to steal, as well as supporting the vigilance noted above. But it must not be strong enough to make sharing break down. And it must be combined with a willingness to punish cheating in others — something which no 'rational self-interest' model can predict or explain, because the process of punishing others is costly to the one who punishes (Frank 1988). In the case of leadership, presumably the earlier predisposition to 'dominate if possible' has not been eradicated; but it has been coupled with the desire and commitment to resist domination too, and to accept an equal relationship as a good outcome. In addition, there must be a willingness to take the initiative, particularly where one is confident of one's knowledge or expertise — but one must also be prepared to listen, and to follow a different line if the group goes that way, or to defer if someone else clearly has superior skill or knowledge. Boehm (1989) similarly models a complex psychology with self-contradictory elements.

An important psychological element underpinning both food-sharing and egalitarian leadership is the process of weighing up one's own position vis-à-vis that of others. If a carcass is being divided up, each individual is extremely sensitive to anything which suggests that he or she is not getting a fair share relative to each other participant. And if one is listening to an expert hunter's proposal on how to get hold of the next carcass, each person is very aware of his own position relative to that of the would-be leader.

It is interesting that apparently the predisposition is not to achieve superiority: not to get more meat than anyone else, nor to dominate the expert. But if anyone else seems to be getting more meat, or becoming too self-assertive, then a trigger is pulled, and someone steps in, to ensure a fair distribution, or to bring the leader back into line. This is not so much keeping up with the Joneses, as making sure that none of the Joneses gets ahead in the first place. Frank (1985) examines a multitude of economic and political effects of this predisposition to measure ourselves against those with whom we feel some affinity.

This tendency could be the outcome of combinations of contradictory dispositions: to get more and at the same time to stop others from getting more; to dominate, and to stop others from dominating. As these contradictory predispositions evolved, one result must have been that behaviour became more and more flexible: there are a host of different possible responses to any social situation. And here there is a likely role for the ability to pass on and absorb a culture — by what is virtually an imprinting process. Culturally defined ways to behave limit the choices that have to be made, helping to resolve what would otherwise be constant psychological dilemmas.

Additionally, with this varied psychology, there is an important role for the ability to think. With so many self-contradictory options there would be a clear adaptive advantage in the ability to hold in mind important features of the environment, particularly the social environment, for long enough so that one's reactions could be resolved into a decision.

Other evidence

The model presented here has not yet been tested on humans in psychological experiments or surveys, although the studies by Cosmides & Tooby on cheater-detection (1992) and by Frank (1988) and Caporael *et al.* (1989) on cooperation, among others, seem to support the model.

One key question which this analysis raises is why, given this evolved psychology, the development of herding and agriculture about 10,000 years ago triggered the creation of big-men, of chiefs, of classes and ultimately of multi-level institutionalized hierarchies. As we have said, there was not time for significant biological evolution to take place: these developments must depend on the same psychology as hunter-gatherer egalitarianism. The answer must lie in the fit between the evolved psychological predispositions and the new environment. Since this is no longer the environment in which humans evolved, the evolved predispositions need not in principle lead to behaviour which is functionally effective. The correlation between the proximate cause of the behaviour and the genetic function of the behaviour may be broken in a radically different environment (Symons 1979, 306–7; Tooby & DeVore 1987, 198). Key aspects of the environment which are different include the availability of stored surpluses; the existence of highly productive assets; the population density; the ability of individuals or small groups to control the surpluses and the assets; and the planning time horizon (Johnson & Earle 1987). Somehow, the counter-dominant tendency became ineffective, although it recurs consistently (Boehm 1993). How that happened would take us far from modelling the early human mind.

Conclusion

The early human mind is likely to have been characterized by psychological dispositions supporting egalitarianism: vigilant food-sharing, informal leadership and counter-dominant behaviour.

David Erdal & Andrew Whiten
Scottish Primate Research Group
School of Psychology
St Andrews University
St Andrews
Fife
KY16 9JU
Scotland

Acknowledgements

With much thanks to Bruce Knauft and Linnda Caporael for stimulating and detailed comments, and to Paul Mellars and Kathleen Gibson for constructive editorial criticism and warm encouragement.

References

Bailey, R.C., 1991. *The Behavioural Ecology of Efe Pygmy Men in the Ituri Forest, Zaire.* (Anthropological Papers 86.) Ann Arbor (MI): University of Michigan Museum of Anthropology.

Bailey, R.C., G. Head, M. Jenike, B. Owen, R. Rechtman & E. Zechenter, 1989. Hunting and gathering in tropical rain forest: is it possible? *American Anthropologist* 91, 59–82.

Balikci, A., 1970. *The Netsilik Eskimo.* New York (NY): Natural History Press.

Bicchieri, M.G. (ed.), 1972. *Hunters and Gatherers Today.* New York (NY): Holt, Rinehart & Winston.

Binford, L.R., 1981. *Bones: Ancient Men and Modern Myths.* New York (NY): Academic Press.

Bleek, D.F., 1928. *The Naron.* Cambridge: Cambridge University Press.

Boehm, C., 1989. Ambivalence and compromise in human nature. *American Anthropologist* 91, 921–39.

Boehm, C., 1993. Egalitarian behaviour and reverse dominance hierarchy. *Current Anthropology* 34, 227–54.

Boehm, C., 1994. Some futher thoughts on reverse dominance hierarchy: reply to Erdal and Whiten. *Current Anthropology* 35, 178–81.

Boesch, C. & H. Boesch, 1989. The hunting behaviour of wild chimpanzees in the Tai National Park. *American Journal of Physical Anthropology* 78, 547–73.

Briggs J.L., 1970. *Never in Anger: a Portrait of an Eskimo Family.* Cambridge (MA): Harvard University Press.

Byrne, R.W. & A. Whiten (eds.), 1988. *Machiavellian Intelligence: Social Expertise and the Evolution of Intellect in Monkeys, Apes and Humans.* Oxford: Oxford University Press.

Byrne, R.W. & A. Whiten, 1992. Cognitive evolution in primates: evidence from tactical deception. *Man* 27, 609–27.

Caporael, L.R., R.M. Dawes, J.M. Orbell & A.J.C. Van de Kragt, 1989. Selfishness examined: cooperation in the absence of egoistic incentives. *Behavioral and Brain Sciences* 12, 683–739.

Cashdan, E.A., 1985. Coping with risk: reciprocity among the Basarwa of Northern Botswana. *Man* 20, 454–74.

Cosmides, L. & J. Tooby. 1992. Cognitive adaptations for social exchange, in *The Adapted Mind: Evolutionary Psychology and the Generation of Culture,* eds. J.H. Barkow, L. Cosmides & J. Tooby. New York (NY): Oxford University Press, 163–228.

Damas, D., 1972. The copper Eskimo, in Bicchieri (ed.), 3–50.

de Waal, F.B.M., 1982. *Chimpanzee Politics.* London: Jonathan Cape.

de Waal, F.B.M., 1992a. Coalitions as part of reciprocal relations in the Arnhem chimpanzee colony, in Harcourt & de Waal (eds.), 233–57.

de Waal, F.B.M., 1992b. Intentional deception in primates. *Evolutionary Anthropology* 1, 86–92.

Endicott, K., 1988. Property, power and conflict among the Batek of Malaysia, in Ingold *et al.* (eds.), vol. 2, 110–27.

Erdal, D. & A. Whiten, 1994. On human egalitarianism: an evolutionary product of Machiavellian status escalation? *Current Anthropology* 35, 175–8.

Foley, R., 1988. Hominids, humans and hunter-gatherers: an evolutionary perspective, in Ingold *et al.* (eds.), vol. 1, 207–21.

Frank R., 1985. *Choosing the Right Pond: Human Behaviour and the Quest for Status.* New York (NY): Oxford University Press.

Frank R., 1988. *Passions within Reason: the Strategic Role of the Emotions.* New York (NY): W.W. Norton.

Gardner, P.M., 1972. The Paliyans, in Bicchieri (ed.), 404–47.

Goodale, J.C., 1971. *Tiwi Wives.* Seattle (WA): University of Washington Press.

Harcourt, A.H., 1988. Alliances in contests and social intelligence, in Byrne & Whiten (eds.), 132–52.

Harcourt, A.H., 1992. Coalitions and alliances: are primates more complex than non-primates?, in Harcourt & de Waal (eds.), 445–71.

Harcourt, A.H. & F.B.M. de Waal, 1992. *Coalitions and Alliances in Humans and other Animals.* Oxford: Oxford University Press.

Hawkes, K., 1993. Why hunter-gatherers work. *Current Anthropology* 34, 341–61.

Hawkes, K., J.F. O'Connell & N.G. Blurton-Jones, 1991. Hunting income patterns among the Hadza; big game, common goods, foraging goals and the evolution of the human diet, in *Foraging Strategies and Natural Diet of Monkeys, Apes and Humans,* eds. A. Whiten & E. Widdowson. Oxford: Clarendon Press, 243–51.

Helm, J., 1972. The Dogrib Indians, in *Hunters and Gatherers Today,* ed. M.G. Bicchieri. New York (NY): Holt, Rinehart and Winston, 51–89.

Hole, F., 1992. Origins of agriculture, in Jones *et al.* (eds.), 373–9

Holmberg, A.R., 1950. *Nomads of the Long Bow: the Siriono of Eastern Bolivia.* (Publication 10.) Washington (DC): Institute of Social Anthropology.

Huntingford, D.W.B., 1955. The economic life of the Dorobo. *Anthropos* 50, 602–34.

Hyde, G.E., 1959. *Indians of the High Plains from the Prehistoric Period to the Coming of the Europeans.* (Civilisation of the American Indians 54.) Norman (OK): University of Oklahoma.

Ingold, T., D. Riches & J. Woodburn (eds.), 1988. *Hunters and Gatherers,* vol. 1: *History, Evolution and Social Change;* vol. 2: *Property , Power and Ideology.* London: Berg.

Isaac, G., 1978. The food sharing behaviour of protohuman hominids. *Scientific American* 238, 90–108.

Johnson, A.W. & T. Earle, 1987. *The Evolution of Human Societies: From Foraging Group to Agrarian State*. Stanford (CA): Stanford University Press.

Jones, S., R. Martin & D. Pilbeam (eds.), 1992. *The Cambridge Encyclopedia of Human Evolution*. Cambridge: Cambridge University Press.

Kaplan, H. & K. Hill, 1985. Food sharing among Ache foragers: tests of explanatory hypotheses. *Current Anthropology* 26, 223–45.

Knauft, B.M., 1991. Violence and sociality in human evolution. *Current Anthropology* 32, 391–428.

Kroeber, A.L., 1939. *Cultural and Natural Areas of Native North America*. Berkeley (CA): University of California Press.

Leacock, E.B., 1954. *The Montagnais Hunting Territory and the Fur Trade*. (Memoir 78.) New York (NY): American Anthropologist.

Leacock, E.B., 1980. Montagnais women and the Jesuit program for colonization, in *Women and Colonization*, eds. M. Etienne & E.B. Leacock. New York (NY): Praeger, 25–42.

Leacock, E.B., 1982. Relations of production in band society, in Leacock & Lee (eds.), 159–70.

Leacock, E.B. & R.B. Lee (eds.), 1982. *Politics and History in Band Societies*. Cambridge: Cambridge University Press.

Leakey, R. & R. Lewin, 1992. *Origins Reconsidered*. London: Little, Brown & Co..

Lee, R.B., 1979. *The !Kung San: Men, Women and Work in a Foraging Society*. Cambridge: Cambridge University Press.

Lee, R.B. & I. DeVore (eds.), 1968. *Man the Hunter*. Chicago (IL): Aldine.

Lovejoy, C.O., 1981. The origin of man. *Science* 211, 341–50.

Marshall, L., 1976 (1961). Sharing, talking and giving: relief of social tension among !Kung Bushmen. Africa 31, 231–49. Reprinted in *Kalahari Hunter-Gatherers: Studies of the !Kung San and their Neighbours*, eds. R.B. Lee & I. DeVore. Cambridge (MA): Harvard University Press, 349–72.

Meggitt, M.J., 1962. *Desert People: a Study of the Walbiri Aborigines of Central Australia*. Chicago (IL): University of Chicago Press.

Morris, B., 1982a. The family, group structuring and trade among S. Indian hunter-gatherers, in Leacock & Lee (eds.), 171–87.

Morris, B., 1982b. *Forest Traders: a Socio-economic Study of the Hill Pandaram*. London: Athlone Press.

Potts, R., 1992. The hominid way of life, in Jones *et al.* (eds.), 325–34.

Provinse, J.H., 1955. The underlying sanctions of Plains Indian culture, in *Social Anthropology of North American Tribes*, ed. F. Eggan. 2nd edition. Chicago (IL): University of Chicago Press, 341–74.

Rasmussen, K., 1931. The Netsilik Eskimos: social life and intellectual culture, in *Report of the Fifth Thule Expedition*, vol. VIII, no. 1–2. Copenhagen: Nordisk Forlag.

Riches, D., 1982. *Northern Nomadic Hunter-Gatherers: a Humanistic Approach*. London: Academic Press.

Riches, D., 1984. Hunting, herding and potlatching: towards a sociological account of prestige. *Man* 19, 234–51.

Rogers, E.S., 1972. The Mistassini Cree, in Bicchieri (ed.), 90–137.

Shepher, J., 1983. *Incest: a Biosocial View*. London: Academic Press.

Smith, E.A., 1988. Risk and uncertainty in the 'original affluent society', in Ingold *et al.* (eds.), vol. 1, 222–51.

Stearman, A.M., 1984. The Yuqui connection: another look at Siriono deculturation. *American Anthropologist* 86, 630–50.

Suttles, W., 1968. Variation in habitat and culture on the north west coast, in *Man in Adaptation*, vol. 2, ed. Y. Cohen. Chicago (IL): Aldine.

Symons, D., 1979. *The Evolution of Human Sexuality*. Oxford: Oxford University Press.

Tanaka, J., 1980. *The San Hunter-Gatherers of the Kalahari: a Study in Ecological Anthropology*. Tokyo: University of Tokyo Press.

Thomas, E.M., 1959. *The Harmless People*. New York (NY): A.A. Knopf.

Tindale, N.B. 1972. The Pitjandjara, in Bicchieri (ed.), 217–68.

Tooby, J. & I. DeVore, 1987. The reconstruction of hominid behavioral evolution through strategic modelling, in *The Evolution of Human Behavior: Primate Models*, ed. W.G. Kinzey. Albany (NY): State University of New York Press, ch. 10, 183–237.

Trivers, R.L., 1971. The evolution of reciprocal altruism. *Quarterly Review of Biology* 46, 35–57.

Turnbull, C., 1965. *Wayward Servants*. London: Eyre and Spottiswoode.

Washburn, S.L. & I. DeVore, 1961. Social behavior of baboons and early man, in *Social Life of Early Man*, ed. S.L. Washburn. Chicago (IL): Aldine, 91–105.

Whiten, A. (ed.), 1991. *Natural Theories of Mind*. Oxford: Basil Blackwell.

Whiten, A., 1993. Evolving a theory of mind: the nature of non-verbal mentalism in other primates, in *Understanding Other Minds*, eds. S. Baron-Cohen, H. Tager-Flusberg & D. Cohen. Oxford: Oxford University Press.

Whiten, A. & R.W. Byrne, 1988. Tactical deception in primates. *Behavioral and Brain Sciences* 11, 233–73.

Wiessner, P., 1982. Risk, reciprocity and social influences on !Kung San economics, in Leacock & Lee (eds.), 61–84

Wilkinson, G.S. 1984. Reciprocal food-sharing in the vampire bat. *Nature* 308, 181–4.

Part IV

Ontogeny and Phylogeny

13
Issues in Mental Development

Christopher Longuet-Higgins

The claim that the mental development of children mirrors the mental evolution of Homo sapiens *appears to rest on Haeckel's controversial principle that 'ontogeny recapitulates phylogeny'. Re-examination of the principle suggests an alternative formulation that correlates the temporal order of expression of an individual's genes with the order in which they were first incorporated into the gene pool of the species. At the cognitive level of enquiry, a comparison of the findings of Piaget and Johnson-Laird suggests that the critical step in the evolution of language was the ability to construct and manipulate mental models — in a word, to think.*

In the early 1970s A.J. Kenny, J.R. Lucas, C.H. Waddington and the writer collaborated in a two-year series of Gifford Lectures on The Phenomenon of Mind. In the first year (Longuet-Higgins *et al.* 1972) we examined The Nature of Mind, and the philosophical issues raised by the enterprise of Artificial Intelligence. In the second year (Longuet-Higgins *et al.* 1973) we turned to a subject of more direct relevance to this symposium, namely The Development of Mind. In an introductory lecture I suggested that one might expect to find connections between the mental development of children, as described by Jean Piaget, and the mental evolution of *Homo sapiens,* as mirrored in the differences between the minds of human beings and apes (Longuet-Higgins *et al.* 1973, 6).

> If it is really true that ontogeny must largely recapitulate phylogeny, then we do not have to wait until the molecular biologist has read our chromosomes before we can begin to understand why our minds develop as they do. We may, instead, be able to gain some enlightenment from studying our evolutionary ancestors and cousins.

What follows is, essentially, a re-examination of those issues.

In the last twenty years the relationships between the minds of children, apes and hominids have been widely discussed by students of human evolution (see, for example, Parker 1977; Gibson 1977; Parker & Gibson 1979; Gibson & Ingold 1993). From the now extensive literature on the subject one may perhaps single out Merlin Donald's recent book *Origins of the Modern Mind* (Donald 1991). A cognitive psychologist by profession, Donald marshals the evidence from disciplines as far apart as archaeology and neuropsychology in support of his claim that the modern mind evolved from the pre-human mind through a series of major adaptations, each leading to the emergence of a new representational system. He posits an intervening layer of cognition between human beings and apes, and suggests that 'if such a vestigial layer of cognition existed, it should have left its mark on the observable structure of modern human cognition'.

Whatever one may feel about the details of Donald's theory, it is difficult to disagree with his presupposition that the close study of any organism is likely to reveal something about its evolutionary past. As for the human mind, we are at last beginning to understand its structure, and the manner in which it develops from early infancy to maturity. It has therefore become more important than ever to examine the claim that we can obtain access to our ancestors' minds by studying what is known about the mental development of our children.

The starting point of the present paper is the axiom that human mentality is the product of an interplay between nature and nurture. We then turn to Haeckel's celebrated claim that 'ontogeny recapitulates phylogeny' and critically examine its biological justification. Though obviously vulnerable to criticism it embodies an important insight that can be put on a firm molecular-genetic basis. The question then arises whether a properly qualified version of the claim would be expected to apply to mental as well as to physical development. Taking the view that it would leads naturally to the Piagetian position that the cognitive development of children affords important clues to human mental evolution. Of particular interest in this connection is the *tour de force* of first language acquisition. If, as Johnson-Laird has suggested, the ability to construct mental models — an important feature of Piaget's 'pre-operational' stage — is actually a prerequisite for language, then this ability — essentially the capacity for coherent thought — could well have been the 'missing link' between the pre-human and the early human mind.

Nature and nurture

It is hardly a matter of controversy that a person's mental furniture is largely determined by their past experience, rather than inherited from their parents. Here the word 'inherited' is used in its biological sense: of a character or trait that is genetically transmitted from parents to offspring, through the physical transfer of DNA. There are, of course, other sorts of inheritance, of no less interest to the anthropologist: people inherit possessions, or titles, or surnames, or beliefs, or prejudices from their forbearers, as well as picking up their parents' language with no apparent effort. But over thousands of generations such non-genetic inheritance is highly vulnerable to disruption and must therefore be a quite unreliable source of evidence about the remote past. By comparison, the inheritance of genetic information is a nearly error-free process, so that a genetically determined character has a good chance of surviving in a population for a considerable time.

The most obvious characters of an individual are physical: height, skin colour and so forth. Mental qualities are altogether less easy to define, and far more subtly related to a person's genetic makeup. A further complication — and this is especially true of our own times — is that people's cognitive faculties are to some extent enhanced or diminished by the cultures and traditions into which they are born. One of the occupational hazards of psychological research is the political sensitivity of issues such as the heritability of intelligence, musical talent or mental instability. But even though human beings do not always develop their full potential, their mental powers must ultimately be limited by the anatomy and physiology of the human brain. In the last resort, therefore, mental development is subject to the same biological constraints as physical development — a proposition we shall take for granted in the paragraphs that follow.

Ontogeny and phylogeny

One of the best known and most widely criticized propositions in developmental biology is Haeckel's 'biogenetic law', to the effect that 'ontogeny recapitulates phylogeny' (Haeckel 1866) — that a developing embryo passes through a sequence of stages in which it resembles, in turn, its evolutionary ancestors. Taken literally, of course, the law is blatantly false: an embryo of a given species could not possibly be mistaken for an animal of a different species, if only because an embryo is not yet capable of independent life. But such objections quite miss the point of Haeckel's epigram — for that is what it is, not just a curious piece of comparative anatomy. It embodies the insight that as soon as Nature has discovered a particular trick she tends to use it over and over again. Once having discovered how to make a vertebrate, say, she will tend to use essentially the same technique for doing so, whether the animal under construction is a fish, a lizard, a mouse, a monkey or a human being. The facts, that recapitulation is not exact in detail, and that it relates only to the immature forms of descendants and ancestors, are mere footnotes to the main thesis.

The history of science demonstrates that appearances can be seriously misleading — that a superficial resemblance does not signify a relationship at a deeper level unless there are other reasons to suspect one. But during the last half century biology has been transformed by the discovery of the molecular mechanisms of biological processes, including the way in which the development of an animal is determined by its genetic makeup — by the DNA it inherits from its parents. This DNA is copied into the chromosomes of every cell in the animal's body, so that each cell will know what to do when its turn comes to divide or differentiate. One may think of all this DNA as constituting a genetic program (Longuet-Higgins 1972) that controls the construction of the animal, and the maintenance of its bodily functions, from the moment of conception. As the program unfolds, more and more of the DNA becomes involved, and ultimately, if everything goes according to plan, the result will be a mature and healthy individual, typical of its species. But that is not the only possibility. There is a small but finite chance that one or both of the DNA strands inherited from the animal's parents contained a mutation — a mistake of some sort — and sooner or later such a mutation is bound to show up as the genetic program unfolds. If this happens fairly late in the process it may not matter, and we shall obtain an individual who is untypical of the species but nonetheless quite able to survive. But if the mistake surfaces early on, its effects will be correspondingly widespread and disruptive, and probably lethal. To take an extreme example, a mutation that corrupted the genetic code, upon which all protein synthesis depends, would instantly abort its unfortunate victim. The upshot of this argument is that those characteristics that distinguish a novel individual are most likely to have arisen from a mutation that was expressed only in the later stages of that creature's development. From time to time such a mutation may well lead to an individual that is fitter than the other members of the species. If that individual survives to reproduce its kind, we chalk up one more success for Darwinian evolution.

Now we are in a position to see why a developing organism might well bear a passing resemblance — or rather, a sequence of passing resemblances — to its evolutionary forbearers. Its DNA will incorporate genes of very different evolutionary ages, but the *genes that control the final stages of the organism's development will tend to be those that were added most recently to the gene pool of the evolving species.* As a result the successive stages of development of the organism will tend to mimic, in the same order, the successive steps in the evolution of the species.

It is therefore not at all unreasonable to expect the mental development of a human child to mirror, however imperfectly, the early mental evolution of *Homo sapiens.*

Genetic epistemology

Perhaps fortunately, most of what is known about the psychological development of children is quite tangential to the study of human mental evolution. An outstanding exception is the work of Jean Piaget and his school. Where Sigmund Freud had focused attention on the influence of nurture on human behaviour, Piaget recognized the crucial importance of nature in the child's cognitive development (Piaget 1926). The phrase 'genetic epistemology' by which he distinguished his research philosophy has come to seem less quaint in the last few years, as others have recognized the importance of his claim that the child's cognitive development runs to a genetically determined schedule. At each stage of increasing sophistication, Piaget maintained, the child's lower-level concepts must be integrated into those at the next level — a process he described as 'structural equilibration'. The term is idiosyncratic but its meaning is easily grasped by anyone who has ever undertaken the development of a complex computer program.

Piaget distinguished four main stages in a person's cognitive development (Piaget & Inhelder 1969), but only the first two stages are relevant to our present concerns. The first two years of the child's life constitute the 'sensorimotor' stage, in which she is mainly preoccupied with controlling her own body. It is during this stage that she becomes aware of the independent existence of the objects around her and of her own separateness from them. There is nothing about the sensorimotor stage that one could regard as specifically human; all primates seem to pass through a stage of this kind, in which they acquire what sensorimotor skills they are going to need in later life (Parker 1976; 1977).

The story is different when one comes to Piaget's second stage — the 'pre-operational' stage (Parker & Gibson 1979). This overlaps with the sensorimotor stage, and lasts until the child is about six or seven. During the pre-operational stage the child learns to manipulate her environment symbolically by means of internal representations of the external world. But — and here *Homo sapiens* is most clearly in evidence — she also begins to master the elements of language — to symbolize objects by words and to order the words mentally, as one must in order to formulate linguistic structures.

The pre-operational stage is clearly where one should look for traces of the 'layer of cognition' that Merlin Donald supposes to intervene between the mind of an ape and that of a human being. It is irresistibly tempting to identify this layer with the ability to build and manipulate mental models of the world, an imaginative faculty that Johnson-Laird has shown to be essential not only for coherent thought but also for the comprehension of language (Johnson-Laird 1983). On this interpretation of the developmental evidence it was a sharpening of our imaginative powers that constituted the penultimate step in our mental evolution, if we think of the advent of language as having been the final step to *Homo sapiens.*

The acquisition of language

All normal children pick up their parents' language without apparent effort. The first modern theory of how this happens was formulated by Noam Chomsky.

In *Aspects of the Theory of Syntax* (1965) Chomsky suggested that we can acquire language only because we come into the world with a certain amount of *general* knowledge about the grammars of natural languages, embodied in what he referred to as linguistic universals. In more formal terms one could speak of a Language Acquisition Device which, given a set of utterances, would return the grammatical rules by which they were generated. These rules, according to the theory, would be syntactic in nature, the business of semantic interpretation being a logically independent task.

If one accepts this as a valid account of human language the question immediately arises how it could possibly have evolved by natural selection. In our joint Gifford lectures (Longuet-Higgins *et al.* 1973) — and he has developed the point in a recent book (Kenny 1989) — Anthony Kenny claimed that the origins of language could not be explained in orthodox Darwinian terms, because, for language to confer any selective advantage on an individual, he or she must already belong to a community that has adopted a number of symbolic conventions. But perhaps this objection is less damaging to a Darwinian view than appears at first sight. Children have no difficulty in inventing languages of their own, and often do so in order to hide their thoughts from grown-ups. As William Wordsworth may have been the first to recognize, the concept of a symbol, whether graphic or phonetic, gives the six-year-old much less pause than it does to philosophers of language:

> Thou best philosopher, who yet dost keep
> Thy heritage, thou eye among the blind. . .

From Chomsky's purely syntactic perspective the evolution of language is little short of a biological miracle. But if we take a broader view and build semantics and pragmatics into the very notion of a language, we may reasonably suppose that it was our ability to construct mental models of the world that permitted us to invent language in the first place, just as it permits the child to acquire it.

Some years ago Richard Power and I set ourselves the task of constructing a semantically-based Language Acquisition Device (Power & Longuet-Higgins 1978). The device took the form of a computer program that would learn from representative examples (such as *vier und zwanzig* : 24) how to name the natural numbers in a range of human languages. We were astonished at the ease with which a system of this sort could acquire even quite elaborate numeral systems, provided it was given the ability to do elementary arithmetic — to add and multiply pairs of numbers together. An approach along the lines laid down in *Aspects* (Chomsky 1965) would have set about the acquisition of a numeral system by attempting to enumerate just those word strings that were used as number names, and no others; but to have done only that — a formidable syntactic problem in itself — would have left unsolved the further problem of assigning these names semantic interpretations. How much easier it is, evidently, to learn a natural language if the relevant cognitive skills are already in place. Who could reasonably have doubted that it would be?

Conclusions

A second look at the relation between the ontogeny and the phylogeny of mind seems to show, not only that the one copies the other (Parker & Gibson 1979) even more closely than could have been expected (Longuet-Higgins *et al.* 1973), but that their resemblance can be given a plausible genetic interpretation: the most novel characters of a species will tend to emerge late in an individual's development, for essentially logistic reasons.

It would be misleading to suggest that cognitive psychology has now reached the stage at which one can draw firm conclusions about the evolution of the human mind by studying the mental development of children. But if the final step to *Homo sapiens* was the invention of language, then the preceding step could well have been the ability to build and manipulate mental models of reality — in a word, to *think*. If that is indeed the case, then it was the imagination no less than the reason that distinguished the early human mind from its forerunners.

Christopher Longuet-Higgins
School of Biological Sciences
University of Sussex
Brighton
BN1 9QG

Acknowledgements

My thanks are due to Paul Mellars for valued encouragement, Kathleen Gibson and Nick Humphrey for pertinent references, and Stuart Sutherland and John Cairns for helpful comments on an earlier draft.

References

Chevalier-Skolnikoff, S. & F.E. Poirier (eds.), 1977. *Primate Biosocial Development*. New York (NY): Garland Press.

Chomsky, N., 1965. *Aspects of the Theory of Syntax*. Cambridge (MA): M.I.T. Press.

Donald, M., 1991. *Origins of the Modern Mind: Three Stages in the Evolution of Culture and Cognition*. Cambridge (MA): Harvard University Press.

Gibson, K.R., 1977. Brain structure and intelligence in macaques and human infants from a Piagetian perspective, in Chevalier-Skolnikoff & Poirier (eds.), 113–57.

Gibson, K.R. & T. Ingold, 1993. *Tools, Language and Cognition in Human Evolution*. Cambridge: Cambridge University Press.

Haeckel, E., 1866. *Generelle Morphologie der Organismen*. Jena: University of Jena.

Johnson-Laird, P.N., 1983. *Mental Models: Towards a Cognitive Science of Language, Inference and Consciousness*. Cambridge: Cambridge University Press.

Kenny, A.J.P., 1989. *The Metaphysics of Mind*. Oxford: Clarendon Press.

Longuet-Higgins, H.C., 1972. What biology is about, in *Towards a Theoretical Biology*, vol. 2: *Sketches*, ed. C.H. Waddington. Edinburgh: Edinburgh University Press, 227–35.

Longuet-Higgins, H.C., A.J.P. Kenny, J.R. Lucas & C.H. Waddington, 1972. *The Nature of Mind*. Edinburgh: Edinburgh University Press.

Longuet-Higgins, H.C., A.J.P. Kenny, J.R. Lucas & C.H. Waddington, 1973. *The Development of Mind*. Edinburgh: Edinburgh University Press.

Parker, S.T., 1976. A Comparative Longitudinal Study of Sensorimotor Development in a Macaque, a Gorilla and a Human Infant from a Piagetian Perspective. Paper presented at the Animal Behaviour Society Conference, Boulder, Colorado.

Parker, S.T., 1977. Piaget's sensorimotor series in an infant macaque: a model for comparing unstereotyped behaviour and intelligence in human and nonhuman primates, in Chevalier-Skolnikoff & Poirier (eds.).

Parker, S.T. & K.R. Gibson, 1979. A model of the evolution of language and intelligence in early hominids. *Behavioural and Brain Sciences* 2, 367–407.

Piaget, J., 1926. *The Language and Thought of the Child*. London: Routledge & Kegan Paul.

Piaget, J. & B. Inhelder, 1969. *The Psychology of the Child*. London: Routledge & Kegan Paul.

14
Development and Evolution of the Symbolic Function: The Role of Working Memory

James Russell

I begin by considering the parallels that can be drawn between the transition towards symbolic behaviour around 40–35,000 BP *and children's acquisition of what J. Piaget called 'the symbolic function' at around 18 months of age. After reviewing the evidence for the stage-wise, 'across-the-board' nature of the transition in children I argue, after Olson (1993), that the essence of symbolic representation consists in the ability to 'hold in mind' a representation that is not caused by what is currently 'held in view'. Evidence is then presented for working memory being the cognitive substrate of the holding-in-mind function and for its localization in the prefrontal cortex. On these bases I then speculate that modern humans' substantially larger working memory capacity formed a cognitive substrate necessary for the evolution of symbolic behaviour and also argue that the evolution of linguistic communication exploited the working memory function.*

The phrase 'genetic epistemology' can be interpreted in a number of ways (Russell 1978; 1979). For Piaget (1970), it meant the study of transitions from lower to higher levels of knowledge — whether these transitions occur in mental development, in the evolution of intelligence or in the historical progression of a science. More controversially, he claimed that the same fundamental principles are at work in all three areas; so that if one wishes to study the acquision of knowledge *per se* the best strategy is to study the most accessible population — children — because, as he put it, 'children are all around us' (Piaget 1970, 10).

Taken to an extreme, this view reduces to recapitulationism; but it also affords a fairly modest and plausible reading. Setting the specific claims of Piagetian theory to one side (e.g. about learning by 'assimilation and accommodation', 'structures-of-the whole', and so forth) it can reasonably be proposed that any knowledge-acquiring population will, handle, say, concrete concepts before it handles abstract ones, will master addition before multiplication, will undergo 'U-shaped' developmental progressions, to give just three examples. Moreover, if there are stage-wise or 'across-the-board' changes during the process of knowledge acquisition it is not absurd to expect that the cognitive underpinnings which are necessary for the transition in one population are similar to those which make it possible in another. In any event, such a watered-down form of 'genetic epistemology' would seem to be defensible.

The possibility I wish to explore is that the cognitive substrate which supports the emergence of what Piaget (1951) called 'the symbolic function' in children of about 18 months of age may be similar to the cognitive substrate which was necessary for the 'explosion' in symbolic behaviour that occurred around 40–35,000 BC (Pfeiffer 1982; Chase & Dibble 1987; White 1989; Mellars 1989). Before proceeding, however, there is a substantial *caveat* to be entered. It is that, at present, we do not actually *possess* any hard data on the cognitive substrate that underpins the development of symbolic thought in children. What we have instead, as will emerge, are *theories* about what the change consists in together with some helpful hints from allied disciplines such as neuropsychology. For this reason, what I will propose can be little more than suggestive. That said, some suggestions are better than others.

The main body of this paper will be a defense of the views that there is indeed an 'across-the-board' shift towards symbolic functioning around the middle of the second year of life and that, following

Olson (1993), we can regard the essence of symbolic behaviour as the ability to 'hold in mind' a representation which diverges from that which is currently 'held in view'. I then argue that *working memory* — a form of information-processing carried forward by the prefrontal cortex — is the cognitive substrate of holding-in-mind. After discussing the importance of working memory to linguistic communication I speculate that working memory may have served as a primary cognitive substrate in the evolution of symbolic behaviour and communication.

I will begin by reviewing what 'symbolic behaviour' means in the case of early man and in the case of toddlers — being very brief in the first case.

A parallel between two kind of symbolic behaviour?[1]

Mellars (1991) has summarized the behavioural changes which occurred between the Middle to the Upper Palaeolithic in Europe within seven categories: (a) a shift from 'flake' to 'blade' forms of stone tool; (b) standardization of the shapes of these tools ('imposed form') together with their increasing variety and complexity; (c) the construction of bone, antler and ivory artefacts; (d) an acceleration of the tempo of technological change; (e) personal ornamentation (e.g. perforated animal teeth and marine shells); (f) naturalistic/representational art; (g) more specialized and systematic forms of social organization (e.g. planned rather than opportunistic hunting).

One may say that either these forms of behaviour are symbolic in a fairly uncontroversial sense, or symbolic processes would seem to make them possible. Specifically, the actions which they require would need to be guided by an ideal form or end-state that is mentally represented during their performance. This is most clearly seen in (a), (b) and (f), because to impose a form on a chunk of material, whether or not the end-product has any practical use, requires a mental representation of that form before it is produced, in addition to the regulation and monitoring of the actions relative to that representation during performance.

What is clearly not the case, however, is that symbolic thought can only be manifested when when an individual *constructs* something. Maintaining in consciousness a representation of a form or of a goal-state irrespective of the current state of reality can obviously be achieved by creatures who have no constructive abilities. This brings me to the case of young children. Their ability to:

- pretend that something is the case,
- imagine the spatial transformations that an occluded object is undergoing,
- imitate what another person did a day or so previously ('deferred' imitation),

would all appear to require representational capacities of this order. Such behaviours are characteristic of 'the symbolic function' whose emergence around 18 months of age was described by Piaget (1951). There is some measure of agreement in the literature that symbolic play (e.g. pretending a toy brick is a cup) normally arrives some time after 18 months of age (Leslie 1987; McCune-Nicolich 1981) although the exact timetable obviously depends upon the definition of 'symbolic'. Search after invisible displacements certainly arrives around this age (Bremner 1988). Deferred imitation is a rather difficult case because it is not easy to discriminate from pretend play (e.g. is the pretend car-driving actually deferred imitation of what mummy did yesterday?). Additionally, Piaget's timetable would seem to be challenged by Meltzoff's (1988) demonstration that infants as young 9 months can remember and imitate the actions performed by an experimenter on a novel object. However this ability can be described more parsimonously as 'remembering what actions a novel object will afford'. Piaget's observations, by contrast, were often of the child's imitation of how another person behaved (rather than what was done to an object) — such as the imitation of a playmate throwing a temper tantrum the previous day (Piaget 1951, 67). If we mean *this* by 'deferred imitation' then there are no strong impediments to saying it arrives some time in the middle of the second year.

But what general cognitive changes underlie the emergence of the symbolic function? One of the main tenets of Piagetian theory is that the first 18 months or so of life (the 'sensorimotor period') are taken up with the child's 'construction' of the concept of a spatio-temporal world that exists independently of her experience of it. The term 'construction' merely implies that this requires the child to be an *agent*, requires her to be able, that is, to be able to change her perceptual inputs at will (Russell 1995, for a defence of this view against some contemporary critics). Although Piaget would not express it quite this way, we can say that a child who is armed with this conception is able to distinguish between the following domains: (1) an independent, refractory physical world; (2) herself as a more or less free, motivated actor within that world; (3) memories and fantasies. During the sensorimotor period, according to the standard theory, these three are confused.

Given this three-way division, the child can engage for the first time in, what Piaget called, 'mental representation', and it is this which is supposed to support symbolic behaviour. Nowadays, the term 'mental representation' has a extremely broad denatation within cognitive psychology, but Piaget meant by it something very particular. He meant the ability to evoke a sensory image of an object or situation when the current situation is otherwise.

Piaget's attempt to pinpoint the cognitive operations which make this achievment possible was not, however, a successful one. 'Interiorized imitation' was the mechanism that was supposed to make Piagetian mental representation possible. Although this idea does some work within Piaget's theory, outside of the theory it is both implausible and untestable.

There have been some recent attempts to redraft or redescribe the phenomenon that Piaget called 'mental representation'. Leslie (1987) redrafted it in a 'computational' style. Writing about pretend play in particular, he proposed that when a child above 18 months of age understands what it means, say, for her mother to pretend that a banana is a telephone (e.g. she puts it to her head like a receiver and talks) the child must be 'decoupling' (a term of art from artificial intelligence) the 'primary' representation of the banana as a banana, and 'copying' this into a 'secondary' representation as 'banana' — in mental quotation marks. The 'banana' representation can then enter into all kinds of counterfactual relations with the result that 'The "banana" is a telephone' is no longer a contradiction — symbolic play becomes possible, in other words. Leslie calls this 'metarepresentation' (after Pylyshyn 1984).

Olson's (1993) recent account of the growth of representational thought, borrowing elements from both Piagetian theory and Leslie's views on pretense, takes as its centrepiece the distinction between the capacity to 'hold in view' and the capacity to 'hold in mind'. This is the terminology I shall borrow, for the following reasons: it implies a more generalizable mechanism than Leslie's 'metapresentation'; it commits us to no particular information-processing account; and it covers more of the symbolic phenomena (e.g. deferred imitation, language, play without object as props). It is a very simple idea: the child is able to experience one thing (people, events, objects) whilst thinking about something else. In a sense, this is hardly moving beyond Piaget's own informal description of what is occurring; but I think it is better to do this — to be modest — than to reach for ideas like 'interiorized imitation' and 'decoupling'. Our goal then becomes that of describing the cognitive substrate for holding-in-mind, the goal of saying what form of information-processing supports it.

But before moving on we need to consider two rather ambiguous cases of representational ability. The first is the child's ability to recognize herself in the mirror, a form of behaviour which clearly has a representational character but which does not obviously possess a holding-in-mind character. The ability is usually tested by surreptitiously making a mark on the child's forehead, placing a mirror before her and observing whether she touches the mark — sometimes called the 'rouge test'. For the child to succeed in doing this, one assumes, she needs to be able to treat the reflection as a 'representation' of her body. On this criterion, the onset of self-recognition comes roughly within the timetable of symbolic functioning, with the earliest onset being at about 15 months and with the majority of infants passing the test at 2 years (Anderson 1984). Perhaps the most defensible view of mirror self-recognition is that it tests for the ability to 'decouple' the visual appearance of one's body from the body itself without testing for the further ability to evoke or manipulate a representation of something not currently present. It is a lower-level ability. Moreover, the response of touching the mark can be regarded as a sensorimotor skill over-and-above that of recognizing one's own reflection; and this would account for the fact that some children do not pass the rouge test until they are nearly in their third year.

The second ambiguous case is that of drawing. One might think that if the symbolic function has an onset of around 18 months then children should begin drawing at that time. But in fact it is rare to see drawing worthy of the name much before about 2-and-a-half years of age (Thomas & Silk 1990). Two-year-old children scribble and they may sometimes produce mandala-like shapes (circles with upright crosses through them) which could conceivably be intended as representations of the human figure. This fact does not, however, present any challenge to the claim that symbolic functioning emerges at around 18 months, because the production of a drawing clearly requires planning, motor and graphic skills over-and-above the necessary representational capacity; and one would reasonably expect the execution of drawings to succeed the representational insight that drawings represent objects. In fact we know very little about toddlers' *understanding* of the representational status of pictures and photographs. Students of mental development sometimes report cases of children under 2 years being fooled by photographs rather as we are by *trompe-l'œil* paintings (e.g. stepping into a picture of some shoes: Perner 1991, 75); but the jury is certainly still out on this question.

Language acquisition and the symbolic function

Piaget also argued that a capacity for 'mental representation' was what made it possible for children to appreciate that things have names. It was supposed to be necessary but not sufficient for language acquisition. On the surface, this looks simply false because children often say their first words at around 12 months of age, and are normally capable of comprehending spoken language as early as 8 months of age (Reznick 1990). We cannot reasonably push back the onset of the symbolic function to 8 months.

The fact that a young child comprehends and produces speech does not, however, imply they grasp the general truth that things and actions have names. Very young children may have their attention appropriately directed by another's speech and they may utter sounds to express a number of communicative functions (Halliday 1975) without any implication that they have this insight. What *would* be evidence for such understanding? Asking the name of an object would certainly be evidence for it; and children do begin doing this after 18 months or so. Indeed this fact was noted by some of the early diary-keepers (e.g. Stern & Stern 1907), and is now usually called 'the nominal insight'. A paradigm of this would be a child checking that the caretaker is attending to the object, pointing to it, then making a sound like 'Da?' whilst making eye-contact with the caretaker. Moreover, a large number of students of language development have recorded a sharp increase in vocabulary size at around 18 months of age, and this has received the label 'vocabulary spurt' or 'naming explosion' (e.g. Bloom 1973; Nelson 1975). One would expect the nominal insight to stimulate the acquisition of new names.

If naming and the other forms of symbolic behaviour which we considered in the previous section all partake of the symbolic function — and thus of 'holding in mind' — then we should expect to find that development in one domain is correlated with development in the others. There is some evidence of this kind (e.g. Bloom *et al.* 1985), and perhaps the clearest evidence of such an association has been produced in the research of Gopnik & Meltzoff. In one of their studies (Gopnik & Meltzoff 1987) the onset of the naming explosion (criterion: more than 10 new words within a 3-week period) was compared with the following measures. The first two could reasonably be regarded as tests of 'symbolic functioning' whilst the third was a control measure. The first test involved search after invisible displacements, namely, the ability to keep track of an object in the experiment's hand as it passed under a number of containers before finally emerging empty. They had to maintain a representation of the temporarily invisible object during the passages of the hand and then infer, on this basis, where it had been dropped — and then search for it. The second test was one of categorization in which 8 objects in two categories (e.g. 4 dolls and 4 triangles) were placed before the subject. In this situation toddlers frequently touch all the members of one category before touching each member of the other, and may spontaneously separate out the two categories. This is not, of course, transparently a measure of symbolic functioning because one could say that what the infants did was evoked by purely visual similarity between the members of each set of 4. But the fact that their behaviour was spontaneous (i.e. they did not have to *learn* to extract the common perceptual features) is at least consistent with their forming some abstract representation of one set as 'dolls' and the other as 'shapes' and then guiding their behaviour by reference to that representation. The third (control) measure was a means–end task in which, for example, a object had be extracted from a tube by the use of a stick. It is very difficult to give a representational gloss to this skill as there was no necessity for keeping in mind the invisible locations of the object; whilst saying that the object was being categorized abstractly as 'the goal' would seem to be otiose. The question was how these three measures would correlate with one another and, more important, how well would each correlate with the onset of the naming explosion.

The results were very clear. There were highly-significant, positive correlations between the onset of the naming explosion and the other symbolic measures: 0.70 with object permanence and 0.78 with categorization. Means–end skill, on the other hand, did not correlate significantly with any of the other three measures, and indeed the correlation between it and the onset of the naming explosion was weakly negative. The correlation between categorization and the object permanence task was positive (0.48), though it fell just short of the conventional level of statistical significance. These are the results one would expect if all three measures are tapping one underlying representational capacity.

An impediment to this conclusion, however, is that correlations like this would also be predicted by those who adopt linguistic determinism — the position that language develops autonomously and that, as it develops, it makes available new modes of mental representation. Is there any evidence for the present, symmetrically opposing view, derived from Piagetian theory, that it is the development of *the symbolic representational capacity itself* which fuels the vocabulary explosion? There is some supporting evidence from an ingenious study by Tomasello & Farrar (1986). These researchers argued that if the causal link was from general representational ability to linguistic ability rather than from linguistic ability to representational ability then infants who are capable of understanding the invisible displacements of objects (so-called 'Stage 6' children) would be more successful in learning a novel term (invented by the experimenters) that refers to any kind of invisible movement than *age-matched* children who do not yet understand invisible displacement ('Stage 5' children). Stage 5 and Stage 6 children learned three novel words — one for invisible displacement (e.g. 'deke' meaning to hide behind one's back), one for visible displacement (e.g. 'dop' meaning to spin on the table) and one monosyllabic English word for an object that was typically unfamiliar to very young children. The Stage 6 children learned all three kinds of word equally well, whereas the Stage 5 children were specifically impaired on the novel word standing for *invisible* displacement. This suggests that it is not linguistic capacity *per se* which enables children to represent the behaviour of invisible objects, but that a representational capacity that is not specifically linguistic supports the acquisition of verbs which refer to actions that bring about invisibility.

But of course the naming explosion is an explosion in the production of *all* kinds of names, not just of those for the actions which cause invisible displacement. Accordingly, we have to give some general account of how holding-in-mind supports lexical development. Olson's way of expressing the link borrows something from the Leslie's (1987) view. He writes:

> [W]e could say that when the child develops the resources to 'hold in mind' the decoupled representation of the visual properties of an apple such that the perceptual system is free to establish a causal relation with the sound of the word 'apple', the child is now in a position to grasp the relation between the name and the thing named. The relation grasped is that expressed by the proposition 'This apple *is called* "apple".' The viewed apple is the subject of the proposition and the name is the predicate. (Olson 1993, 7)

Or we may say that the child can represent apples in their absence and thus can appreciate the linkage between the set of perceptual properties that is 'applehood' and an arbitrary sign. There is all the difference in the world between this and forming an *association* between apples and the sound of the word 'apple'. Indeed it is difficult to imagine how a child would ever come to ask the name of an unfamiliar object if he or she were only capable of forming associative links between things and sounds. In other words, the so-called 'naming insight' must be grounded in an appreciation of what it means for a sign to stand for a referent.

It is now time to list the elements of the symbolic function that may be said to emerge after about 18 months of age:

- search for objects after their invisible displacement;
- pretend play and deferred imitation (though often difficult to distinguish);
- the categorization of objects by touching and sorting;
- the nominal insight and vocabulary spurt.

And here are the unclear cases:

- understanding the representational status of drawings and pictures (the evidence is unclear);
- mirror self-recognition (which does not seem to require holding-in-mind).

Holding-in-mind and the brain

Turning from behaviour to the brain, what can be said about the neural substrate of holding-in-mind? It would be easy and neat if one could point to a growth spurt occurring in the human brain around 18 months after birth. But clear evidence here is hard to come by, partly because so much depends upon what one takes 'growth' to mean — volume, connectivity, synaptic density and dendritic development (Huttenlocher 1990), synaptic 'pruning' (Huttenlocher 1984), myelination (Konner 1991), EEG coherence and phase (Thatcher 1991). One can look for growth phenomena that are *consistent* at least with the timetable that has just been sketched. For example Gibson (1981) reports that between 15 and 24 months of age layers IV, V, and VI of the association areas begin to myelinate; and these are areas which, because they involve cortico–cortico projections (as well as connections to the mid-brain), one would reasonably expect to be necessary for developing high-level representations.

An alternative to looking for correlations between brain maturation and symbolic behaviour is to proceed in a more *a priori* fashion. Here we might ask whether there is some form of information-processing which is carried forward by a particular neural structure and which could reasonably be said to support holding-in-mind. Such a form of information-processing is 'working memory'. Working memory is not a very precise term, and an extremely broad and flexible conception of it will be used in this paper. But perhaps the best way of distinguishing it from related phenomena such as short-term memory span is by calling it 'memory for the future' or 'prospective memory' — keeping a representation (e.g. a set of data, a rule, a route) active when engaged in doing something else until the time comes to report or act on the information. In experimental psychology, working memory is assumed to have an 'executive component', meaning that the individual must not merely hold a set of data in a passive, rehearsal 'loop' but must do this while also performing some mental procedure 'on-line', such as counting (see Baddeley 1986, ch. 9).

Clearly, my use of the term is much broader than this: I am assuming that as long as the individual is maintaining in mind a representation to some further end and as long as the representation is not causally determined by the current perceptual input then we can speak of 'working memory'. Although this is too broad a definition to be of use to the experimental psychologist, it has an heuristic value in the present discussion.

Where is working memory localized? The prefrontal cortex (PFC) would appear to be the obvious candidate, given the fact that it has an 'executive' character (see above reference to Baddeley). However the story is complicated by the fact that a subcortical structure called the hippocampus also plays a role in some forms of working memory. The hippocampus is implicated, for example, in the ability to perform so-called 'delayed-response tasks' in which, say, a monkey sees a food bait placed at one of two locations, a screen is then dropped to prevent responding, after which the animal must search for the bait having had to 'hold in mind' the fact that the food is, say, on the right rather than the left. The location of the bait changes from trial to trial. Lesions of the PFC, however, impair performance on these tasks *even at very short delays*; whereas the hippocampus is primarily implicated in the ability to search after longer delays (Squire 1992) and in specifically spatial tasks (O'Keefe 1993). The conservative conclusion is that both the hippocampus and PFC have a role to play in working memory tasks as primitive as delayed-response, but that the PFC is crucial to the *immediate*, on-line representational function of working memory.

But what of the more cognitively 'rich' form of working memory that one would expect to underpin symbolic functioning? Can we more confidently say that the PFC is primarily responsible? I will briefly consider the evidence.

We can usefully distinguish between two kinds of model of prefrontal functioning in humans — 'mixed' and 'unitary' models. Mixed models hold that PFC performs a number of functions with no common denominator between them and bearing only a family resemblance to one another — planning, self-regulation and self-monitoring, maintenance of behavioural 'set', instigation and inhibition of behaviour, the spatio-temporal segmentation and the organization of events. (There is no one theorist associated with the mixed theory, though many neuropsychologists treat it as a good conservative assumption.) The unitary theories, by contrast, claim that we can posit a central function for the PFC. One of these[2] emphasizes the role of working memory. Goldman-Rakic's (1987) model pictures the PFC as receiving representations from other centres in the brain and as keeping these activated for long enough to control behaviour by reference to them rather than by reference to what is currently being experienced. This is referred to as 'on-line' or 'representational' memory. One may say that representational memory guides behaviour via 'mental models' (Johnson-Laird 1983; Halford 1993) of reality rather than in terms of whatever perception is telling us about the current state of reality.

Depite the fact that Goldman-Rakic's model is derived principally from monkey work of the kind just described, while mixed models are derived from human neuropsychology (studies of brain-damaged subjects), there is reason to believe that the representational memory model describes performance on tests of human prefrontal functioning more successfully. Daigneault, Braün, and Whitaker (1992) conducted a systematic comparison of the mixed theory against Goldman-Rakic's unitary theory, with a group of 259 normal adults. They did this by presenting the subjects with a battery of 'frontal' tests which afforded 11 kinds of kinds of error altogether. Because the two approaches predict different error patterns across tasks, the incidence of these error-types could then be set beside the five functions (named in the previous paragraph) which the mixed theory posit and also set beside four functions derived *a priori* from the Goldman-Rakic model — regulation of manual responses by verbal memory, regulation of verbal responses by verbal memory, regulation of manual responses by visuo-spatial memory, regulation of simple responses by representational memory on both visual and verbal tasks. Without going into statistical details, the latter was the better fit.

There do not appear, then, to be any obvious impediments to the conclusion the PFC is the neural structure primarily responsible for working memory and thus for the holding-in-mind that supports symbolic behaviour. It may indeed be the case that the primary function of the prefrontal cortex is that of holding-in-mind, and that the *regulation* of behaviour, with which the PFC has traditionally been associated, is achieved by virtue of this capacity.

Working memory, evolution and communication

It is now time to take stock. *If* the development of symbolic functioning in the child occurs all of a piece, and *if* much the same can also be said of the emergence of symbolic functioning in modern humans, and *if* working memory can be said to be the cognitive substrate for the holding-in-mind which is necessary for symbolic functioning in children, then we can reasonably speculate that the evolution of working memory formed a major cognitive substrate for the emergence of symbolic functioning in modern humans. That is to say, it may make sense to see symbolic functioning as natural, eventual outgrowth of an increase in mental *processing efficiency* (Case 1984, for application of this notion in cognitive development).

This is not of course to suggest that working memory was *sufficient* for the emergence of symbolic functioning. Neither is it to deny that social pressures — perhaps of the kind described by Humphrey (1984) — and the invention of signing systems had major roles to play. But one can say that every advance in social functioning, such as the ability to manipulate the behaviour of others or to achieve strategic cooperation, requires an information-processing substrate. Perhaps the evolution of this substrate — working memory — provided the necessary cognitive grounding for social evolution. A related point is that the 'mind-reading' capacities whose evolution is currently being hotly debated by primatalogists (e.g. Povinelli 1993; Whiten 1993) can only occur in creatures which have the right kind of *first-person* experience of their own mental functions, and that one central aspect of this experience is regulating one's behaviour in terms of mental models of reality — a point which also bears making in the context of ontogeny (Russell 1995; in press).

The obvious question then concerns the evolutionary pressures contributing to the growth of working memory itself. It is not hard to produce plausible scenarios here. Just about every interaction with the physical, animal and human world will be facilitated by the ability to hold in mind information that is independent of what is currently being held in view. Evolution will accordingly select the individual with the greater working memory capacity. My aim in this section, however, is not to speculate about the selective pressures, but to suggest how holding-in-mind can facilitate the development of communcation between conspecifics. Holding-in-mind gives the individual far more than access to arbitrary signs.

The process on which I want to concentrate is that of *predication*. Making a statement means saying something about something and so inevitably requires the communicator to establish both an 'argument' (a referring expression) and a 'predicator'. Moreover, the successful use of a referring expression can require a 'predicate' (e.g. the *green* one): predicates qualify the referring expression and allow the referent to be identified by the addressee. I will discuss first the cognitive demands of successful predication and second the pragmatic demands. In both cases, holding-in-mind has a crucial part to play.

(a) Predication and the 'generality constraint'

Let us consider what the capacity for qualifying the argument 'a' with the predicate 'F' in the referring expression 'Fa' consists in. If we say that somebody can think 'the white car', for example, what fundamental cognitive capacity do we thereby attribute to her? The philosopher Gareth Evans (1982) has argued that it makes no sense to attribute to somebody a thought such as Fa unless she also has the capacity to think Fb or Fc or Fd, and so forth, as well as Ga or Ha or Ka, and so forth. In other words, the capacity to think 'the white car' entails a near-infinite potential capacity for thinking 'the black car' or 'the big car' or 'the white cat', or the 'the white house'. Evans called this 'the generality constraint'. That is to say, a capacity for successful predication entails what philosophers call an 'holistic' or 'structured' (Peacocke 1989) capacity, one aspect of which is the ability to combine arguments and predicates *at will*.

Where does holding-in-mind come into this? In a situation in which I wish to refer to the white car, my cognitive system is, as it were, considering but excluding a very wide range of other predicates simply because a great number of other things will be true of the car in addition to its white colour. This does not mean that in the very mental act of characterizing it as white I am also privately and explicitly characterizing it to myself as a Renault or as second-hand. But it does imply that the thought about the car as white is not *solely* a thought about it as white in the moment I refer to it. A number of further predicates are held in abeyance — held in mind. If you say back to me, for example, 'You mean the Renault' I do not have to, as it were, 'think again'. More important, I can also hold in mind predicates which *cannot* truly be applied to it — 'red', 'a Fiat' and so forth — and must do just this whenever I negate, deny and deceive. As Bennett (1964) has argued, the denial of a statement would appear to be a necessary feature of rational communication; and it is difficult to imagine how denial would be possible if the generality constraint were not exercised in predication.

What this implies about communcation is that a capacity for holding-in-mind can be seen to underly our capacity for the free attachment of arguments to predicates and predicates to arguments.

(b) Predication and relevance

I now consider pragmatic constraints upon *which* predicates to use in a given situation. This will clearly depend upon a number of factors, but it will certainly depend upon the number and kind of potential, competing referents additional to the target referent. Returning to the example of the white car, in one case there may be three cars, all of which are white, two of which are large and one of which is small. You wish to refer to the last-mentioned item. In this case, clearly, 'the white car' will not secure reference but 'the small car' will secure it. In a second situation, the same target item is alongside two other cars again, but this time they are equally as small as the target, but black in colour. In this case 'the small car' will not secure reference but 'the white car' will secure it.

This second example is taken from Olson, who writes: 'If there is some means available for preserving the fact that these two construals [i.e. as "small" and as "white"] are of the *same* [car], the conditions are set for *comparing the content of what is held in mind with that held in view*.' (1993, 6, my italics) Now this is, in fact, a *different* application of the holding-in-mind concept from that discussed under (a) because it is evident from the quotation that Olson takes the *relevant predicate* to be that which is held in mind whilst the percept of the item (e.g. as a small, white, car-shaped thing) is that which is held in view. Unlike Olson, I believe it is implausible to think of percepts in this holistic way and would prefer to say — as I did in (a) — that numerous potential predicates are held in mind as the relevant one is selected, under pragmatic constraints.

But the way in which the holding-in-mind concept is put to work here is not crucial for my present purpose, which is to make the point that holding-in-mind — the information-processing face of which is working memory — is one of the cognitive substrates for successful predication.

Some conclusions — and a possible impediment to them

This is what I have been arguing.

- It is reasonable to search for a common cognitive substrate for the emergence of symbolizing in the developing child and in evolving humans.
- Both the archaeological and developmental evidence suggest that the emergence of symbolizing has a stage-like character.
- Symbolic functioning involves holding a representation in mind which differs from the percept held in view (Olson 1993) and the information-processing substrate of this capacity is working memory, a capacity which is carried forward by the prefrontal cortex.
- The selective advantage of a large working memory is clear.
- A capacity for holding-in-mind will not only support symbolic behaviour but will also support the linguistic function of predication.

This leads then to the following conclusion:

- Working memory is a good candidate for being one of the principal non-social and non-linguistic factors in the evolution of human symbolic intelligence.

I must stress that what is being proposed here is a substrate and not a trigger. A relatively large working memory capacity will not *of itself* produce symbolic behaviour and predication, as I said before. Indeed it is impossible to imagine how it could. Moreover, we know that anatomically modern populations were present around 90–100,000 BP (Stringer & Andrews 1988) which is considerably before 40,000 BP at which the onset of symbolic behaviour in Europe is generally timed. The cognitive substrate may have existed, therefore, around 50,000 years before it came to be recruited to symbolic ends; and one can plausibly regard the progression within that time span as being an essentially social one.

I want to turn finally to an apparent difficulty with the claim that cognitive changes around 90–100,000 BP provided the necessary cognitive substrate for symbolic and linguistic functioning. Evidence that non-human primates have symbolic capacities that are as sophisticated as that possessed by the human child in her second year would undermine the claim that working memory capacity was one of the main factors that separated modern humans from Neanderthals. For saying that working memory is the substrate of symbolic functioning would thereby entail the impausible claim that apes have a superior working memory capacity to Neanderthals. This is not the place to discuss such vexed questions as whether apes' tool use is evidence of full-blown symbolic functioning (Chevalier-Skolnikoff 1989; Gibson 1988; 1989), whether they show symbolic play (Mitchell 1991; Smith 1982) or what is the significance of chimpanzees' ability to pass mirror self-recognition tests (Gallup 1982). What I will do instead is to confront the *strongest* evidence which has yet emerged for symbolic behaviour in the ape and then show why this does not undermine the present thesis.

Savage-Rumbaugh *et al.* (1993) have reported that the language comprehension capacities of an 8-year-old bonobo chimp (*Pan paniscus*), which had been raised in a linguistic environment similar to that enjoyed by humans, was not only comparable with the comprehension capacities of a normal 2-year-old human child but superior to them in some respects.[3] If we can place the evolutionary divergence between *Homo* and *Pan* at around 4–6 million years ago (Sibley & Ahlquist 1984) then this would suggest that the cognitive substrate for language which is at least as sophisticated as that of a 2-year-old existed long before modern humans — with their dramatially larger prefrontal brains — emerged. The argument then runs that what prevented chimpanzees from evolving speech is that they did not, unlike *Homo*, evolve the bipedalism which lead to the reorientation of the laryngeal tract necessary for vocalization.

But there is a major flaw in the argument. It relies upon the assumption that language comprehension and language production are sides of the same coin. The chimpanzee excels in the comprehension of commands from trainers; and so it is tempting to think that if only he could speak he would be talking to us like a 2-year-old. But there are reasons for thinking that the cognitive requirements for speech production far exceed those of comprehension.

First, in comprehending commands the animal had its attention focused by heard *sentences*; but in producing commands, or any other kind of speech act, and individual has to focus the attention of the *addressee* in an appropriate and relevant manner (*cf.* my discussion of predication above). This requires not only a conception of someone's being informed and uninformed about a fact, but also the conception of informing another through a symbolic means. There is, in fact, evidence that chimpanzees are able to discriminate appropriately between an individual who was informed about an event versus one who was not informed about it (Povinelli *et al.* 1990); but there is no evidence that they are able to direct symbolized information in an appropriate way, and certainly no evidence that they do this with anything approaching the communicative repertoir of a very young child. I am, in fact, in broad agreement with Tomasello, Kruger and Ratner's (1993) sceptical interpretation of the ape 'theory of mind' evidence.

Second, and more important, language production involves choosing something to say from an infinity of things which could be said. It is not just a matter of obtaining goods, services and succour from others, as Chomsky (1959) pointed out many years ago. Moreover, this choice must be determined to some degree by an awareness of current interests, state of mind and cognitive capacities of the addressee — something at which very young children excel (Dunn 1991). There is no analogue to this in the comprehension of instructions, which *needs to involve no conception of the producer of sentences as a cognitive entity or even as an agent*. It would seem to be a mistake to regard the comprehension of instructions as evidence of the kind of symbolic functioning which has been discussed here.

Notes

1. It will become evident that, while my approach is 'Piagetian' in the sense of being concerned with global rather than modular developmental changes (for a good discussion of the latter see Robson Brown 1993), it clearly diverges from previous parallels between human evolution and the Piagetian stages. Writers on this subject who take a Piagetian tack tend to argue that the transition to the Upper Palaeolithic marks a transition to 'formal operational' intelligence, achieved in human development around adolescence (Gibson 1985; Parker 1985; Wynn 1991). As a developmental psychologist, my view is that — whatever Piaget and Piagetians (e.g. Furth 1966) thought on the subject — 'operational' thought is difficult, if not impossible, to assess without verbal evidence; all Piaget's procedures were verbal. But rather than take a firm stand on this matter, I will leave open the question of whether the hypothesized transition towards symbolic functioning brought in its immediate wake some

form of operational thinking — in a way in which it does *not* in human development. (My scepticism here relates to the scepticism I shall voice towards the end of the paper about what recent experiments have demonstrated about ape linguistic competence and ape 'theory of mind'.)

2. Another unitary model of PFC functioning is owing to Norman and Shallice (see Shallice 1988, ch. 14). They proposed a 'supervisory attentional system' which oversees the functioning of routinely scheduled systems and generates action appropriate to novel circumstances. But the very notion of over-seeing implies the on-line maintenance of models of current activity and the ability of hold alternative action schemes in abeyance as the current action is being carried out.
3. Here are a few examples of what the *bonobo* (called Kanzi) could do. He was able to develop 'creative' solutions to instructions. For example he once took an object outside in the rain when told to 'make X wet'. He was also sensitive to word-order, interpreting 'put the egg in the juice' and 'put the juice in the egg' appropriately, for example. Moreover, the objects available did not normally suggest how the instruction was to be interpreted. He was able, for example, to carry out 'Feed your ball some tomato'. (The ball did have a schematic face on it however.)

Acknowledgements

I am grateful to Paul Mellars for commissioning this paper.

James Russell
Department of Experimental Psychology
University of Cambridge
Downing Street
Cambridge
CB2 3EB

References

Anderson, J.R., 1984. The development of self-recognition: a review. *Developmental Psychobiology* 17, 35–49.

Baddeley, A., 1986. *Working Memory*. Oxford: Oxford Unversity Press.

Bennett, J., 1964. *Rationality*. London: Routledge & Kegan Paul.

Bloom, L., 1973. *One Word at a Time: the Use of Single-word Utterances before Syntax*. Cambridge (MA): M.I.T. Press.

Bloom, L., K. Lifter & J.M. Broughton, 1985. The convergence of early cognition and language in the second year of life: problems in conceptualisation and measurement, in *Children's Single-word Speech*, ed. M. Barrett. London: John Wiley, 34–67.

Bremner, J.G., 1988. *Infancy*. Oxford: Basil Blackwell.

Butterworth, G. (ed.), 1985. *Evolution and Developmental Psychology*. Brighton: Harvester.

Case, R., 1984. The process of stage-transition: a neo-Piagetian view, in *Mechanisms of Cognitive Development*, ed. R.J. Sternberg. New York (NY): Freeman, 56–76.

Chase, P.G. & H.L. Dibble, 1987. Middle Palaeolithic symbolism: a review of current evidence and interpretations. *Journal of Anthropological Archaeology* 6, 263–93.

Chevalier-Skolnikoff, S., 1989. Spontaneous tool use and sensorimotor intelligence in *Cebus* compared with other monkey and apes. *Behavioural and Brain Sciences* 12, 561–627.

Chomksy, N., 1959. Review of 'Verbal Behaviour' by B.F. Skinner. *Language* 35, 26–58.

Daigneault, S., C.M.J. Braün & H.A. Whitaker, 1992. An empirical test of two opposing theoretical models of prefrontal function. *Brain and Cognition* 19, 48–71.

Dunn, J., 1991. Understanding others: evidence from naturalistic studies of children, in *Natural Theories of Mind*, ed. A. Whiten. Oxford: Basil Blackwell, 89–99.

Evans, G., 1982. *The Varieties of Reference*. Oxford: Oxford University Press.

Furth, H.G., 1966. *Thinking without Language*. New York (NY): Collier Macmillan (Free Press).

Gallup, G.G., 1982. Self-awareness and the emergence of mind in primates. *American Journal of Primatology* 2, 237–48.

Gibson, K.R., 1981. Comparative neuro-ontogeny: its implications for the development of human intelligence, in *Infancy and Epistemology*, ed. G.E. Butterworth. Brighton: Harvester, 52–82.

Gibson, K.R., 1985. Has the evolution of intellegence stagnated since Neanderthal Man? in Butterworth (ed.).

Gibson, K.R., 1988. Brain size and the evolution of language, in *The Genesis of Language: a Different Kind of Evidence*, ed. M. Landsberg. New York (NY): Mounton de Gruyer, 122–43.

Gibson, K.R., 1989. Tool use in cebus monkeys: moving from orthodox to neo-Piagetian analyses. (Commentary on Chavalier-Skolnikoff). *Brain and Behavioural Sciences* 12, 598–9.

Goldman-Rakic, P.S., 1987. Circuitry of primate prefrontal cortex and regulation of behaviour by representational memory, in *Handbook of Physiology: the Nervous System*, vol. 5, ed. F. Plum. New York (NY): Oxford University Press, 373–417.

Gopnik, A. & A. Melzoff, 1987. The development of categorization in the second year of life and its relation to other cognitive and linguistic developments. *Child Development* 58, 1523–31.

Halford, G.S., 1993. *Children's Understanding: the Development of Mental Models*. Hillsdale (NJ): Lawrence Erlbaum.

Halliday, M.A.K., 1975. *Learning How to Mean*. London: Edward Arnold.

Humphrey, N., 1984. *Consciousness Regained: Chapters in the Development of Mind*. Oxford: Oxford University Press.

Huttenlocher, P.R., 1984. Synapse elimination and plasticity in developing human cerebral cortex. *American Journal of Mental Disorders* 88, 488–96.

Huttenlocher, P.R., 1990. Morphometric study of the human cerebral cortex development. *Neuropsychologia* 28, 517–27.

Johnson-Laird, P.N., 1983. *Mental Models*. Cambridge: Cambridge University Press.

Konner, M., 1991. Universals of behavioural development in relation to brain myelination, in *Brain Maturation and Cognitive Development: Comparative and Cross-Cultural Perspectives*, eds. K.R. Gibson & A.C. Peterson. New York (NY): Aldine de Gruyter, 181–222.

Leslie, A.M., 1987. Pretence and representation in infancy: the origins of 'theory of mind'. *Psychological Review* 94, 84–106.

McCune-Nicolich, L., 1981. Toward symbolic function: structure of early use of pretend games and potential parallels with language. *Child Development* 52, 785–97.

Mellars, P.A., 1989. Major isssues in the emergence of modern humans. *Current Anthropology* 30, 349–85.

Mellars, P.A., 1991. Cognitive changes and the emergence of modern humans in Europe. *Cambridge Archaeological Journal* 1(1), 63–76.

Meltzoff, A.N., 1988. Infant imitation and memory: nine-month-olds in immediate and deferred tests. *Child Development* 59, 217–25.

Mitchell, R.W., 1991. Bateson's concepts of 'metacommunication' in play. *New Ideas in Psychology* 9, 73–87.

Nelson, K., 1975. The nominal shift in semantic–syntactic development. *Cognitive Psychology* 7, 461–79.

O'Keefe, J., 1993. The Hippocampus and Spatial Memory. Paper presented at 'The Zangwill Club'. Department of Experimental Psychology, Cambridge University, October, 1993.

Olson, D.R., 1993. The development of representations: the origins of mental life. *Canadian Psychology* 34, 1–14.

Parker, S., 1985. Higher intelligence as adaptation for social and technological strategies in early *Homo sapiens*, in Butterworth (ed.).

Peacocke, C., 1989. *Transcendental Arguments in the Theory of Content*. Oxford: Clarendon Press.

Perner, J., 1991. *Understanding the Representational Mind*. Cambridge (MA): M.I.T. Press.

Pfeiffer, J.E., 1982. *The Creative Explosion: an Enquiry into the Origins of Art and Religion*. New York (NY): Harper & Row.

Piaget, J., 1951. *Play, Dreams and Imitation in Childhood*. London: Routledge & Kegan Paul.

Piaget, J., 1970. *Genetic Epistemology*. New York (NY): Columbia University Press.

Povinelli, D.J., 1993. Reconstructing the evolution of mind. *American Psychologist* 48, 152–63.

Povinelli, D.J., K.E. Nelson & S.T. Boysen, 1990. Inferences about guessing and knowing by chimpanzees (*Pan trologytes*). *Journal of Comparative Psychology* 104, 203–10.

Pylyshyn, Z., 1984. *Computation and Cognition: Toward a Foundation for Cognitive Science*. Cambridge (MA): M.I.T. Press.

Reznick, J.S., 1990. Visual preference as a test of infant word comprehension. *Applied Psycholinguistics* 11, 145–65.

Robson Brown, K., 1993. An alternative approach to cognition in lower Palaeolithic: the modular approach. *Cambridge Archaeological Journal* 3(2), 231–45.

Russell, J., 1978. *The Aquisition of Knowledge*. London: Macmillan Press.

Russell, J., 1979. The status of genetic epistemology. *Journal for the Theory of Social Behaviour* 9, 53–70.

Russell, J., 1995. 'At two with Nature': agency and the development of self-world dualism, in *The Body and the Self*, ed. J. Bermúdez, A.J. Marcel & N. Eilan. Cambridge (MA): M.I.T. Press.

Russell, J., in press. *Agency: Its Role in Mental Development*. Hove: Lawrence Erlbaum.

Savage-Rumbaugh, E.S., J. Murphy, R.A. Sevcik, K.E. Brakke, S.L. Williams & D.M. Rumbaugh, 1993. *Language Comprehension in Ape and Child*. (Monographs of the Society for Research in Child Development 58, Nos. 3–4.).

Shallice, T., 1988. *From Neuropsychology to Mental Structure*. Cambridge: Cambridge University Press.

Sibley, C.G. & J.E. Ahlquist, 1984. The phylogeny of hominoid primates as indicated by DNA–DNA hybridization. *Journal of Molecular Evolution* 20, 2–15.

Smith, P.K., 1982. Does play matter? Functional and evolutionary aspects of animal and human play. *Behavioural and Brain Sciences* 5, 139–55.

Squire, L.R., 1992. Memory in the hippocampus: a synthesis from findings with rats, monkeys and humans. *Psychological Review* 99, 195–231.

Stern, C. & W. Stern, 1907. *Die Kindersprache*. Leipzig: Barth.

Stringer, C.B. & P. Andrews, 1988. Genetic and fossil evidence for the origin of modern humans. *Science* 239, 1263–8.

Thatcher, R.W., 1991. Maturation of the human frontal lobes: physiological evidence for staging. *Developmental Neuropsychology* 7, 397–419.

Thomas, G.V. & A.M.J. Silk, 1990. *An Introduction to the Psychology of Children's Drawings*. London: Harvester-Wheatsheaf.

Tomasello, M. & M.J. Farrar, 1986. Object permanence and relational words: a lexical training study. *Journal of Child Language* 13, 495–505.

Tomasello, M., A.C. Kruger & H.H. Ratner, 1993. Cultural learning. *Behavioural and Brain Sciences* 16, 495–552.

White, R., 1989. Production complexity and standardization in Early Aurignacian bead and pendant manufacture: evolutionary implications, in *The Human Revolution: Behavioural and Biological Perspectives on the Origins of Modern Humans*, eds. P. Mellars & C.B. Stringer. Edinburgh: Edinburgh University Press, 366–90.

Whiten, A., 1993. Evolving theories of mind: the nature of non-verbal mentalism in other primates, in *Understanding Other Minds*, eds. S. Baron-Cohen, H. Tager-Flusberg, D. Cohen, & F. Volkmar. Oxford: Oxford University Press, 367–96.

Wynn, T., 1991. Archaeological evidence for modern intelligence, in *The Origins of Human Behaviour*, ed. R. Foley. London: Unwin Hyman & World Archaeology.

Part V

Interpreting the Archaeological Record

15

Comparing Chimpanzee and Early Hominid Techniques: Some Contributions to Cultural and Cognitive Questions

Frédéric Joulian

This paper presents ethological data on the patterns of tool-use by West African chimpanzees and illustrates their relevance to an understanding of the earliest hominid stone tool techniques. First, I will outline the theoretical and methodological framework. I will then explain the main results of my doctoral dissertation concerning the existence of technical traditions among chimpanzees, and address the problem of comparing chimpanzee and prehistoric hominid tools. To illustrate this problem I will present two techniques, one from Ivory Coast chimpanzees, the other from Plio-pleistocene hominids. Finally, using these techniques, I shall attempt to extract information concerning the levels of knowledge and intelligence of the tool-users.

Theoretical and methodological framework

The theoretical and methodological framework employed in this study is what I will describe as an 'ethoarchaeological' one. Briefly, by 'ethoarchaeology' I mean recognizing and dealing with archaeological questions (e.g. the shape and function of tools, spatial patterning of assemblages, characterization of diets, etc.), and exploring them in a different universe, that of one of our closest relatives in the animal world, the chimpanzees.

In addition to its phylogenetic proximity to man, the common chimpanzee is the non-human primate which makes the greatest use of tools, almost on a daily basis. Chimpanzee tools play an essential role in food extraction. Some (in particular stone hammers and anvils) are identical to prehistoric tools, especially to the Plio-pleistocene examples (see below). Thus chimpanzees provide the only example in the animal world of activities and artefacts analogous to those of prehistoric humans, although set within a different anatomical and behavioural framework.

As a result of these differences, we have to keep in mind that we are comparing two different evolutionary histories — the hominid one, and the 'panid' one. Early hominid evolution took place between 8 million years ago (the evolutionary division between panid and hominid lines) and 1.5 million years — the period which I refer to here as the Plio-pleistocene (Coppens 1994). The time span of chimpanzee evolution of course extends up to the present day.

My own perspective is that chimpanzee behaviour can provide us with a valuable model to help understand some of the behaviour of Plio-pleistocene hominids, but not in terms of a directly explanatory model, as most scholars usually argue (Tanner 1987). In other words, I am not transferring a behavioural model directly from one species to another, but simply attempting to define some of the basic anthropological issues with regard to the nature of the archaeological record, as well as to the ethological record.

Nut-cracking traditions

My research has been concerned specifically with the nut-cracking activities of Ivory Coast chimpanzees. Results of a number of earlier studies of the same topic have already been published (e.g. Savage & Wyman 1844; Struhsaker & Hunkeler 1971; Sugiyama & Koman 1979; Boesch & Boesch 1983; Anderson *et al.* 1983; Whitesides 1985; Kortlandt & Holzhaus 1987) and I have summarized the available ethological data in Figure 15.1 and Tables 15.1 and 15.2 (*cf.* Joulian 1993; 1994b; 1995).

Figure 15.1. *Chimpanzee nut-cracking sites published prior to the present study.*

Table 15.1. *Nuts: present and cracked, by site.*

Site	Country	Environment	Nut							Field study			Authors
			Adans.	Co.	Pd.	Pi.	Det.	Sac.	El. g.	Long-t.	Short-t.	Survey	
Mt Assirik	Senegal	Woodland–savanna	C *	Abs	Abs	Abs	P ?	Abs	P ?		x		Bermejo *et al.* 1989
Tiwaï Island	Sierra Leone	Primary forest	Abs	P ?	P ?	P ?	C	P ?	P ?			x	Whitesides 1985
Sapo forest	Liberia	Primary forest	Abs	C	C	C	P ?	C	P ?		x		Anderson *et al.* 1983
Beatty's site	Liberia	Primary forest							C			x	Beatty 1951
Cap Palmas	Liberia	Primary forest	Abs	C	P ?	P ?	P ?	P ?	M			x	Savage & Wyman 1844
Mt Canton	Guinea	Prim./second. for.	P ?	C	P ?	P ?	P ?	Abs	P			x	Kortlandt *et al.* 1987
Bossou	Guinea	Second. forest	P ?	Abs	Abs	M	Abs *	Abs	C	x			Sugiyama 1979; 1989
Taï	Ivory Coast	Primary forest	Abs	C	C	C	C	C	P	x			Boesch *et al.* 1983; 1990
Sakré	Ivory Coast	Primary forest	Abs	C	C	P	P	P	P		x		Struhsaker *et al.* 1971

C Cracked
M Nut eaten but not cracked (direct observation or found in faeces)
P Nut present (not cracked) recorded on the chimpanzee territory
P ? Probable presence (following the range of the species)
Abs Nut absent
No observation
* - At Mt Assirik Adansonia nut is perhaps cracked directly on a stone-anvil without hammer.
* - At Bossou *Detarium senegalense* is absent but *Detarium microcarpum*, (relatively close species) is present and consumed.

Adans. *Adansonia digitata*
Co. *Coula edulis*
Pd. *Panda oleosa*
Pi. *Parinari excelsa*
Det. *Detarium senegalense*
Sac. *Sacoglottis gabonensis*
El. g. *Elaeis guinensis*

Long-t. Long-term study
Short-t. Short-term study
Survey Survey or short-term study

Table 15.2. *Nut-cracking sites: wooden and stone tools (published sites).*

SITES Country		Mt Assirik (Senegal)	Tiwaï Island (Sierra Leone)	Beatty's site (Liberia)	Sapo forest (Liberia)	Mt Canton (Guinea)	Bossou (Guinea)	Sakré (Ivory Coast)	Taï * (Ivory Coast)
Raw material	Stone	Yes	Yes	Yes	Yes	Yes	Yes	Yes	Yes
	Wood	Yes ?	Yes	?	Yes	?	Yes	Yes	Yes
ANVILS									
Adansonia	Stone	Yes							
	Wood	No							
Coula	Stone				Yes (15)	Yes (8)		No	Yes (13)
	Wood				Yes (3)	No ?		Yes (21 ?)	Yes (735)
Detarium	Stone		Yes						Yes (1)
	Wood		No						Yes (124)
Elaeis	Stone			Yes			Yes (29)		
	Wood			No ?			No		
Panda	Stone				Yes (4)			No	Yes (441)
	Wood				Yes (1)			Yes (5)	Yes (27)
Parinari	Stone				Yes (10)				Yes (77)
	Wood				Yes (4)				Yes (2)
Sacoglottis	Stone				Yes (10)				No
	Wood				No				Yes (14)
HAMMERS									
Adansonia	Stone	Yes							
	Wood	No							
Coula	Stone				Yes (8 ?)	Yes (7)		Yes (2)	Yes (44)
	Wood				No	No ?		Yes (>20)	Yes (513)
Detarium	Stone		Yes (16)						Yes (7)
	Wood		No						Yes (44)
Elaeis	Stone			Yes			Yes		
	Wood			No ?			No		
Panda	Stone				?			Yes (5)	Yes (70)
	Wood				?			No	Yes (8)
Parinari	Stone				Yes (8 ?)				Yes (12)
	Wood				No				Yes (19)
Sacoglottis	Stone				?				No
	Wood				?				Yes (2)

? No precision
( ) Anvil or hammer's numbers
* 1983s data

All of these earlier observations have pointed to the existence of cultural differences in nut-cracking activities (McGrew 1985; 1992; Sugiyama 1993), but as yet the variations have only been described from a qualitative viewpoint — mainly because of the heterogeneity of the observation conditions and recording methods employed. In 1988–89 I conducted fieldwork with the aim of recording quantitative differences in these nut-cracking activities, in association with the research project of C. Boesch on

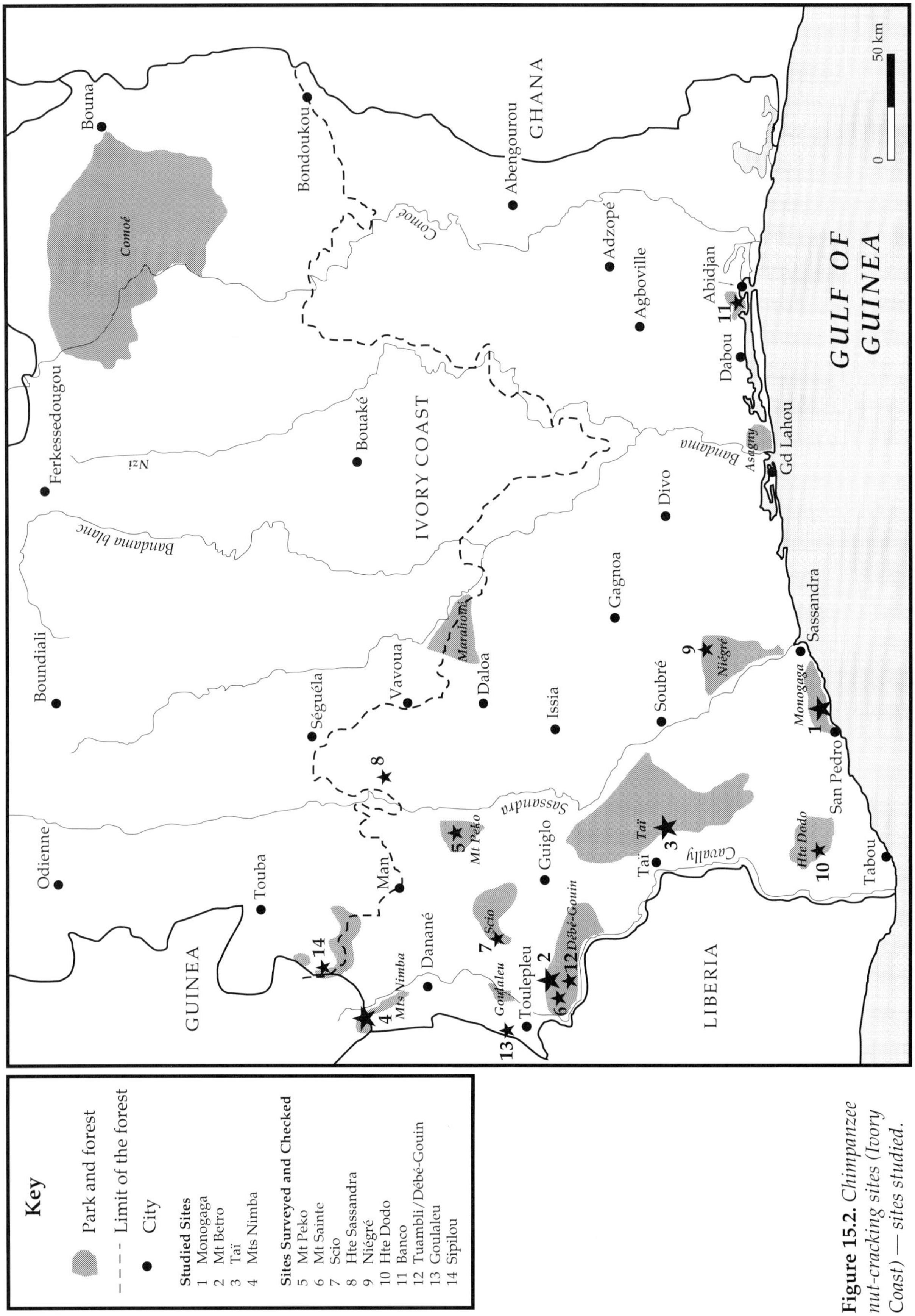

Figure 15.2. *Chimpanzee nut-cracking sites (Ivory Coast) — sites studied.*

'Traditions in West African chimpanzees' (Boesch *et al.* 1994). By using the same methodology at a range of different sites we were able to observe and document both the traces of nut-cracking and the activities of the chimpanzees themselves, from both an environmental and ethological perspective.

To make this possible, we employed a number of ecological transects and documented three main features: (a) the chimpanzee themselves (their presence, nests, density, etc.); (b) the distribution of nut-bearing trees and their productivity (*Panda oleosa, Parinari excelsa, Coula edulis, Detarium senegalense, Sacoglottis gabonensis, Elaeis guinensis*); and (c) finally the availability of suitable raw material for nut-cracking activities (i.e. potential anvils and hammers, in wood or stone). We also mapped the working areas. Analysis of raw material availability has allowed us to understand chimpanzee choices and to make comparisons between the different sites.

The four sites studied in detail were excavated, mapped and recorded following archaeological techniques that we adapted to the rainforest conditions. In this way we were able to collect data directly comparable to that available from the hominid archaeological record. In this section I will focus on one specific topic which is fundamental to prehistory: demonstrating the existence of traditional differences in chimpanzee tool-use.

New sites, surveyed and studied

The primary objective of the field studies — to identify the maximum range of variation over a wide geographical area (but one offering common environmental characteristics) — led us to choose the following sites (see Fig. 15.2):

Mt Nimba	(detailed study)
Mt Betro	(detailed study)
Taï–Nipla	(detailed study)
Taï–Audrenisrou	(detailed study)
Monogaga	(detailed study)
Ht Sassandra classified forest	(rapid study/survey)
Mt Sainte classified forest	(rapid study/survey)
Scio classified forest	(rapid study/survey)
Banco National Park	(rapid study/survey)
Mt Peko National Park	(survey)
Pékanhoubli–Toulépleu	(survey)
Goulaleu–Toulépleu	(survey)
Niégré classified forest	(survey)
Hte Dodo classified forest	(survey)

(For additional information and other sites, see Boesch *et al.* 1994.)

As indicated above, fourteen sites have been surveyed, and four have been studied in detail (Taï–Audrenisrou, Taï–Nipla, Monogaga, Mts Nimba, Mt Betro). The results have allowed us to document significant variations in both the techniques employed and the choices made by the chimpanzees which are apparently independent of the environment (see Tables 15.3 & 15.4).

The choices made by West African chimpanzees can be observed at different levels. Adopting McGrew's approach (1992) the following examples are considered:

First level: to eat a specific nut or not; Taï chimpanzees neither eat nor crack the *Elaeis guinensis* nut, Bossou chimpanzees do.

Second level: to crack the nut or not. Monogaga chimpanzees eat the *Parinari excelsa* nut (as indicated by freshly chewed exocarp wadges in my observations) but they do not crack it to get at the kernels as is done at Taï (Boesch & Boesch 1983) or at Sapo (Anderson *et al.* 1983).

Third level: to crack with a hammer or not. At Mt Assirik the *Adansonia digitata* nut is probably cracked without a hammer — chimpanzees are reported to hold the nut in the hand and pound it on a stone anvil. This indirect report (Bermejo *et al.* 1989), however, remains problematic and is discussed further by McGrew (1992). The large *Strychnos* nut in Gombe (Goodall 1986) is large enough (grapefruit size) to be cracked without a hammer, but is ignored by the Taï chimpanzees.

Fourth level: to crack with one or other specific form of tool. Sapo forest chimpanzees (in Liberia) crack the nut only with stone hammers (Anderson *et al.* 1983) while Taï groups use either stone or wooden hammers.

Fifth level: the precise mode of cracking the nut — for example, to crack on the ground or crack in the tree, employing a particular operational sequence or a particular gesture. Bossou chimpanzees always crack the nut holding the hammer in one hand, while Taï chimpanzees can grip the hammer with one, two (Sugiyama *et al.* 1993) or three hands.

The comparative study of Panda nut cracking in Taï, Monogaga, Mts Nimba and Mt Betro revealed that although most of the variations of the activity could be explained by environmental factors, some of them cannot (Joulian 1993).

Table 15.3. *Nuts (Ivory Coast): present and cracked, by site.*

Site	Environment	Nut: Co.	Pd.	Pi.	Det.	Sac.	El. g.	Irv.	Kl.
Monogaga	Primary forest	C	C	M	P ?	P	P	C ?	P
Mt Betro	Prim. for. degraded	P	C	P	P	P ?	P	(P)	(P)
Taï	Primary forest	C	C	C	C	C	P	P	P
Nimba	Primary forest	P	C	M ?	P	Abs	P	(P)	(P)
Mt Peko	Second. forest	(P)	P	(P)	(P)	Abs	P		
Mt Sainte	Second. forest	P	P	P	(P)	P ?	P	P	C ?
Scio	Prim. for. degraded	P	C	P	P	P	P	(P)	(P)
Hte Sassandra	Prim. for. degraded	(P)	(P)	P	(P)	P	(P)	P	C ? M ?
Niégré	Primary forest	P	(P)	P	(P)	(P)	(P)	(P)	(P)
Hte Dodo	Prim. for. degraded	P	P	P	(P)	P	(P)	(P)	(P)
Banco	Primary forest	C	(P)	P	(P)	P	(P)	(P)	Abs
Tuambli	Prim. for. degraded	(P)	(P)	P	(P)	P ?	P	(P)	(P)
Goulaleu	Primary forest	P	(P)	(P)	(P)	P ?	P	(P)	(P)
Sipilou	Prim. for. degraded	P	P ?	(P)	(P)	P ?	P	(P)	(P)

C	Cracked	Co.	*Coula edulis*
C ?	Cracking to check, perhaps cracking without hammer	Pd.	*Panda oleosa*
M	Nut eaten but not cracked (direct observation or found in faeces)	Pi.	*Parinari excelsa*
P	Nut present (not cracked) recorded on the chimpanzee territory	Det.	*Detarium senegalense*
(P)	Probable presence (following the range of the species)	Sac.	*Sacoglottis gabonensis*
P ?	Dubious presence	El. g.	*Elaeis guinensis*
Abs	Absence	Irv.	*Irvengia gabonensis*
	No observation	Kl.	*Klainedoxa gabonensis*

Table 15.4. *Nut-cracking sites (Ivory Coast): tools (qualitative and quantitative data).*

SITES		Monogaga	Mt Betro	Taï–Nipla	Nimba	Scio	Mt Sainte	Hte Sassandra	Banco
Available raw material	Stone	Yes	Yes +	Yes	Yes ++	Yes +	Yes	Yes	Yes --
	Wood	Yes	Yes	Yes +	Yes	Yes	Yes	Yes	Yes -
ANVILS									
Coula	Stone	Yes	No ?	Yes	No				(iron) Yes
	Wood	Yes	No ?	Yes	No				Yes
Panda	Stone	Yes	Yes	Yes	Yes	Yes			
	Wood	Yes	Yes	Yes	No	No ?			
Parinari	Stone	No	No	Yes	No				No
	Wood	No	No	Yes	No				No
Detarium	Stone		No ?	Yes	No				
	Wood		No ?	Yes	No				
Sacoglottis	Stone			?					No
	Wood			Yes					No
Elaeis	Stone			No	No				
	Wood			No	No				
Irvingia	Stone	Yes ?		No					
	Wood	No ?		No					
Klainedoxa	Stone			No			No ?	No ?	
	Wood			No ?			Yes ?	Yes ?	
HAMMERS									
Coula	Stone	Yes	No ?	Yes	No				Yes
	Wood	No	No ?	Yes	No				No
Panda	Stone	Yes	Yes	Yes	Yes	Yes			
	Wood	No	No	Yes	No	No ?			
Parinari	Stone	No	No	Yes	No				No
	Wood	No	No	Yes	No				No
Detarium	Stone		No ?	Yes	No				
	Wood		No ?	Yes	No				
Sacoglottis	Stone			Yes ?					No ?
	Wood			Yes ?					No
Elaeis	Stone			No	No				
	Wood			No	No				
Irvingia	Stone	?		No					
	Wood	?		No					
Klainedoxa	Stone			No ?			Yes ?	?	
	Wood						Yes ?	?	

empty cell = tree absent (*cf.* Table 15.3) or information requiring confirmation.

Because it can be studied more quickly and extensively we started by exploring the fourth level of variation — that relating to the morphological traits of the stone hammers and anvils. This revealed four main patterns:

(a) the hammers are larger at Mt Betro than at Taï. The availability of hammers (in term of size) is however similar at the two sites, but Mt Betro chimpanzees appear to prefer larger implements;

(b) 'cupules' (pits) are more abundant in the Mt Betro hammers (up to 5 per hammer) compared to the hammers from the three other sites;

(c) there is a more intensive use of the hammers on two sides at Mt Betro (52 per cent) compared to Taï or Monogaga (26 per cent). This and the preceding features could of course be linked to a longer or more intensive use of the hammers at the Mt Betro site, but on Mt Betro there is plenty of good quality stone. Thus, the more extensive use of the tools must reflect a specific technique (see Joulian 1995);

(d) one type of anvil (which I have called 'enclume-cupule' (pit-anvil) exists at Monogaga and Mt Betro, but is rarely found at Taï.

The above examples express clear variations in the choices and technical habits in nut-cracking behaviour between different sites. Even if the precise meanings of these variations are still not fully understood, the first essential task must be to document and record the evidence of such characteristics — an especially urgent task at present, in view of the endangered nature of the West African chimpanzee populations.

In the context of modern human societies we could ask various questions about, say, the ways of cutting tomatoes (for example in Provence as opposed to Paris) or about the ways in which knives are handled in Africa and in Europe. Why, for example, do West African people peel oranges by pushing the knife *away* from the body while Europeans peel oranges by pulling the knife *toward* the body? Such variations can sometimes be explained by historical or sociological factors, but most of these reasons remain unexplained. We are now beginning to observe similar technical variations in chimpanzee behaviour, but are still a long way from reconstructing the cultural history of chimpanzees. These patterns, so variable or apparently of so little functional significance, are nevertheless present, and demonstrate the existence in chimpanzees of a whole set of qualities which are far removed from the behaviourally deterministic patterns in which they were generally interpreted a few decades ago.

The aim of the preceding was to illustrate the existence of separate traditions or technical variants between different chimpanzee groups. We must now seek to understand what such variations mean. For human societies, very diverse explanations for such variations can be invoked, relating for example to the expression of separate group identities, to the perpetuation of knowledge, or social cohesion. Are such explanations relevant to the kind of traditional variations I have documented, and more generally for chimpanzees? I will pursue this question further in the final section.

Analogous tools

If we compare chimpanzee stone tools to the oldest (Plio-pleistocene) stone tools we can observe identical 'anvils', or 'utilized materials'. Hammers or anvils from Olduvai gorge (Leakey 1971; 1994) or from Melka Konturé (Chavaillon 1986), show similar features (use-wear, shape, etc.) to those used by modern Ivory Coast chimpanzees. Figure 15.3 illustrates such similarity and immediately raises more questions than it answers. For example, use-wear can occur on one, two, or more sides of the tools. Let us assume that these variations, like many others, are not obviously functional, but can express the choices of a social group, either animal or human. A chimpanzee, anatomically and perhaps behaviourally very different from *Homo habilis*, *Australopithecus* and *Paranthropus* (the three potential craftsmen), can produce analogous tools to theirs. This observation has many implications for understanding early hominid lifestyles, but primarily reflects the limitations of our current interpretative frameworks.

A chimpanzee and a pre-Acheulean culture?

Chimpanzee tools provide the opportunity to reconsider the question of 'tradition' from less anthropomorphic bases. But can we — as some ethologists do — speak of 'chimpanzee material culture'? To tackle this question I adopted an explicitly anthropological position, and used an anthropologist's definition of culture to study the social and material productions of the chimpanzees. I will not elaborate on the question of culture here (*cf.* Joulian 1993; 1994b; 1995). Instead I will give two simple definitions of the notions of 'culture' and 'tradition'. By 'culture', I choose from among the scores of anthropological definitions, the very succinct but discriminating one of Maurice Godelier (1992): *'la culture c'est l'ensemble des normes idéelles pour exister'* ('culture is the whole

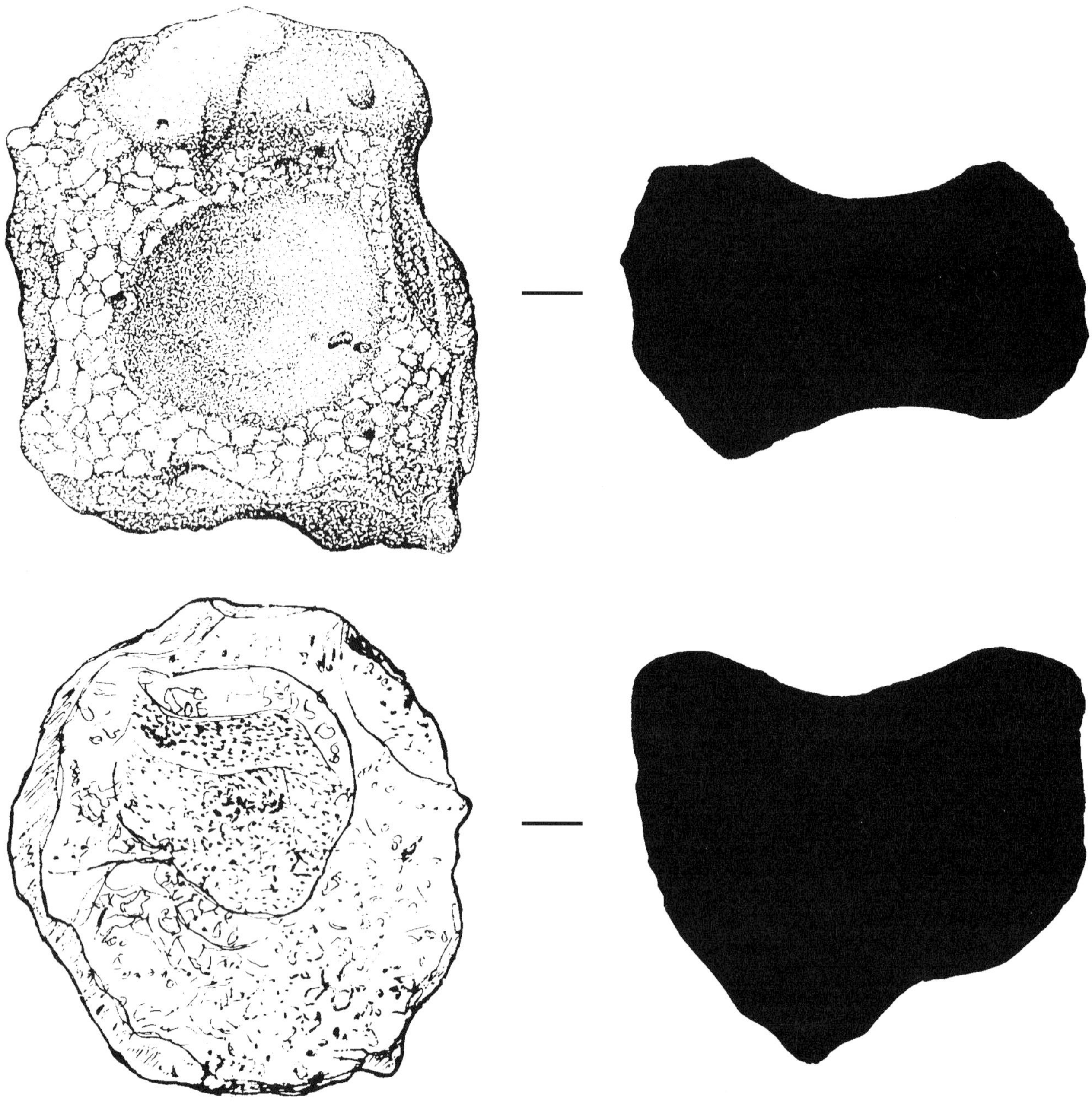

Figure 15.3. *Chimpanzee and prehistoric tools.*

set of ideational rules necessary to exist') and Tylor's classic definition (1871): 'Culture or civilisation . . . is that complex whole which includes knowledge, belief, art, law, morals, custom, and any other capabilities and habits acquired by man as a member of society.'

By 'tradition' I mean a part of culture stabilized and transmitted at least over one generation. Tradition shares the same objects and the same characteristics as culture (*cf.* Table 15.5), but only partially; culture comprises the whole set of elements. I prefer to use the word 'tradition' to describe the nut-cracking techniques discussed here, because such techniques represent only one particular aspect of the chimpanzee's behaviour. The other point to emphasize is that even if a tradition (technical or otherwise) has the same characteristics as culture, it does not belong to the same dimension, and consequently cannot produce the same phenomena (historical accumulation, system functioning, etc.). It should be emphasized that this conception differs clearly from that usually employed by ethologists or psychologists. Nishida (1987), Tomasello (1990) or Premack & Premack (1994), for example, think of tradition mainly in term of information transmission and neglect the phenomenological impact of objects upon the world.

If we turn to the pre-Acheulean materials and ask whether they are 'cultural' and whether Plio-pleistocene societies are 'cultural' we face a enormous problem. Possible answers are very diverse and the methodical and analytical approaches to this question made by Stiles (1979) or Isaac (1984) indicated that Oldowan tool variability expresses neither ethnic nor cultural differentiation, but only raw material differentiation.

My aim, of course, is to identify and compare (following anthropological concepts) how far elements of 'culture' can be identified in chimpanzees and in pre-Acheulean hominids. Table 15.5, which summarizes 33 characteristics, provides the strongest possible basis for a comparison of the two situations in terms of culture, because here culture is being viewed not simply as a means of separating animals from humans, and because culture is being expressed in the broadest possible terms — in other words, both by its objects and its dynamic properties (in contrast to the conceptions of some prehistorians (e.g. Toth 1987) or psychologists (e.g. Tomasello 1990)).

Without entering into the details of these comparisons (see Joulian 1993; 1994a) we can emphasize two main points. Viewed in terms of these comparisons, chimpanzee and Plio-pleistocene societies are: (1) broadly comparable and, (2) similar in overall complexity. They do not differ in nature. These conclusions are akin those of Wynn & McGrew (1989) who concluded that Oldowan tools and toolmaking procedures 'can be all found in the repertoire of living apes'. This kind of purely typological comparison is, however, of limited value, and we need much more detailed analysis, especially on the nature of underlying activities — the only way in which we can gain insight into the associated technical knowledge which lay behind tool production, and, to some degree, into prehistoric cognitive capacities.

In the next section, I will present a new framework to tackle these topics, and suggest new paths of analysis.

Table 15.5. *Comparison of chimpanzee and early hominid 'culture'.*

	Chimpanzee (*Pan troglodytes*)	Ethological bias	Plio-pleistocene hominids	Archaeological bias	Modern humans
OBJECTS					
ideas	No	+	No	+	Yes
knowledge	Yes	–	Yes ?	–	Yes
values	No	–	No ?	–	Yes
norms, rules	Yes	–	Yes ?	+	Yes
icons	No	+	No	+	Yes
concrete objects	Yes	–	Yes ?	–	Yes
(basic choices)	? or Yes	? or –	? or No	? or –	Yes
(secondary choices)	? or Yes	? or –	?	?	Yes
QUALITIES					
produced	Yes	–	Yes	–	Yes
transmitted	Yes	–	Yes	–	Yes
learned	Yes	–	Yes	–	Yes
social	Yes	–	Yes	–	Yes
shared, integrated	Yes	–	Yes ?	+	Yes
cross-generational	Yes	–	Yes	–	Yes
durable	Yes	–	Yes	–	Yes
individual	Yes	–	Yes ?	–	Yes
specific, autonomors	Yes	–	No ?	–	Yes
relational	Yes ?	–	?	+	Yes
diffused	No ?	+	No	+ ?	Yes
delimited	Yes	–	No	–	Yes
symbolic, 'ideal'	?	+	No ?	+	Yes
concrete	Yes	–	Yes ?	–	Yes
'patterning' - normative	Yes	–	No ?	–	Yes
identity, discriminant	?	–	?	+	Yes
standardized	Yes	–	Yes	–	Yes
style	No	–	No	–	Yes
changeable	?	–	?	+	Yes
innovative	Yes	–	Yes ?	–	Yes
cumulative	?	+	Yes	+	Yes
historical	Yes ?	+	Yes	–	Yes
obsolete	Yes ?	–	No	+	Yes
systemic	?	?	?	?	Yes
adaptive	Yes	–	Yes	–	Yes
non-adaptative	Yes	–	No	–	Yes
arbitrary	No ?	–	No ?	+	Yes

Yes = 25 + = 6 Yes = 17 + = 12

Studying techniques and intelligence today

Why did I choose to work on objects or techniques, and not on social aspects? To begin with, the framework I use is essentially 'empirico-inductive' and not 'hypothetico-deductive' — in other words I approach the study from the bottom-up rather than from the top-down.

As already mentioned, there are certain artefacts and technical procedures which are more or less identical in both early hominids and chimpanzees. For the earliest hominids we only have access to discarded artefacts and palaeo-environmental data, while for chimpanzees we can observe both the objects and the behaviour with which they are associated.

In general, chimpanzees give us a unique access to possible interpretations of archaeological objects which I have called 'models of understanding'. This corresponds to what I would describe as a first level of comparison — in fact the only level of interpretation which we can falsify in our present state of knowledge.

In this situation how should one work with the data? The approach I have adopted is essentially empirical, and conforms to the analyses of cultural technology developed by Leroi-Gourhan (1971; 1973), Robert Cresswell (1990) or Pierre Lemonnier (1983; 1990; 1993).

As the principle means of analysis, this approach uses the notion of *chaîne opératoire* (operational chain) and *système technique* (technical system) in order to describe techniques and their place in the social world.

Briefly, the idea of the *chaîne opératoire* is to provide a continuous description of a technical activity, including the different elements which comprise it, from the initial acquisition of the raw material to the discard of the tool. In other words the life of the object is followed and described throughout its functioning cycle. This description constitutes a basis upon which further analysis can be carried out. The concept of *système technique* involves considering techniques as social productions, and we usually do this by splitting the system into three main components: the tools, the processes, and the knowledge involved.

I should like to concentrate now on the characterization of the technical knowledge of chimpanzees and pre-Acheulean hominids. A comparison between the technical knowledge involved in the use of similar tools by chimpanzees on the one hand and pre-Acheulean hominids on the other allows us to raise some questions concerning the cognitive capacities of our early ancestors. But before proceeding, why do we need to ask questions concerning the intelligence of our ancestors in such a specific way?

The reason is that the attempts to model and interpret technological behaviour and knowledge that have been made by non-prehistorians do not take sufficient account of the nature and limits of the available evidence, and do not try to integrate assessments from current research in prehistoric technology. For example, many of the typologies developed for the Oldowan artefacts — such as those of Mary Leakey (1969; 1971) or Jean Chavaillon (1979) — are actually based on formal distinctions which are extremely arbitrary. Despite these shortcomings they have been used by some researchers (for example Wynn 1981; 1989) to reconstruct the mental capacities of early hominids. What we require is a technological approach which tries to integrate such stone artefacts within the context of their production and use.

For example, on the basis of some of my own preliminary observations on material from Bed I at Olduvai, some of the 'anvils' described by Mary Leakey could very well be either hammerstones or alternatively simply by-products of bipolar knapping. The same problem arises with some of the modified tools — most notably with choppers and analogous forms. According to Mary Leakey's classification these can be attributed to various categories on the basis of shape or to so-called functional categories (utilized material, hammer, polyhedron, nucleus, or chopper, etc.).

How can one conduct a cognitive analysis on such a weak basis? How can one evaluate mental capacities from such relatively unstandardized artefacts, to which we can attribute very different functions? Once we recognize these limitations we must proceed very carefully.

Case study : description and analysis of two *chaînes opératoires*

In order to form an impression of the complexity of some of the activities of chimpanzees and early hominids I shall now describe two *chaînes opératoires* — the extraction of *Panda oleosa* nuts by chimpanzees and the knapping of a chopper by Plio-pleistocene hominids. The following descriptions synthesize a wide range of information. There are presented in terms of the notions of 'tools', 'processes' and 'knowledge' as defined above. (For other examples see a comparative analysis of termite fishing techniques by chimpanzees and modern humans (Joulian & Roulon-Doko 1994) or the 'anthropotechnologic' approach developed by Philippe Geslin (1994; 1995).)

To these notions I have added two key concepts from ergonomy: that of 'task' and 'activity'. Together, these enable us to describe both the behaviours and the objects involved. There are of course limitations: the two operational chains correspond to different tasks and activities, and we do not have access to the same parts of the two sequences. Nevertheless it would seem that gestures involved in cracking nuts are related in many ways (notably the motor control of movement) to the percussion gesture involved in knapping stone. In any event, this is the most comparable technical activity we can find to make an empirical comparison of hominid and chimpanzee techniques.

(A) *Chaîne opératoire* of extraction of the *Panda oleosa* nut
(personal observations, and Boesch & Boesch 1983; 1984; 1990; 1991; 1992)

Technique:
- feeding technique, *Panda oleosa* nut cracking (Ivory Coast, Taï region, Taï–Audrenisrou group)

Product: Nut (*Panda oleosa*)
- ripens from December to March
- extreme hardness (1.5 Tonne pressure required to open it, therefore very powerful hit needed)
- trilobate nut (kernels can easily be smashed, therefore very precise motor control needed)

Environment:
- evergreen forest, tropical humid climate

Actors:
Chimpanzees *Pan troglodytes* (entire group)
- females are more successful than males
- only adults and sub-adults succeed in cracking nuts

Social interactions:
- individual and collective social activities

TOOLS
Hammers:
stone: - chosen (according to weight and mechanical properties).
wood: - chosen (according to dimensions, weight and mechanical properties).
- modified (reduction in size for some *c.* 6.7 per cent)

Anvils:
stone: - chosen (or not, according the availability)
wood: - chosen (or not, according the availability)
- cupules (pits) cleaned of the nutshell when it is necessary

Twig:
wood: - chosen
- modified (reduction in size for some)

TECHNICAL PROCESSES (see Table 15.6)

KNOWLEDGE (knowledge, know-how)

REPRESENTATION:

of the project:
- plan, concept: kernel to get at (mental image?)

of the tools:
- according to their physical and mechanical properties
- according to the availability of the materials, the chimpanzees would prefer stone hammers to wooden hammers (even the hardest ones).
- choice of the anvil is less important if there are

Table 15.6. Panda oleosa *nut-cracking operational chain.*

TASKS			ELEMENTS OF THE ACTIVITY							
Main task	Phases	Sub-tasks	Tools	Place	Actors (activity)	Main postures	Prehension	Continuous Discont.	Moving	Time duration ex:
Obtain the nut kernels for eating	Phase I: Gather the nuts	• Choice of the tree where collect • Look for a hammer/ • Carry to the anvil • Gather the nuts/ • Carry nut to the anvil	hands feet mouth	tree and around the tree	males, females individual, collective	knuckle-walking, bipedal	hands feet mouth	D D C	Yes Yes/No Yes Yes/No Yes/No	several minutes depending on distance to the tree
	Phase II: Open the nuts	• Actor's positioning • Place the nut on the anvil • Hold the hammer (glance) • Hit (several hits) • Leave the hammer (open-eat) • Replace the nut • Hit again • Leave the hammer	hands feet anvil hammer	tree trunk and around the tree (× 10 m)	individual male mother//infant	sitting	hands feet fingers fingers	C C/D C/D C C/D C/D C/D	Yes if several anvils No if action on same anvil	few seconds to 1 min./nut whole sequence few min. to 1 hour or more
	Phase III: Feed on the nut	• Direct comsumption/ • Indirect comsumption looking for a twig/ modification of the twig extraction of the almond feeding	lips teeth fingers twig	tree and around the tree	individual male mother//infant	sitting	hands fingers	C/D C/D C/D C/D C/D C/D	Yes if several anvils No if action on same anvil	seconds more if sharing

stone hammers. When materials are available, chimpanzees prefer granite or gneiss rather than quartzite; quartzite is preferred to laterite.
- choice of tools is biased, favouring some characteristics (weight, hardness, material) and requires no more than the use of analogical representations for these tools (*cf.* Knowledge: representation of techniques, pp. 185–7).
- bias in terms of shape has not yet been demonstrated.
- some choices are not dictated by genetic or environmental factors, but by traditional factors (*cf.* Nut-cracking traditions, pp. 173–9). These tools reflect collective uses and collective representations. In many respects these correspond to the 'types' we recognize in prehistoric tools.
- traditional choices have been demonstrated for termite fishing probes (McGrew & Tutin 1979; McGrew 1992).

of the activity:
- from different perspectives of space and time.
- anticipation of the season: for some nuts (*Coula edulis*), female chimpanzees climb trees, check whether the nuts are ripe and crack them in the trees.
- anticipation of location: chimpanzees can travel long distances across the forest toward an invisible *Panda* tree, to exploit it.
- stone tool transportation: distances can exceed hundreds of metres in Taï–Audrenisrou group. For the Mt Betro and Taï–Nipla groups, chimpanzees might choose inselbergs (isolated outcrops of granite in the forest) because stone availability for anvil and hammer stones is greater in these locations (Joulian 1993; 1995).

of actions:
- sequence of gestures (or processes): anvil + nut + hammer
- positioning (and re-positioning) of the nut on the anvil; pit depression is preferred; pounding; extraction of the almond; consumption.
- different gestures
  - positioning the nut
  - holding the hammer (with two hands)
  - striking (speed of percussion)
  - controlling (motor control of the hammer rebound)
  - extracting: direct extraction of the almond extraction using another tool (small twig)
- knowledge susceptible to modification (behaviour flexibility)
- complicated know-how, requiring a long time to learn (around 10 years to learn to crack *Panda* nuts efficiently)
- 'teaching': two examples could be cited:
(a) a mother showing her offspring how to position a *Panda* nut correctly; or
(b) another mother demonstrating the correct way of holding a hammer (Boesch 1991).

This indicates that representations of gestures and processes actually exist and that they are not uniquely linked to motor patterns or know-how. Such observations demonstrate that knowledge and know-how are not only acquired by imitation,

Table 15.7. *Oldowan chopper-knapping operational chain.*

TASKS			ELEMENTS OF THE ACTIVITY							
Main task	Phases	Sub-tasks	Tools	Place	Actors (activity)	Main postures	Prehension	Continuous Discont.	Moving	Time duration ex :
Producing a useful cutting edge	Phase I : Getting the raw material	• Look for the raw material • Look for a hammer • Carrying it to a workshop	hands	outcrops riverbanks	collective/ individual ?	upright	hands, arms, containers ?	C/D C/D C/D	Yes/No Yes/No Yes/No	few minutes to a few hours or more (?)
	Phase II : Knapping the tool	• Positioning of the craftsman • Positioning of the raw material, and hammerstone • Hit-strike (chopper debitage) • Turning of the nucleus • Hit-retouch (retouch) • Hit-flake (flake)	hands hammer (pebble) hammer	diverse, lakeshore, riverbank, floodplain . . .	individual	sitting/ on the knee/ upright	manual	C C C C/D C/D C/D C/D	No No No Yes/No Yes/No Yes/No Yes/No	few second to 5–10 minutes
	Phase III : Use of the tool	• Use of the chopper • Use of the flake wood-working plant-working (saw, cut, scrape, . . .) butchery-working (cut, dismember, . . .)	chopper or flake	diverse, lakeshore, riverbank, floodplain . . .	collective/ individual ? collective (big game)	sitting/ on the knee/ upright/ bent	manual	D & C/D	Yes/No	few minutes to several hours or years ?

observational learning or other modes of transmission, but also by more active learning, and express a general representation of know-how. The next question (as yet unanswered) is to determine if such representations are actually collective ones? The existence of technical traditions (discussed in Nutcracking traditions, pp. 173–9) related to evidence of 'teaching' would lead us to answer this question in the affirmative.

(B) *'Chaîne opératoire'* of the knapping of a pre-Acheulean stone tool:

Technique:
shaping of a lithic tool

Product:
quartz, quartzite, etc.

Availability:
raw materials obtained from no more than 2–3 km (i.e. at Omo or West Turkana sites)

TOOLS
Hammers:
stone (hard pebbles, etc.)

TECHNICAL PROCESSES (see Table 15.7)

KNOWLEDGE

REPRESENTATION:

of the project:
- cutting edge

of the tool:
- of the raw material: different qualities and shapes different availability and accessibility
- of the tool (hammer-stone): different shapes, sizes and weight

of actions:
- sequence of gestures where mechanical constraints play an important role at a specific moment during the positioning of the struck pebble and hammer-stone (minimum of 90 degree angulation) with the aim of producing a conchoidal fracture.
- motor control of the speed
- the mastering of these two factors indicates an operational complexity in the processes, but cannot necessarily be interpreted in terms of algorithmical representation.

Comment
For the Oldowan period (before 2 million years) Hélène Roche noted the high percentage of tools (*c.* 70 per cent) and 'little chunked flakes broken, like products of the flaking'. She described a weakness in the production of the oldest tools. 'Technical behaviour seems to be very stereotyped and rigid, with poor adaptation to the raw material. This is due to the weakness of the conceptual schema and to a certain lack of skill. If the aim is to obtain good cutting flakes, one may say that the success ratio is quite low' (Roche 1989).

Knowledge: representation of techniques

Let us now consider the third component of the technical system — knowledge. Usually prehistorians apply, on the one hand, notions of 'concept' and 'method' (which correspond to the representations of craftsmen's tasks) and on the other hand, notions of 'conceptual scheme' and 'know-how', with the aim of reflecting the modalities and variations in the technical chain (Boëda 1991; Karlin 1991; Pelegrin 1991). Such analytical concepts are not totally appropriate to describe both living behaviours (chimpanzee ethology) and reconstructed behaviours (prehistoric technologies), so I would prefer to employ the broad concept of 'representation'. The term 'representation' has the dual advantage of relating both to individual cognitive factors (the mental representation and mental image of the psychologists: Kosslyn 1980; Vauclair 1992), and to social factors, as in 'the social representations of techniques' (Lemonnier 1991; 1993).

The internal representations reflect different levels of information processing according to the things represented (tools, processes or, in our case, concepts). These representations exist '. . . only by the types of behaviors which establish them and make them function' (Bresson 1987). In this case behaviour and material productions can be as valid a guide to the representation of animal and pre-hominid techniques as language is for modern humans.

There are two questions to keep in mind: (1) What is being represented (elements of the environment, animate/inanimate objects, actions, physical appearance, physical principles, artefact types)? (2) What are the modes of representation and mental manipulation of techniques, depending on whether the representations deal with objects or series of gestures or processes? (The structure of the preceding *chaînes opératoires* deals partially with these questions.)

To address these two questions I shall make use of the distinction established by Bresson (1987) between 'analogical' and 'non-analogical' systems of representation. An analogical system establishes a resemblance between the representation and the represented (for example, the physical model or the mental map). In the second system, the 'non-analogical' or algorithmical, the resemblance between representation and represented is not retained, and representation is carried out through an arbitrary system of code — as for example language.

According to Bresson '. . . there is no analogical representation which can refer to actions, operations, orders or transformations'. Furthermore he gives us an essential indication when he writes: '. . . in the analogical mode it is possible to represent only configurations or states or eventually like in the cinema, succession of states'.

Following these ideas I consider that tools, products, raw material, etc. can be referred to analogical representation, for example to mental images. On the other hand, complex processes (which necessitate long-term apprenticeships) refer to algorithmical representation — in other words to a syntax-like code and to the transmission of knowledge.

This leads us to question the relevant type of code in the technical activities of hominids: Were they using language as a medium or not?

What can be said is that a linguistic medium is not necessarily implied in the most complicated (but still relatively simple) *chaîne opératoire* of pre-Acheulean hominids. The *chaînes opératoires* that we have described do not necessarily imply representation of an algorithmical nature. This does not mean that chimpanzees or the earliest hominids were incapable of a language; they could have had the potential. But we have to look for other dimensions beyond the problematic linguistic one (imitation learning and sociality is one of them).

Most if not all of the technical actions described above for chimpanzees or for the pre-Acheulean can be interpreted in terms of the mental image notion. Mental images not only allowed the anticipation of real movements (for example, mental rotations in chimpanzee actions: Brésard 1985) but also allowed certain forms of computation (e.g. analogical computation of interactions) and consequently the ability to anticipate some actions to be carried out in the future (as for example in stone-knapping).

Let us turn to the comparative chimpanzee/pre-hominid *chaînes opératoires* I have proposed above and ask ourselves whether these complex percussion gestures which exist among living chimpanzees resemble those which were undertaken by early hominids, between 2.0 and 2.5 million years ago? It would seem that the gestures involved in cracking nuts are closely related in many ways (notably in the motor control of movement) to the percussion gesture of prehistoric stone-knappers.

What could we say about these prehistoric gestures? How did they evolve from simple to complex? In the case of pre-Acheulean tools, can we assume that we are dealing with complex technical behaviours illustrating fully 'integrated' capacities (in Leroi-Gourhan's sense) — very far removed from those of the chimpanzees — and that the earliest hominid tools only needed further elaboration to generate fully 'cultural' Palaeolithic assemblages? We can no longer accept this traditional view.

Do the actions of 'hitting', like a chimpanzee, or 'removing matter' (e.g. reducing stone), like an *Australopithecus*, correspond to very different psycho-motor schemas? And could we argue, as many technologists do (e.g. Pelegrin 1993; Pigeot 1991) that mastering a platform and mastering the conchoidal fracture are highly significant characteristics which prove early hominid abilities to be superior to those of chimpanzees?

This paper has challenged such statements. We have shown that traditional variations exist in both nut-cracking activities and associated tools, and that one can establish the close proximity of chimpanzee culture to the Plio-pleistocene hominid culture. We can also argue that traditional variations in Plio-pleistocene lithic assemblages have not yet been demonstrated. While Plio-pleistocene industries reveal some tools analogous to those of chimpanzees, what seems to be most significant in this context is that the activities of Plio-pleistocene hominids and chimpanzees can actually be compared and are in many respects surprisingly comparable (even if the specific tasks and functionings are different). Similar activity stuctures are identifiable in each group and fill the so-called 'gaps' that many prehistorians still draw between apes and hominids.

The precise differences have still to be clarified and defined; perhaps we should play closer attention to the length and organization of the *chaînes opératoires*. The greatest complexity can perhaps be seen in the length of such operational sequences, or in the way the sequences are integrated with non-technical activities, and especially with social relations. The degree of integration of technical and non-technical activities in chimpanzees (and other primates: *cf.* Strum 1994) and in prehistoric societies is probably one of the best ways of estimating cognitive capacities. We are, however, only just beginning to investigate these questions.

To conclude, I have presented a comparative study of chimpanzee and pre-Acheulean techniques. This study involves the elaboration of specific concepts with which to study techniques (*chaînes opératoires*, technical system, task, activity, representation). These concepts can be applied to the empirical study of the behaviour of living chimpanzees with an 'etho-archaeological' approach, and can be brought to bear on cognitive questions. It seems to me that *there are no major differences in qualitative terms* between what the chimpanzee do when they cracks nuts and what pre-Acheulean hominids did when they knapped a chopper or produced a flake.

Looking for clear-cut differences between apes and early hominids no longer seems to be essential. These dichotomy-seeking approaches have revealed their limitations. What we need now are good tools — both concepts and methods — to compare and appreciate the similarities as well the differences, and perhaps to better estimate the 'cognitivo-socialo-cultural sphere' which mobilizes modern prehistory.

Acknowledgements

The fieldwork was encouraged by Professor Yves Coppens and supported by a grant from the Simone and Cino Del Duca Foundation and a Ministère de l'Education contribution. I am very indebted to the CSRS in Abidjan and the Taï Chimpanzee Project directed by Christophe Boesch. I should also like to thank the Eaux et Forêts and the Institut d'Ecologie Tropicale agents who made the survey and the study possible.

Frédéric Joulian
Laboratoire de Paléoanthropologie et de Préhistoire
Collège de France
11 Place Marcelin Berthelot
75231 Paris
Cédex 05
France

References

Anderson, J.R., E.A. Williamson & J. Carter, 1983. Chimpanzee *Pan troglodytes verus* of Sapo forest, Liberia, density nests, tools, and meat eating. *Primates* 24(4), 594–601.

Beatty, H., 1951 A note on the behavior of the chimpanzee. *Journal of Mammalogy* 32(1), 118.

Bermejo, M., G. Illera & J. Sabater-Pi, 1989. New observations on the tool-behavior of chimpanzees from Mt Assirik (Sénégal, West Africa). *Primates* 30(1), 65–73.

Berthelet, A. & J. Chavaillon (eds.), 1993. *The Use of Tools by Human and Non-human Primates*. Oxford: Oxford University Press.

Beyries, S., 1993. Peut-on déterminer la fonction des outils du paléolithique le plus ancien?, in Berthelet & Chavaillon (eds.), 225–36.

Boëda, E., 1991. Approche de la variabilité des systèmes de production lithique des industries du paléolithique inférieur et moyen: chronique d'une variabilité attendue. *Techniques et Culture* 17–18, 37–79.

Boesch, C., 1991. Teaching among wild chimpanzees. *Animal Behaviour* 41, 530–32.

Boesch, C. & H. Boesch, 1983. Optimisation of nut-cracking with natural hammers by wild chimpanzees. *Animal Behaviour* 83, 265–86.

Boesch, C. & H. Boesch, 1984. Possible causes of sex differences in the use of natural hammer by wild chimpanzees. *Journal of Human Evolution* 5, 415–40.

Boesch, C. & H. Boesch, 1990. Tool-use and tool-making in wild chimpanzees. *Folia Primatologica* 54, 86–99.

Boesch, C. & H. Boesch, 1991. Les chimpanzés et l'outil. *La Recherche* 233, 724–31.

Boesch, C. & H. Boesch, 1992. Transmission aspects of tool-use in wild chimpanzees, in *Tools, Language and Cognition in Human Evolution*, eds. K.R. Gibson & T. Ingold. Cambridge: Cambridge University Press, 171–83.

Boesch, C., P. Marchesi, N. Marchesi, B. Fruth & F. Joulian, 1994. Is nut cracking in wild chimpanzees a cultural behaviour? *Journal of Human Evolution* 26, 325–38.

Brésard, B., 1985. Ethologie cognitive des anthropoïdes: quelques orientations actuelles. *Annales de la Fondation Fyssen* 1, 45–53.

Bresson, F., 1987. Les fonctions de représentation et de communication, in *Psychologie, Encyclopédie de la Pléiade*, eds. J. Piaget, P. Mounoud & J-P. Bronckart. Paris: Nrf, Gallimard, 933–82.

Chavaillon, J., 1979. Essai pour une typologie du matériel de percussion. *Bulletin de la Société Préhistorique Française* 76(8) 230–33.

Chavaillon, J., 1986. Premiers outils et vie en société, in *Définition et origines de l'homme, Table Ronde Internationale du CNRS*, ed. M. Sakka. Paris: CNRS, 309–15.

Coppens, Y., 1994. Une histoire de l'origine des hominidés. *Pour la Science* 201, 64–71.

Cresswell, R., 1990. 'A new technology' revisited. *Archaeological Review from Cambridge Technology in the Humanities* 9(1) 40–54.

Geslin, P., 1994. *Ethnologie des techniques. Architecture cérémonielle Papago au Mexique*. Paris: L'Harmattan.

Geslin, P., 1995. La parenté cachée: regard d'un ethnographe sur l'anthropotechnologie. *Le Travail Humain* n° spécial en hommage à Alain Wisner, 15 p.

Godelier, M., 1992. Lecture, in 'Culture Matérielle' Seminar. Université de Paris I, Sorbonne, Paris.

Goodall, J., 1986. *The Chimpanzees of Gombe. Patterns of Behavior*. Cambridge (MA) & London: The Belknap Press of Harvard University Press.

Isaac, G., 1984. The Archaeology of human origins: studies of the lower pleistocene in East Africa, 1971–1981, in *Advances in World Archaeology*, vol. 3, eds. F. Wendorf & A.E. Close. Orlando (FL): Academic Press, 1–87.

Joulian, F., 1993. Application de l'éthologie des chimpanzés Ouest-Africains au comportement des premiers hominidés: le problème de la culture. Thèse de doctorat, Université de Paris I Sorbonne (dir. J. Chavaillon).

Joulian, F., 1994a. Peut-on parler d'un système technique chimpanzé? Primatologie et archéologie comparées. in Latour & Lemonnier (eds.), 45–64.

Joulian, F., 1994b. Culture and material culture in chimpanzees and early hominids. in *Current Primatology*, vol. II: *Social development, Learning and Behaviour. Selected Proceedings of the XIVth Congress of the International Primatological Society*, eds. J.J. Roeder, B. Thierry, J.R. Anderson & N. Herrenschmidt. Strasbourg: Univ. Louis Pasteur, 397–404.

Joulian, F., 1995. Mise en évidence de différences traditionnelles dans le cassage des noix chez les chimpanzés (*Pan troglodytes*) de la Côte d'Ivoire, implications paléoanthropologiques. *Journal des Africanistes* 65(2).

Joulian, F. & P. Roulon-Doko, 1994. Comparaison d'une activité technique chez les hommes et chez les chimpanzés: la collecte des termites. *Techniques et Culture* 23–4, 29–62.

Karlin, C., 1991. Connaissances et savoir-faire: comment analyser un processus technique en préhistoire, in *Technologia y cadenas operativas liticas*, eds. M. Mora, X. Terradas, A. Parpal & C. Plana. (Treballs de arqueologia 1.) Barcelona: Publicacion de la universitat autonoma de Barcelona, 99–124.

Keeley, L. & N. Toth, 1981. Microwear polishes on early stone tools from Koobi-Fora, Kénya. *Nature* 293, 464–5.

Kortlandt, A. & E. Holzhaus, 1987. New data on the use of stone-tools by chimpanzees in Guinea and Liberia. *Primates* 28(4), 473–96.

Kosslyn, S.M., 1980. Les images mentales. *La Recherche* 108, 158–64.

Latour, B. & P. Lemonnier (eds.), 1994. *De la Préhistoire aux missiles balistiques: l'intelligence sociale des techniques.* Paris: La Découverte.

Leakey, M., 1971. *Olduvai Gorge*, vol. 3. Cambridge: Cambridge University Press.

Lemonnier, P., 1983. L'étude des systèmes techniques, une urgence en technologie culturelle. *Techniques et Culture* 1, 11–34.

Lemonnier, P., 1986. The study of material culture today: toward an anthropology of technical systems. *Journal of Anthropological Archaeology* 5, 147–86.

Lemonnier, P., 1990. Topsy turvy techniques. Remarks on the social representation of techniques. *Archaeological Review from Cambridge, Technology in the Humanities* 9(1), 27–37.

Lemonnier, P., 1991. De la culture matérielle à la culture? Ethnologie des techniques et préhistoire, in *25 ans d'études technologiques en préhistoire. Bilan et perspectives. Actes des XIè rencontres internationales d'archéologie et d'histoire d'Antibes*, eds. CRA du CNRS. Juan les Pins: APDCA, 15–20.

Lemonnier, P. (ed.), 1993. *Technological Choices: Transformation in Material Cultures since the Neolithic.* London: Routledge.

Leroi-Gourhan, A., 1971. *Evolution et techniques*, tome 1: *L'homme et la matière*. Paris: Albin-Michel.

Leroi-Gourhan, A., 1973. *Evolution et techniques*, tome 2: *Milieu et techniques*. Paris: Albin-Michel.

McGrew, W.C., 1985. Chimpanzee and the oil palm. *Social Biology and Human Affairs* 50(1), 7–25.

McGrew, W.C., 1992. *Chimpanzee Material Culture: Implications for Human Evolution*. Cambridge: Cambridge University Press.

McGrew, W.C. & C. Tutin, 1979. Evidence for social custom in wild chimpanzee? *Man* 13, 234–51.

Nishida, T., 1987. Local traditions and cultural transmission, in *Primates Societies*, eds. B.B. Smuts & D.L. Cheney. Chicago (IL): Chicago University Press, 462–74.

Pelegrin, J., 1991. Prehistoric lithic technology: some aspects of research. *Archaeological Review from Cambridge, Technology in the Humanities* 9(1), 116–25.

Pelegrin, J., 1993. Connaissance et savoir faire à travers l'analyse d'un processus technique en préhistoire, in Berthelet & Chavaillon (eds.).

Pigeot, N., 1991. Réflexions sur l'histoire technique de l'homme: de l'évolution cognitive à l'évolution culturelle. *Paleo* 3, 167–200.

Premack, D. & A.J. Premack, 1994. Why animals have neither culture nor history, in *Companion Encyclopedia of Anthropology*, ed. T. Ingold. London: Routledge, 350–65.

Roche, H., 1989. Technological evolution in early hominids. *Ossa* 14, 97–8 (and unpublished paper, Stockholm symposium 1987, 8 p.).

Savage, T.S. & J. Wyman, 1843/44. Observations of external characters and habits of *Troglodytes niger*, Geoff. and its organisation. *Boston Journal of Natural History* 4, 362–76, 381–6.

Stiles, D., 1979. Paleolithic culture and cultural change: experimentation in theory and method. *Current Anthropology* 20, 1–22.

Struhsaker, T.T. & P. Hunkeler, 1971. Evidence of tool-using by chimpanzees in the Ivory-Coast. *Folia Primatologica* 15, 212–19.

Strum, S.C., 1994. Une société complexe sans culture matérielle: le cas des babouins, in Latour & Lemonnier (eds.), 27–44.

Sugiyama, Y., 1989. Local variation of tool and tool behavior among wild chimpanzee populations, in *Behavioral Studies of Wild Chimpanzees at Bossou, Guinea*, ed. Y. Sugiyama. Inuyama: Kyoto University Primate Research Institute, 1–15.

Sugiyama, Y., 1993. Local variation of tool and tool behavior among wild chimpanzees populations, in

Berthelet & Chavaillon (eds.), 175–90.

Sugiyama, Y. & J. Koman, 1979. Tool-using and making behavior in wild chimpanzees at Bossou, Guinea. *Primates* 20, 513–24.

Sugiyama, Y., T. Fushimi, O. Sakura & T. Matsuzawa, 1993. Hand preference and tool-use in wild chimpanzees. *Primates* 34(2), 151–9.

Tanner, N.M., 1987. The chimpanzee model revisited and the gathering hypothesis, in *The Evolution of Human Behavior: Primate Models*, ed. W.G. Kinzey. New York (NY): SUNY Press, 3–27.

Tomasello, M., 1990. Cultural transmission in the tool and communicatory signaling of chimpanzees?, in *Language and Intelligence in Monkeys and Apes Comparative Developmental Perspectives*, eds. S.T. Parker & K.R. Gibson. New York (NY): Cambridge University Press, 274–311.

Toth, N., 1987. La naissance de la culture. *Pour la Science* 116, 55–66.

Tylor, E.B., 1871. *Primitive Culture*. Boston: Estes & Lauriat.

Vauclair, J., 1992. *L'intelligence animale*. Paris: Seuil.

Whitesides, G.H., 1985. Nut-cracking by wild chimpanzees in Sierra-Leone. *Primates* 26, 91–4.

Wynn, T., 1981. The intelligence of Oldowan hominids. *Journal Human Evolution* 10, 529–41.

Wynn, T., 1989. *The Evolution of Spatial Competence*. (Illinois Studies in Anthropology 17.) Urbana & Chicago (IL): University of Illinois Press.

Wynn, T. & W.C. McGrew, 1989. An ape's view of the Oldowan. *Man* 24, 383–98.

16

Mental Abilities of Early *Homo*: Elements of Constraint and Choice in Rule Systems

J.A.J. Gowlett

Our most direct evidence for early human behaviour, apart from that of fossils, is that of technology in the form of stone artefacts. Not only do these provide several kinds of functional and economic information, but the steps of their manufacture can reveal pathways of decision-making or problem-solving, thus giving insights into early human abilities. The extent of the elements of design which underlie the artefacts is debated by archaeologists. It is not always easy to determine which aspects are chosen (i.e. could be varied arbitrarily), and which are constraints directly or indirectly imposed by the raw materials and functions. The paper explores this problem, concentrating on artefacts of the Acheulean tradition, especially handaxes or bifaces of about one million years ago. Comparisons are made with design forms and levels of skill practised in later times and the study of size-transformations of designs is emphasized. It is concluded that there is an essential technological continuity through the last million years which transcends labels such as 'presapiens' and 'sapiens'. Modern design is no less affected by constraints, although modern designers have the benefit of recourse to theory.

Introduction: the central place of archaeology and data?

> 'It seems undeniable that the prime responsibility of any discipline must be to maintain its factual basis in good order. It is equally true that the superstructure of insight and understanding is what really interests most of the participant scholars and the rest of intellectual humanity.' (Isaac 1971)

My aim is that this paper should be grounded in archaeological data. There are two fundamental approaches possible in aiming to model the early human mind: that of comparative study based on living organisms of the present; and that of gaining data from the past, in the form of hominid remains, or in the form of archaeological evidence such as stone tools. Much can be achieved from the present, but the lesson, indeed the rationale, of archaeology is that past data cannot be bypassed.

The particular aim of this paper is to take archaeological evidence — in the time range of *Homo erectus* and early *Homo sapiens* — and to examine the question of 'constraint or choice' as revealed in the decisions underlying early artefact manufacture. Such activities of the hominids can provide insight into the extent of their mental processing and their ability to impose and follow cultural rules. Because views about this have developed rapidly, I deliberately reuse the structure of my paper of 1984 to examine the progress of the last ten years. Again I would like to provide specific examples of hard evidence in early hominid behaviour.

Why use the archaeological record? Implicitly the reasons are given above: general comparative modelling could not tell us whether technology has a history of five million years or 500,000 years. Only archaeology is available to tell us that the 'hard' technology of stone flaking began around 2.5 million years ago, and that it has developed markedly if not uniformly until the past few thousand years.

Such technology is not a perfect index to past abilities. It is well known that many abilities may be held without being expressed (*cf.* Mellars & Stringer

1989). Thus direct correlation of archaeological evidence with actual abilities is dangerous, but it must always be true that cultural evidence entails abilities *sufficient to sustain it*. That is, it sets minimum levels of abilities.

There is a basic question underlying this work. How are we to understand how human beings have become so different from apes over a few million years? To approach that problem entails coming to grips with rates of change, and continuity. It is easier to emphasize discontinuity, to draw a line under the abilities of modern humans. But modern apes exhibit formidable abilities with only a third as much brain as we have. Their minds and ours are seated in brains which possess vast number of neurones and synapses. If early hominids possessed only half as many as we do, they would still have exceptional mental abilities by any yardstick except that of modern humans.

Progress of recent years

What are the major advances of the last ten years in data and ideas? In hominization the issues of bipedalism and then of brain enlargement/tool-making remain central to the subject. There have been fewer major interdisciplinary field projects and hominid finds (until very recently), but many new interpretations. It has been confirmed that the early hominids were fully bipedal, but evidence as to the great length and strength of the arms is just one factor arguing that the australopithecines were more apelike than was once thought (Beynon & Wood 1987; Johanson *et al.* 1987; Wood 1992). For the earlier australopithecines this impression is reinforced by the large canines of the newly discovered *Australopithecus ramidus* (White *et al.* 1994). Archaeological evidence appears at about 2.5 Ma (Harris 1983; Ronen 1991), and a date of 2.4 Ma has been claimed for earliest *Homo* from a site at Chemeron in Kenya (Hill *et al.* 1992). Whether or not this classification of the Chemeron hominid finds support, the increasing acceptance that there were three species of early *Homo* by 1.7–1.8 Ma argues for the genus emerging at a considerably earlier date.

In this context *Homo erectus* appears suddenly, much like our earlier image of the striding Cro-magnons. The finds give the impression of a being of great physical competence — with all of the stature of modern humans, and far greater strength — though with intellectual endowment open to debate. The new dates from southeast Asia of 1.8 Ma for early *Homo* (Swisher *et al.* 1994) provide near-proof of powerful hominids who had already extended across much of the Old World by the time that stone tools become well-established in the archaeological record. In short, *Homo erectus* had more than two-thirds of the cranial capacity of modern humans; had a more formidable body; had effective technology; and spread widely around the world more than 1.5 million years ago. This hardly gives us licence to assume a primitive adaptation, and urges the need to test for its complexity. But some workers still feel the danger is of seeing too much complexity — the hominids who had everything but the stockmarket. In the face of such varied opinion one aim of archaeology must be to find more objective, generally agreed, tests of its evidence.

Hypotheses of recent years and their problems

The relationship between biological and cultural evolution remains one of the great challenges to our understanding. In ten years there has been a perceptible cycle of ideas. Technology, which had been out of favour, has re-emerged as a respectable aspect of cultural behaviour for study, especially in a social context (e.g. Sinclair & Schlanger 1990). This is helpful because the primary evidence of early archaeology is largely technological.

The interpretative ideas favoured by a past generation, however, such as the hunting hypothesis (e.g. Ardrey 1976; Washburn & Lancaster 1968), have now fallen by the wayside. The food-sharing model supported by le Gros Clark (1967) and Isaac (1976) has fared little better. Schick (1987) has recast the idea into a 'favoured places' hypothesis, aiming to decouple the observed evidence of stone and bone scatters from higher levels of interpretation. But others have moved to seeing home bases as inherently implausible, until far later times, since it is alleged the lakeside areas would be too dangerous for early hominids as the haunts of carnivores (Binford 1981; Potts 1992). Likewise, hunting is impossible to prove, and often difficult to envisage in the case of early small hominids and fearsomely large prey. This has led to scavenging models, and to a search for criteria that may be recognizable in the archaeological record (Blumenschine & Selvaggio 1988; 1991). Many of the new ideas have come close to being assumed by default, because the old ideas could not be proven. They have received surprisingly little testing in their own right.

Other more general models have survived. The positive feedback model of Tobias (1981; 1991) entails interdependence between brain, hands and eyes, on the one hand, and technological culture on the other. Tobias sees language as a principal facilitator in this interaction. Sociobiologists point out that cultural activity is likely to be favoured only if it helps at a genetic level. One approach, seemingly compatible with that of Tobias is the gene-culture coevolution of E.O. Wilson (e.g. 1983). There are some difficulties in having technology as the prime mover in such systems. If human evolution is technology-driven, it becomes hard to explain that technology is far more used by some societies than others, and by some individuals more than others in a given society. There is scope for other mechanisms, such as the social intelligence or competition models advanced by Humphrey (1976) or Alexander (1979; 1989). It has been suggested that as language developed and cultural content increased, so group boundaries would become reinforced. This possibility was considered by Isaac (1972) who felt that low levels of linguistic ability in the earlier Palaeolithic might be consonant with the relative stability and lack of geographical variation in artefact traditions.

In summary, technology is a part of culture; it has a social context of learning and usually of practice; and is linked with brain evolution. But of course it is not the whole story. The problem is that it is not a one-to-one index. Nevertheless, recent assertions that early artefacts do not require advanced cognitive abilities may result from a failure to analyze the actual evidence to the full. Technology remains the most useful index of past abilities which we have.

More than an alternative approach; mental abilities

Amongst the cadre of hypotheses, psychological approaches can be seen as merely another alternative. Ten years ago one had to argue that they should be admissible. But there is near universal agreement that human evolution, in the end, is characterized by astonishing developments of mind, and so there has been steady progress towards cognitive studies. Certainly if our obligation is to analyze the whole pattern of a species or family's evolution, then the psychological aspects must be covered.

Preoccupation with this aspect of human evolution goes back beyond the modern 'cognitive' archaeology. It is interesting, rather than contentious, to reflect that Palaeolithic archaeology as so often, arrived there first (e.g. Schmidt 1936; Oakley 1981). The problem of application lies in the variety of the psychological approaches employable, contrasted with the narrowness of the data base which can be applied to them.

Nevertheless, among those actually working on the early material, as opposed to commentators, there appears to have been a convergence in recent years, regardless of the psychological theory invoked. In Isaac's work, psychology was implicit; Wynn refers specifically to Piaget; Toth employs experimental criteria (e.g. Isaac 1977; 1986; Wynn 1979; 1981; 1985; Toth 1985). I have attempted to follow a routine-based approach, which I trace back ultimately to Craik (1943), Wiener (1954), and which was developed by Miller *et al.* (1960). Its analysis of planning has been taken further by e.g. Sacerdoti (1977) or Hoc (1988). In psychology such routines have been employed by for example Annett (1969) and Connolly (1979). French archaeologists have tended to follow Leroi-Gourhan in the documentation of 'operational chains' (e.g. Pelegrin *et al.* 1988; Boeda *et al.* 1990). There are differences of interpretation: some tend to equate early hominid activity with that of apes, others put it at a higher level; some see the record as 'steppy', other as a gradient; some emphasize describing the chain, others the decision points. There are, however, underlying levels of agreement both about the high levels of ability in apes and hominids, and about the legitimacy of the approach. There is also a common concern with how hominids structured activity through space and time, and how they handled form in the environment.

The literature nevertheless shows an extreme divergence in views expressed — chiefly disagreement with archaeologists whose work is based in later periods, and who see early hominids as inherently primitive or limited in their abilities (Binford 1989; Dibble 1989; Davidson & Noble 1993).

If the views range from those who see non-cultural programmed activity, to those who see rules and design content in early artefacts, how do we proceed? I would argue that a balanced and structured approach to assessing possible complexity is called for. If we assume that complexity does not exist, we are unlikely to test for it. Equally, if we assume that apparent complexity is real complexity, we may fail to determine when it is the product of simple underlying factors. In my view one must test for complexity: you cannot find pattern in material if it is not there. What explanation we then accept is another matter.

In 1984 it seemed fair to conclude consideration of mental abilities by emphasizing parallels between the structure of language and tool manufacture. Holloway (1969) had pointed out similarities of 'grammar' between the two. Much recent work has explored these issues (e.g Gibson 1993; papers in Gibson & Ingold 1993). Gibson argues for an integrated pattern of interpretation by the brain. Wynn (1993) sees differences, noting that individuals vary much more in technological proficiency and participation than they do in lingustic competence. Like Graves (1994) I am unwilling to attribute language to early humans on the basis of archaeological evidence. It may be that eventually we will recognize in the manufacture of sophisticated artefacts abilities or packages of abilities which must be linked with language — but we are not yet in any position to do so. The extent of chimpanzee cultural activity now recognized might have led us to think, *a priori*, that they would need language as we know it: but the chimpanzees do not, although they plainly possess and express mental abilities that are symbolic in scope (Passingham 1981; Savage-Rumbaugh 1986; McGrew 1993). It does seem relevant that in modern humans concentration on a complex technical process requires suppression of language (stop talking . . . I can't think). This seems to indicate a bottleneck in the brain processes. The suppression occurs because otherwise there is a loss of concentration, and that may be because two competing systems are attempting to use the same neural circuitry. This would concur with Deacon's observation that language uses much pre-existing brain circuitry (Deacon 1992). This argument takes us back to the work of Holloway (1969) who postulated that increase in brain size in *Homo* had much to do with achieving redundancy, in the interests of improving concentration. There may in fact be two great demands: (1) for concentration, or the filtering of the extraneous from the vast inputs which enter the senses; (2) parallel processing, which allows many different task requirements to be monitored and sequenced.

Case studies

Navigation in space and time

Human beings have a more extended sense of orientation in time than other creatures, and look further ahead. A basic assumption can be made that the transport of artefacts across a landscape will tell us something both about territoriality (Isaac 1981), and about foresight and planning depth. Although these points seem to have been accepted by Binford for the later Palaeolithic, he appears to reject Lower Palaeolithic evidence (Binford 1989).

There is no doubt, however, that long-distance transport of artefact material occurs on early sites. Hay (1976) points out that almost every large artefact concentration includes material that has been transported a long distance — over 10 km. Much more extensive documentation can now be provided. It may be enough to cite here the lower Palaeolithic site of Gadeb, in Ethiopia, where several bifaces appeared to have been transported more than 100 km (Clark 1980). Similarly at Kilombe (a site discussed below) two obsidian bifaces appear among many hundreds made from local lavas, and the implication is of long-distance transport. Then on the later Lower Palaeolithic site of Arago in France about one third of artefacts were of exotic materials, often transported 30 km (Wilson 1988). Later instances of such transport in the Middle Palaeolithic have been summarized by Otte (1990).

The interpretation of such cases is not altogether straightforward. How do we distinguish between tool-in-hand cases, and systematic transport; or exchange systems; or non-purposive tool migration, as suggested by McGrew (1992; 1993)?; an idea to which Sept (1992) gives some support.

Modelling through computer simulations has the potential to provide a realistic view (e.g. Steele 1994); equally, the existing archaeological record should be explored. If the hypothesis of 'incremental artefact migration' is to gain support, then we would expect sites in a given area to contain mixtures of artefacts according to their distances from prime sources. In fact, Bed II at Olduvai provides an example of two neighbouring Acheulean sites each provisioned almost exclusively from one source — lava at EF-HR, and quartz at DK-EE (Hay 1976; Leakey 1971). On some Bed I sites the early hominids have demonstrated a preference for flakes of quartz and quartzite, but for core tools of lava, indicating not only choice of the most appropriate materials, but transport of selected elements. The lava core tools are found without the lava flakes generated in knapping them, implying that the cores had a use in their own right. Although we have only a limited number of early sites, their raw material evidence points consistently in the same direction. Any incremental movement model would produce a landscape littered with artefacts from several sources, and sites ought to yield them in proportion to distance, but the picture is far more complicated. Archaeologists working on the early sites have consistently concluded that the makers knew the properties of stones and sought them out (e.g. Leakey 1971; Isaac 1977; Schick & Toth 1993).

Stone tool making — analysis of a skill

Our comparative framework for the study of stone

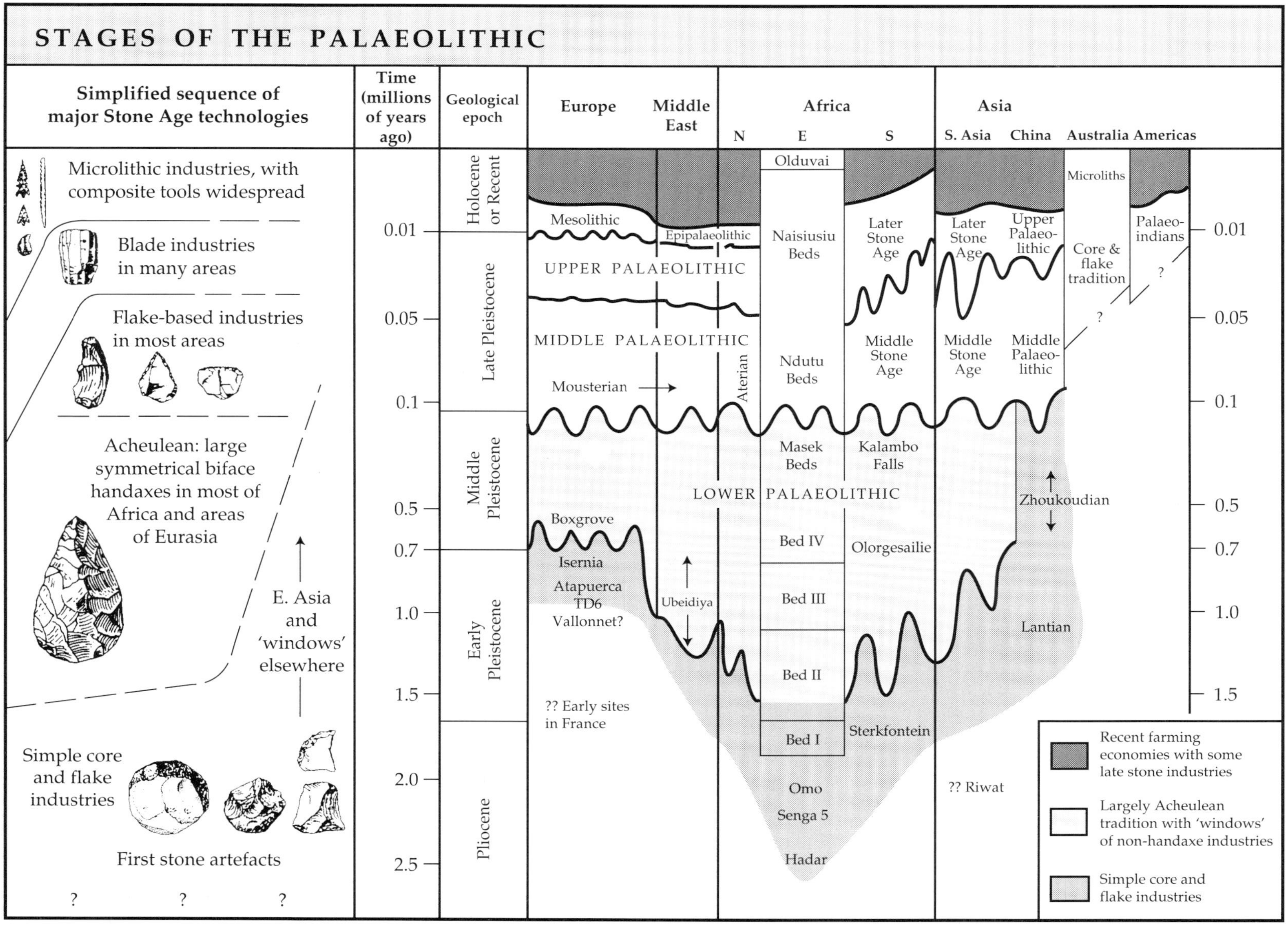

Figure 16.1. *A time chart showing the major archaeological phases of the Pleistocene.*

tool making has improved greatly over the past few years, partly because of experimental studies using human subjects, partly because the use of stone objects by chimpanzees, both in the wild and in captivity, has now been firmly documented (Boesch 1993; McGrew 1993; Toth *et al.* 1993; Schick & Toth 1993 all provide bibliographies). Ironically, tools have traditionally been used to accentuate human distinctiveness ('man the toolmaker'), in the face of arguments by others who regarded them as irrelevant — but their importance is now less contested precisely because other primates do use them. Early artefacts provide information in the following ways at least:

1. as indicators of function and economic practices;
2. as 'visiting cards' — indicators of behaviour in a geographic frame (which also has temporal implications);
3. as indices of problem-solving and the imposition of design form;
4. as indicators of style considerations;
5. as indicators of symbolic activity, but only in certain circumstances.

Of these the first two are long established; (3) is the subject of much new work; (4) and (5) are the subject of highly interesting discussions, which show the difficulties of disentangling style and function (Sackett 1982; 1986), and of demonstrating the existence of symbolic intention in early times (Chase & Dibble 1987; Chase 1991; Duff *et al.* 1992).

In terms of artefact complexity — closely linked with design form — it is still useful to make a basic division of the record into the early Oldowan artefacts (*c.* 2.5–1.6 Ma); those of the long-lived Acheulean tradition (*c.* 1.5–0.15 Ma); then those of later periods.

The artefacts of all these periods are linked by the most basic elements of process, however much they may differ in elaboration. The essence of the process is that flakes are struck in sequence from a core — often a natural cobble in the earliest industries. The maker must be able to maintain a correct edge angle on the core, strike with the required force at the right point, and judge where to strike so as to maintain the viability of the core. If there is an additional purpose, that of shaping a core tool, the maker will need to draw on a learned *instruction set* — otherwise no set form can emerge. As I have remarked earlier, single cases may tell us little about design norms, and this is why good samples of material are needed from known excavated contexts.

Two aspects tend to emerge from analyses of early stone artefacts. I labelled these as a 'mental template of form' and 'procedural template' (Gowlett 1984), but have since attempted to use a more neutral vocabulary employing the term 'instruction set'. Nevertheless, the distinctions of procedure and form appear time and again in different schools of thought. In using the *operational chains* of Leroi-Gourhan, authors distinguish them as *'savoir-faire'* and *'connaissances'*. It is possible that the language conceals some more substantial differences. Although I have adopted Leroi-Gourhan's term 'operational chain' my own inspiration comes from the routine-based analyses of cognitive psychology (for a modern example see Connolly & Dalgleish 1989). Such work, perhaps stemming from that of Miller *et al.* (1960), has always maintained a distinction between procedural and declarative aspects. This has been reiterated in recent work on cognitive psychology (e.g. Hoc 1988).

A question now worth addressing is whether this distinction stems from modern habits of human thinking, or whether it actually represents a more fundamental separation in the brain. Logically the two aspects should be distinct in artefact manufacture and use. The procedural aspect is necessary for manufacture — a controlling routine, which involves regular tests to determine whether the manufacture is proceeding according to plan. Once the tool is made, however, the procedural routine becomes irrelevant: the tool is ready for use in its finished form, and is recognized by that finished form (i.e. when it is to be used again after being put down). Some authors have argued that the concept of a finished tool is a fallacy, maintained by western science inured in the ways of mass production (Dibble 1989; Davidson 1991; Davidson & Noble 1993). But the concept of using an unfinished tool seems a greater fallacy. Ethnographic records show that tool-makers are often extremely choosy about which tools they will or will not use, emphasizing that a tool must be 'correct' (i.e. finished). Equally, we resharpen a pencil to use it again when it is blunt, but this does not mean that the pencil was unfinished the first time that it came into use. One must concede that a stone tool, like a pencil, may have provision for further working — the kind of trimming and re-use envisaged by Dibble. Whether such provision for re-trimming is itself intentional is another question.

It seems inevitable that the maker must have some mental recourse to the 'finished tool', or he could not make it. He could not even assess whether it was finished or satisfactory without a reference for comparison. This possession of an internal referent probably presents one of the strongest cases for our saying that there is a symbolic element in early artefacts; but it would seem that we have to be careful about the definition of symbol which we use, and to distinguish between an internal and an externally projected symbolism. I hope to return to this subject elsewhere.

Two critiques can be applied to the early artefact material, which could be held to diminish its importance:

(1) *apes can make stone tools.* Wynn & McGrew (1989) have argued that this places ape and early hominid abilities on a similar plane. We would then question whether very early stone tools could tell us much about hominization; they might be so much within the scope of the apes that they are fully in the overlap zone. This topic has been excellently explored recently by Toth *et al.* (1993). It had long been known that apes used stones in the wild, and could be induced to use them in captivity, but there had been no examples of extensive flaking. The bonobo Kanzi has altered this, but 'his skill in flaking stone still contrasts sharply to that of Oldowan hominids as manifested in the archaeological record' (Toth *et al.* 1993, 89). Those who require a Rubicon between ape and hominid abilities may be satisfied by this, but presumably there was an earlier moment when hominid stone tool flaking abilities were more similar to those of the chimpanzee. What seems plain is that from about 2.5 Ma early hominids possessed cultural abilities of a technical kind which surpass those of modern apes — otherwise we would not have archaeological sites. The chances of chimpanzees accumulating equivalent sites are, as Wrangham (1992) has put it, 'vanishingly small'.

(2) *there is a lack of formal design content.* The cultural content of the Oldowan industries can be debated, and there is no real consensus to the extent of design form. Mary Leakey's (1971) typology implicitly saw all core tools as designed end products, thus separately named, and there was no category of cores as such. One could argue that the scheme was intended purely descriptively, without reference to function. Nevertheless, Toth's studies have delivered a new balance, through emphasizing the importance of flakes and flake tools at a functional level (Toth 1985). Thus many 'core tools' may have been used primarily to make flakes, without design content. In contrast, Roche (1980) has seen statistical evidence of patterns in the choppers, e.g. from FLK, at Olduvai, which suggest that they do have an element of design. Toth has argued that such form as there is stems largely from the maker's need to continue flaking in a path of least resistance. He suggests that edge angle may be one criterion for distinguishing cores from core tools and that in the latter function requires that it be less than 80 degrees. Nevertheless, he and Schick argue for the use of core tools, and also for a high level of intelligence in the tool-maker (Schick & Toth 1993). A 'chopper' from Chesowanja provides a rare case of direct support for Toth's interpretations: it retains a delicate feather of stone on its edge, about 2 mm thick. This accidental and incidental product of flaking proves that this particular specimen was never used as a tool (Gowlett *et al.* 1981, fig. 3a).

From these differing viewpoints one can extract perhaps a minimum view that even in the earliest industries there is an instruction set, a procedural template, behind core-making or shaping a core tool, regardless of whether it is the flakes or the core which are about to be used. The instruction set includes concepts of form, at least so that the maker will not be 'boxed in', losing the opportunity to strike further flakes. However one interprets the intent, the artefacts allow a mapping of hominid activities through space and time that is not possible for other species.

I revise here my own diagram of process in the Oldowan (Gowlett 1984), published before Toth's paper of 1985, partly to take into account some of his and Schick's (1991) points, and partly because my original version seems graphically to take too firm a view of 'types'. I was taught from the beginning to regard these as 'classes' — artefacts of classification, rather than 'types' known to the makers (this was one of the lessons drummed into all his students by Charles McBurney), but since we still use names with apparent functional connotations this is constantly misleading to those who are new to archaeological terminology. The new version of the diagram attempts to portray a form field (Fig. 16.2). The degree to which these classes may intergrade without having separate modalities remains largely untested, although it seems likely that at Olduvai DK intermediate forms between chopper and discoids are rare (Gowlett 1990; *cf.* Roche 1980). Toth and Schick have also demonstrated the extent to which the Oldowan, like the later Acheulean, involves preworking on a quarry site, and then the introduction of select blanks only to the use site.

Most authors feel that there is an increase in design form when the Acheulean succeeds the Oldowan about 1.5 million years ago (e.g. Isaac 1969; 1986; Gowlett 1986; Toth & Schick 1993; Wynn 1993). The main new element is the classic Acheulean biface, often made on a blank consisting of a large flake. The record is not precise enough to show us whether the idea of the bilaterally symmetrical biface emerged before or after the idea of striking such large flakes.

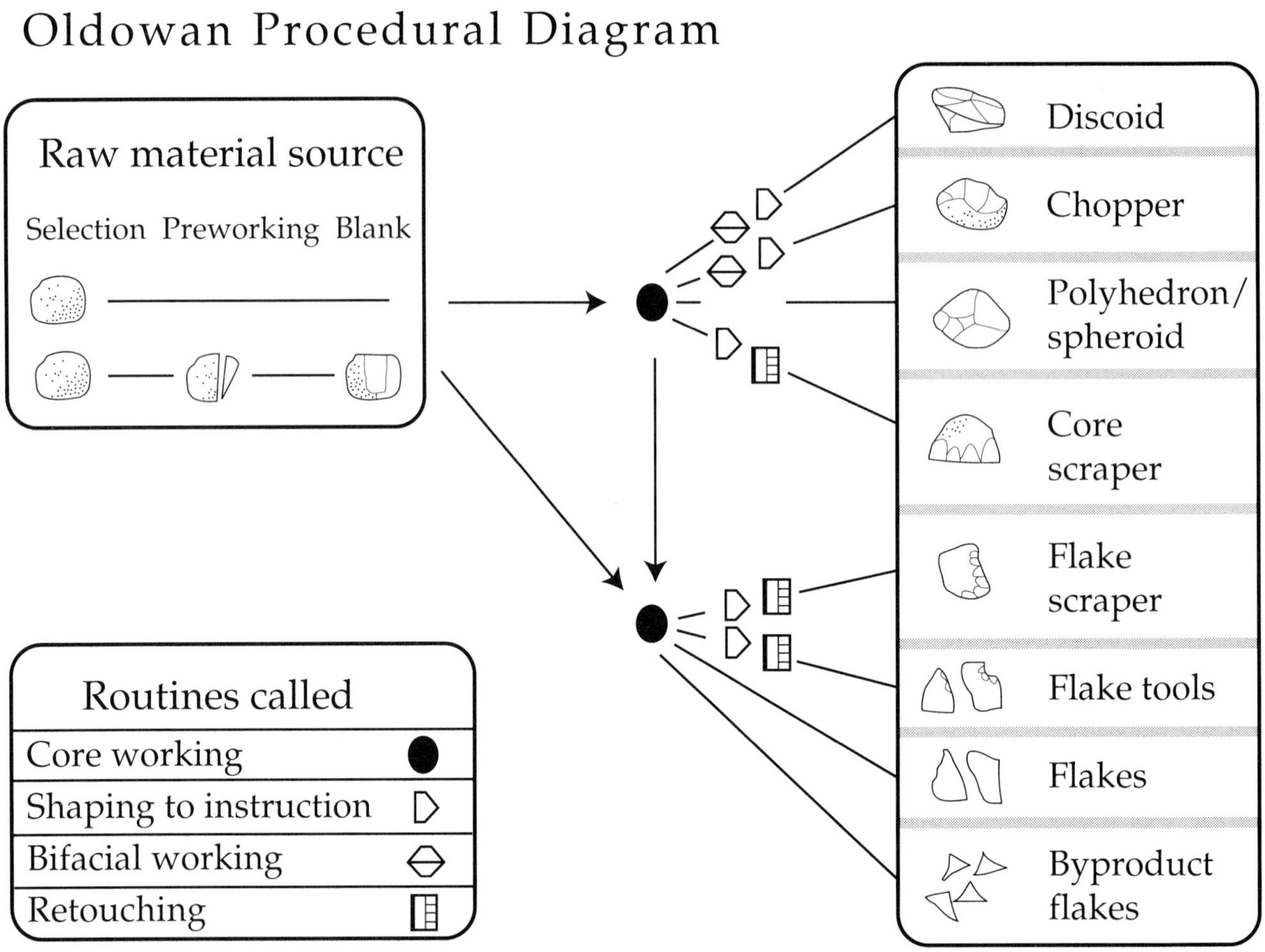

Figure 16.2. *A procedural diagram for the Oldowan. This outlines the main processes involved in the manufacture of early industries between about 2.0–1.5 Ma* (cf. *Gowlett 1984 for an earlier version).*

Throughout the Acheulean large cobbles used as blanks provided an alternative route to a similar end-product. This is an important feature in my theme of 'constraint or choice'. On occasion, at least, bifaces were also made of bone (Leakey 1971; Mania 1995), demonstrating that the makers were not constrained by *material* to produce a certain form.

It seems inescapable the biface entails a more complex procedural template for the Acheulean than for the Oldowan (Gowlett 1984). The existence of alternative routines for biface manufacture (Fig. 16.3) adds to the overall complexity: with certain exceptions (Belfer-Cohen & Goren-Inbar 1994) we have scarcely started documenting *differences* of routine in different Acheulean contexts.

Previously I have suggested that the discoid may be a pointer towards the emergence of Acheulean (Gowlett 1986) an idea weighed by Toth (1990) in his discussion of symmetry in early artefacts. I felt that the presence of all-round bifacial working on Oldowan discoids was significant, because technically a classic biface could be regarded as little more than an elongated discoid (Gowlett 1984; 1986). On consideration, however, I feel one should emphasize that the discoid 'establishes' some of the technical routines involved in a biface, but that it may have no functional overlap with bifaces at all. It is also evident that the biface represents more than the addition of a long axis: several other form-defining rules appear to be present (Crompton & Gowlett 1993).

Thomas Wynn (1985) has argued for a less strong contrast between the Oldowan and early Acheulean, than between the early and later Acheulean. The question is partly bedevilled by 'noise' in that some early assemblages on good raw materials are better made than others which are far later. It also seems likely that some specimens, such as the lava series from Kariandusi are more asymmetrical

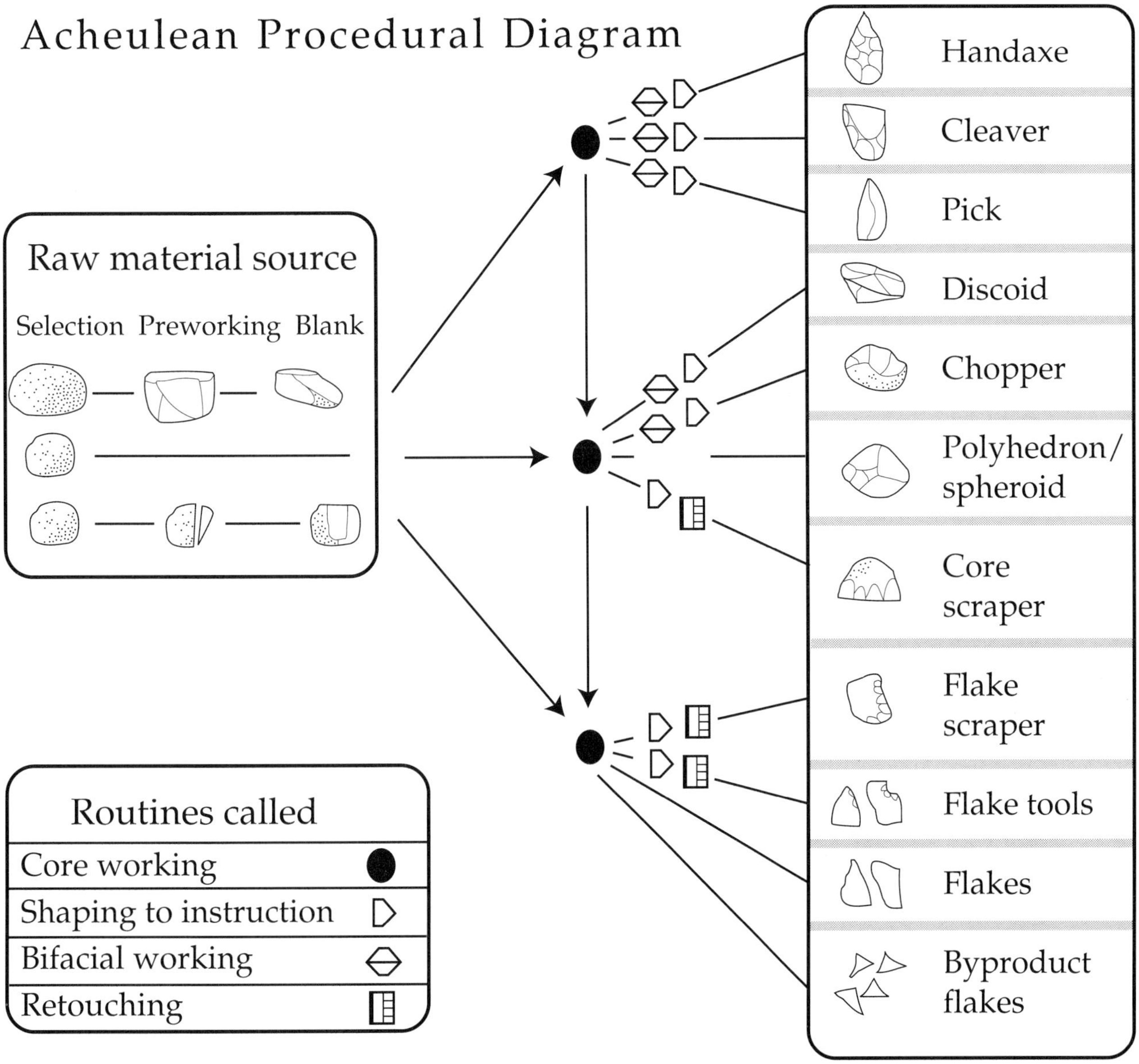

Figure 16.3. *A procedural diagram for the Acheulean. This applies particularly to the 'classic' African sites in the time range* c. *1.4 to 0.3 Ma, in which large cobbles were used as cores to provide biface blanks. Procedures in other industries may have been somewhat different* (cf. *Gowlett 1984 for an earlier version).*

than other contemporary series such as Kilombe (Gowlett & Crompton 1994). It is hard to see functional needs being entirely responsible for such marked asymmetries in particular assemblages: there may well also be a stylistic element (*cf.* Sackett 1982; Isaac 1977). If we can take into account enough representative assemblages, it should be possible to construct tests of Wynn's hypothesis objectively, and to determine whether there is any overall increase in symmetricality in Acheulean assemblages through time.

Other skills: fire control and animal butchery

Our understanding of early routines would be much improved if we could incorporate documentation of skills other than stone-working. Perhaps the most disappointing aspect of the last ten years is that little progress has been made in the understanding of fire-use, although good summaries have been given (Clark & Harris 1985; James 1989). Routines of butchery have been well-studied in ethnographic examples, and compared with natural sequences of carcass disintegration (e.g. Hill & Behrensmeyer 1985). It is not easy, however, to project this work into the early hominid past. Cutmarks on bone are a potentially poor diagnostic, in that hominids achieving early access into a rich carcass probably would produce very few marks on bone. Equally, carnivore chewing of bones will often yield very few signs of chewing (Brain 1981). Of course both cutmarks and toothmarks have been found on early bone assemblages, but the diversity of opinions expressed suggests that the firm causal relationships sought by Binford (1981) have not yet been achieved (for a review, see McBrearty 1989). More progress has been made in our knowledge of wood and its use, through discoveries of this material on several sites (Bosinski 1992; Belitzky *et al.* 1991), sometimes with signs of human intervention. Thus a documentation of routines is gradually being assembled.

Foundations for the origins of mathematics and art

Human beings are generally accepted to have two sets of symbolic abilities: those of language and of number. Logically these two can be seen to interconnect, but it seems likely that the brain tends to handle them separately. Mathematics does of course represent far more than number, one important area being that of functions. Functions provide a link with art, in that both involve transformations. It would seem, for example, that we have no way of defining a curve except through a function which is often represented graphically in terms of two (or more) orthogonal axes. In looking at human operation of such 'knowledge' we are considering some of the brain's most advanced abilities. It is unlikely that these would have simple roots. It is vital to distinguish between the developed disciplines which we now see, which have a maximum time depth of about 5000 years, and the underlying abilities on which they are based. These may depend on preconditions and precursors which have a much longer history. I have tried to emphasize this distinction in my own work, but have still found it subject to misunderstanding (e.g. Davidson 1991).

Clarification may be needed about the kinds of precursor abilities which we are looking for. Language — and especially its origins — is notoriously difficult to trace, and the same might be said about numerical abilities. On the other hand, transforming abilities — which are only one facet of the later disciplines — have a good chance of leaving a trace in the archaeological record.

I will try to sketch out a general theory of the assumptions. First, a fundamental characteristic of cultural behaviour appears to be the imposition of arbitrary form (Geertz 1964; Holloway 1969). Arguably this form must always be drawn from the environment in some way, whether learnt culturally from another individual, or perceived and analyzed *ab initio*. It is analyzed and then reimposed, in similar or perhaps different form. We can be sure the manufacture of any object involves great numbers of transformations and other calculations by the brain. Evolution seems to be very good at using existing machinery for other purposes when making new departures (Deacon 1992 points this out in the case of language). In the case of imposing form on material, it would seem that perception, particularly of moving objects that change scale, places demands on the brain somewhat similar to those of tool-making. Seeing a particular object, near or far, a creature needs to be able to interpret it as one and the same.

In a sense then, transformations employed in manufacturing objects are an externally imposed projection of perceptive abilities common to many animals. To analyze the outside world any creature must be able to analyze its properties. To make a shaped artefact entails reassembling such analyzed characteristics in the outside world. It also, however, requires their formalization in an instruction set which must involve rules or grammar (*cf.* Chippindale 1986; Hassan 1988; Wynn 1993).

Bifaces of the long-lasting Acheulean tradition provide some of our best documentation of such activities by early *Homo*. Some Acheulean sites preserve large numbers of bifaces which can be studied as sets. Examples are Ubeidiya in Israel (Bar-Yosef & Goren-Inbar 1992), Olorgesailie (Isaac 1977), Kariandusi and Kilombe in Kenya (Gowlett 1991; Crompton & Gowlett 1993; Gowlett & Crompton 1994; Wynn & Tierson 1990), or Warren Hill in England (Roe 1981).

Form and rules at Kilombe can be used as an example of early human abilities expressed in stone. As the next section draws extensively on the site evidence, it may be useful to recapitulate that the site is in the central Rift Valley of Kenya, probably 0.7–1.0 Ma old, and that as with the rest of the Acheulean, we have little idea of the actual use of the large numbers of bifaces (Gowlett 1991).

My earlier work examined the geometry of artefact sets, especially the handaxes or bifaces (Gowlett 1982; 1984). I found remarkably high correlations between dimensions and observed that lines fitted to the data, when extended, passed through the origins of the two axes, or close to this point. As the bifaces were made in a range of lengths from about 8 cm to 25 cm, this provides a good example of transformation. I used it to argue that the makers were able to use a geometrically accurate sense of proportion (Fig. 16.4).

Critics have argued that this is merely a result of functional needs, rather than the manufacture of finished artefacts to a preset plan (Dibble 1989; Davidson & Noble 1993). But this more deterministic view also presents problems — as noted by other observers (e.g. Ingold 1993; Hayden 1993). Its arguments fail on at least three grounds in my view; in each of these it is less parsimonious in explanation than the alternative view which admits cultural control of planning and rule systems by early *Homo*.

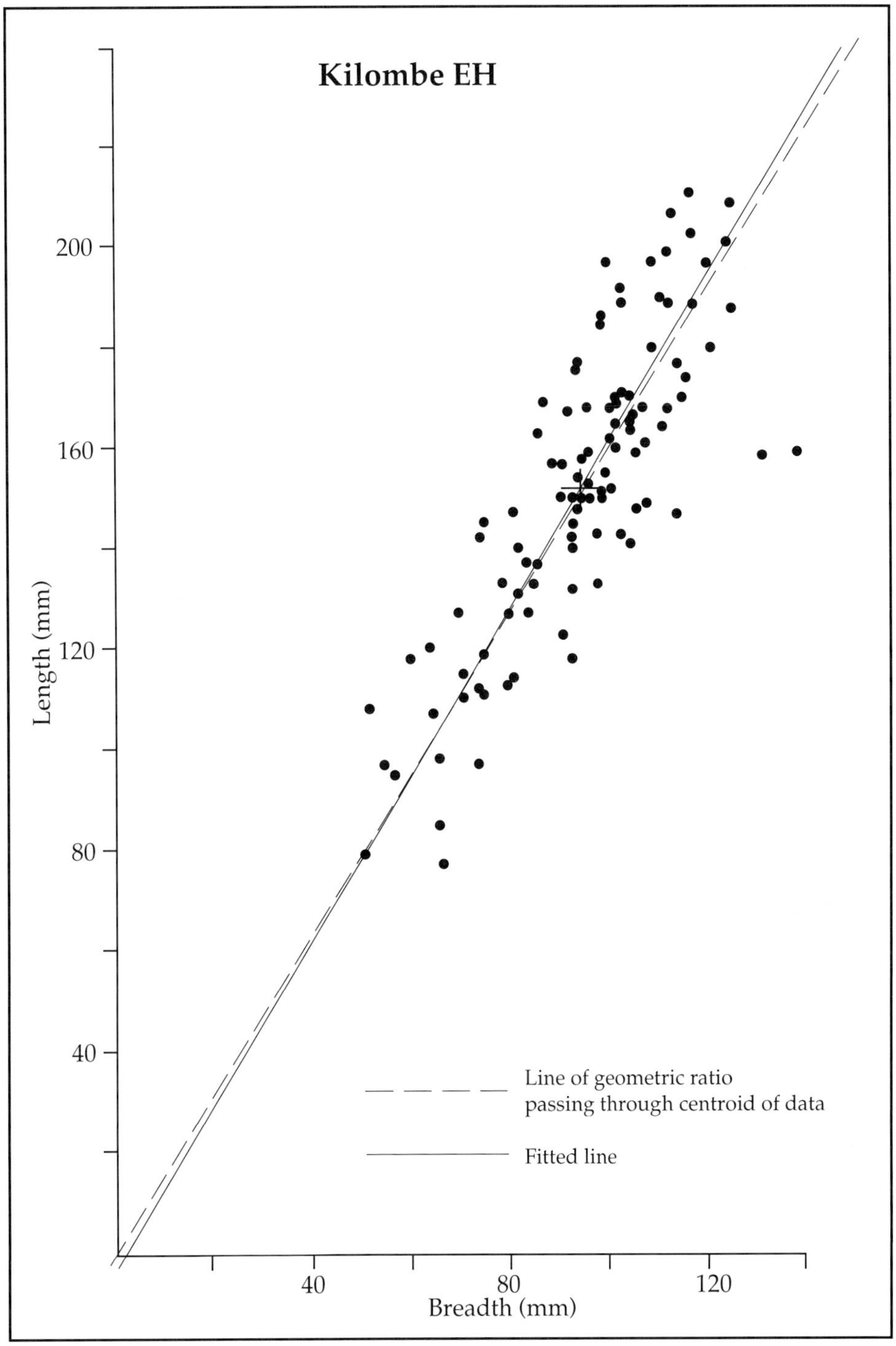

Figure 16.4. *The Breadth/Length diagram for bifaces from Kilombe EH.*

1. The deterministic view implies — at least according to some authors — that *Homo erectus* operates as an automaton, not through learned cultural behaviour; as a fundamental characteristic of the higher primates is increased plasticity of behaviour, it seems exceedingly unlikely that ancestors of the species which has the most labile behaviour of all (*Homo sapiens*), would have reverted to patterns of behaviour under direct genetic control, and then have abandoned this characteristic without trace.
2. It regards bifaces as cores made to yield flakes, rather than as artefacts planned to serve functions in their own right. This view ignores the result, that the flakes produced would largely be in the wrong size spectrum: they would not correspond with the modal size of flakes which we know early artefact makers were regularly seeking. Secondly, bifaces are a wasteful and inefficient form compared with cores geared to flake production (and such cores are after all frequently preserved). The shape of a biface is

difficult to produce, and a completely unwarranted burden on a knapper whose goal is to produce flakes for other purposes. Graves (1994) completes the argument by noting that soft-hammer technique is restricted to shaping bifaces, and not used for flake production.

3. Some authors make the implication that function imposes its need directly. That is, an artefact assumes its shape not because the maker wishes to make it that way in order to serve a function, but because the interaction with the function leaves no alternative. But the archaeological and modern records show that there are almost always alternatives (hence Sackett's (1986) use of the term isochrestic). On occasion artefacts may be shaped through application (i.e. 'bedded in' to a task) but it seems fair to say that in general they are constructed in anticipation of a function, not through the function (*cf.* Ingold 1993).

The deterministic views have posed a legitimate challenge, but they do not seem to be held by any workers who have actually studied large collections of early Acheulean material, and it would be unfortunate if a red herring diverted attention from other questions of research interest. The Acheulean is not as constant as alleged. There is in reality much variation in biface sets (*cf.* Roe 1964; 1968; Isaac 1977; 1984; Wynn & Tierson 1990; Crompton & Gowlett 1993; Gowlett & Crompton 1994). The point of challenge is to discover what accounts for that variation and what sets limits on it. We need to explore the data with the aim of understanding better both how a population of tool-makers achieves standardization, and at the same time how and why it permits variability. I use some aspects of the Kilombe data for this purpose.

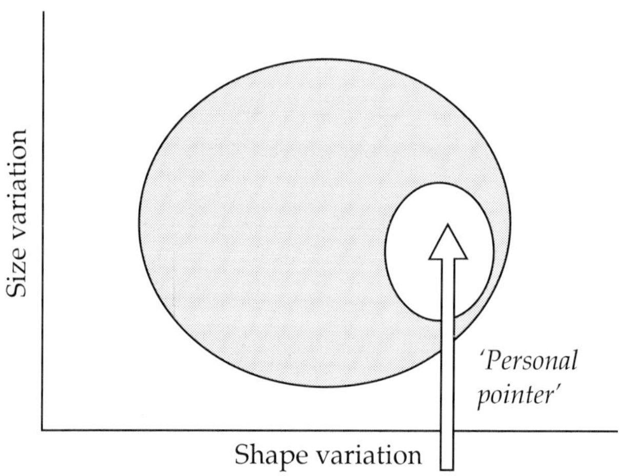

Figure 16.5. *The personal pointer: each time a biface is made it involves choices which amount to positioning a selection pointer within a field of variation. An individual might well operate within a subset of the range used by the group ('the industry').*

Kilombe offers a great advantage in that one may study the total output from one surface, rather than a few perhaps unrelated specimens (Gowlett 1991). Through observing the total output of a past human group in a given area we can make generalizations. But conversely this introduces the interesting problem that we know virtually nothing of the relationship between an *individual's* production and the *group* production — and hence very little about the causes underlying the total field of variation. We do not know, for example, whether small bifaces were made by small individuals; or whether size differences are more likely to be task-related. Nor is it easy to see whether we have subdesigns in the artefacts, as opposed to a continuous field of variation (see below).

It seems likely, however, that each individual knapper operates a subset of the total instruction set in the cultural memory of the group. When making a new biface the knapper exerts certain choices, which amount to moving a personal pointer to a position within the field of shape/size variation (Fig. 16.5). Of course there may be complicating factors, such as the tractability of the raw material, and whether the site is a factory or use site.

(1) Geometric proportion — a subject for further study

For the purposes of this exploration, I have concentrated on the variables Length and Breadth, which help to define planform of bifaces. At Kilombe correlations of length and breadth are in the range *c.* 0.80–0.95. Thus the whole assemblages allow fitting of lines that are quite consistent, but within an assemblage

Table 16.1. *Correlation coefficients for biface dimensions.*

	Length and Breadth R	Thickness on L and B Multiple R
Kilombe		
EH	0.83	0.59
AH	0.93	0.57
AC	0.77	0.56
AD	0.82	0.59
SAH	0.83	0.81
KZ	0.85	0.58
Kilombe		
All	0.83	0.56
Kariandusi		
Upper (Obsidian)	0.77	0.77
Lower (Lava)	0.59	0.29
Kapthurin	0.77	0.63
Casablanca		
STIC	0.66	0.46
Cunette D2	0.88	0.65

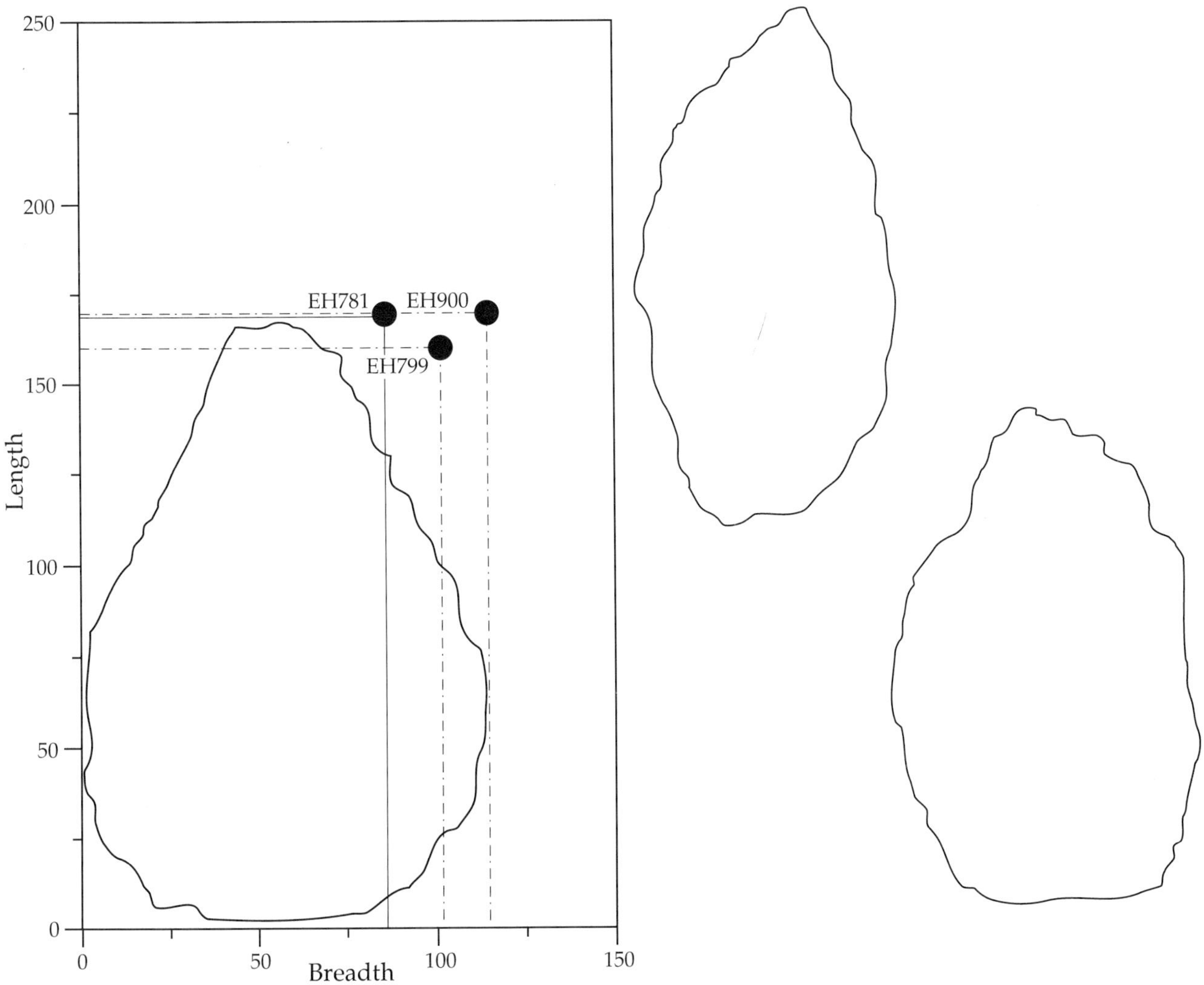

Figure 16.6. *Three bifaces selected to illustrate shape differences at Kilombe. The specimens are of similar length, and demonstrate the range of variation in breadth/length relations which was acceptable on the site.*

two bifaces may be of quite obviously different proportions (Fig. 16.6): and such variation can be said to be regularly acceptable to the makers.

Nevertheless, all Kilombe assemblages consistently have a mean ratio of *c.* 0.61 for breadth/length, suggesting a group norm which did not vary: although the total site production may represent hundreds of years.

Other sites present different ratios of breadth/length. One may also note that only some Acheulean biface assemblages present very high correlations of the dimensions (Table 16.1). I would take this to confirm that the artefacts are not universally 'programmed', either in the makers' minds or by functional constraint. Less well-made assemblages seem to present lower correlations of dimensions (e.g. Kariandusi lava set, or STIC at Casablanca). It may be that sites close to raw material will tend to have lower correlations, compared with assemblages which have been transported far afield, in the sense that only satisfactory blanks or finished specimens will justify the effort of being carried so far.

(2) Allometric adjustment

When archaeologists utilize a ratio, as they often do, they are in effect assuming a geometric relationship between the two variables. Yet the existence of such a relationship in early material could not be taken for granted. Exploration of it has given us a good deal of information about early transforming abilities, as I have tried to show. But one can go further, and begin to examine residuals — in this case departures from geometric proportion.

A good example is provided by allometric

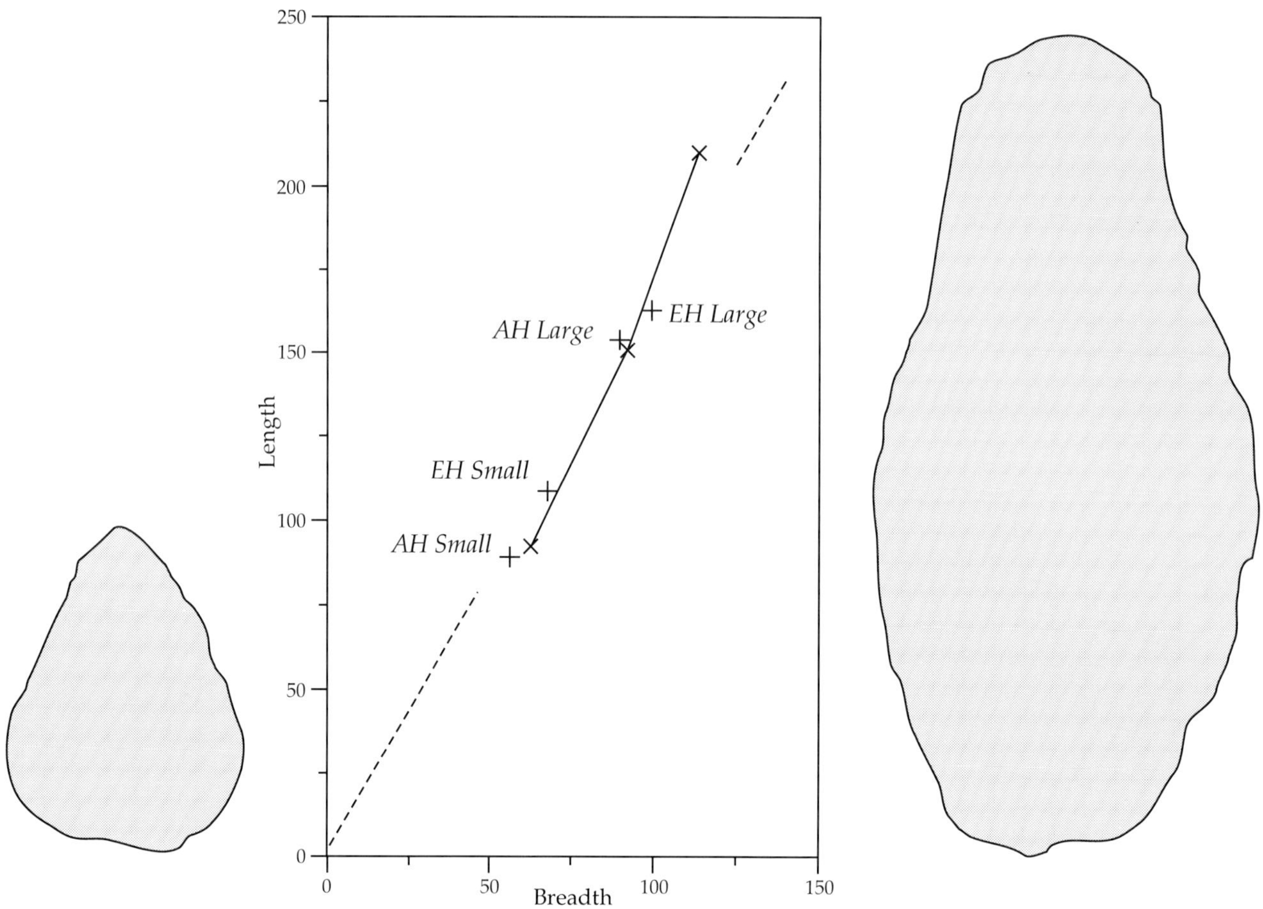

Figure 16.7. *Allometry in breadth/length relations at Kilombe. The diagram shows the adjustment to breadth/length relations made for different sizes of biface. Departures from a geometrically constant ratio of breadth/length are most apparent in very small and very large specimens, typical examples of which are illustrated. (For details of groups see Crompton & Gowlett 1993.)*

(size-related) adjustments, which can be seen to affect the breadth/length relationship, as well as relationships between other variables (Crompton & Gowlett 1993). The effect of the adjustment is that very long bifaces are made somewhat narrow, and very short ones somewhat broad. Those in between depart very little from geometric proportions (Fig. 16.7).

Crompton & Gowlett (1993) found that the butt zone is conservative both in shape and size, and considered that this may be linked with the need for hand-hold, i.e. that hand size does not vary as much as biface size. This work approached the problem of allometry from a multivariate analysis, but it need not imply that the early hominids themselves could handle several variables at the same time — something which modern humans find difficult. A simpler explanation would be that they had *ad hoc* rules involving assessment of up to three variables. Thus the relatively fixed relationship between length and breadth might be overridden if weight was tending to become excessive. Weight rises as the cube of linear dimension, so a biface 20 cm long would be eight times the weight of a biface 10 cm long, if geometric proportions were observed exactly. In practice at Kilombe the larger biface is consistently made only five to six times heavier, implying a very significant adjustment.

(3) Filters

In addition to such adjustments it can be observed that there are also filters or cut-offs limiting potential variation: for example, a biface cannot be more than so small, or so large. This can also be illustrated from length and breadth measurements. Plots of flakes (as opposed to bifaces) in Acheulean industries exhibit a fan-shaped distribution (Fig. 16.8). This is the direct product of knapping, without subsequent selection. A similar range of shapes is produced

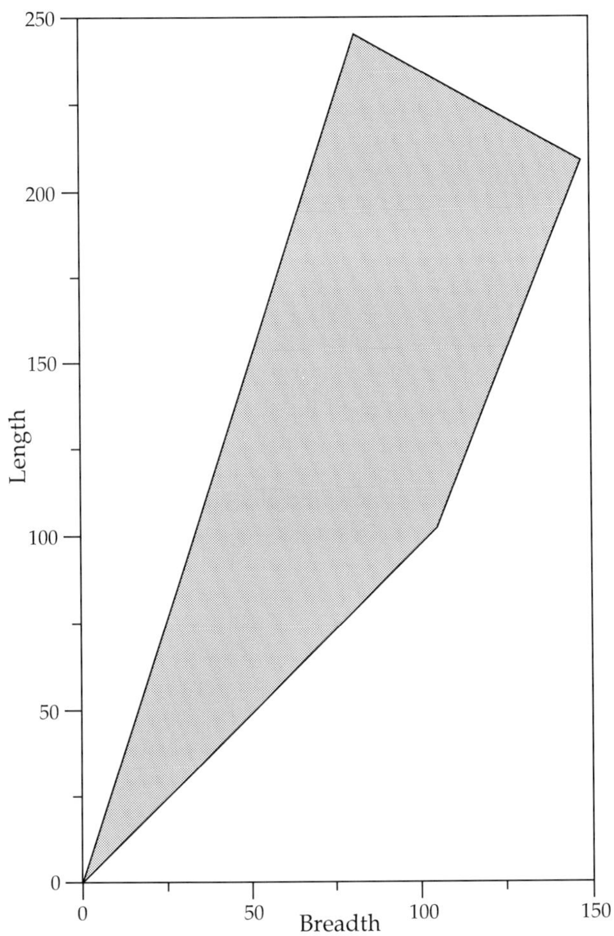

Figure 16.8. *The fan-shaped distribution: small flakes and long flakes from Kilombe and other sites were made in approximately the same range of breadth/length relations. Hence, the plot shows a fan-shaped distribution. One corner of this becomes cut off, because very large broad flakes were not struck.*

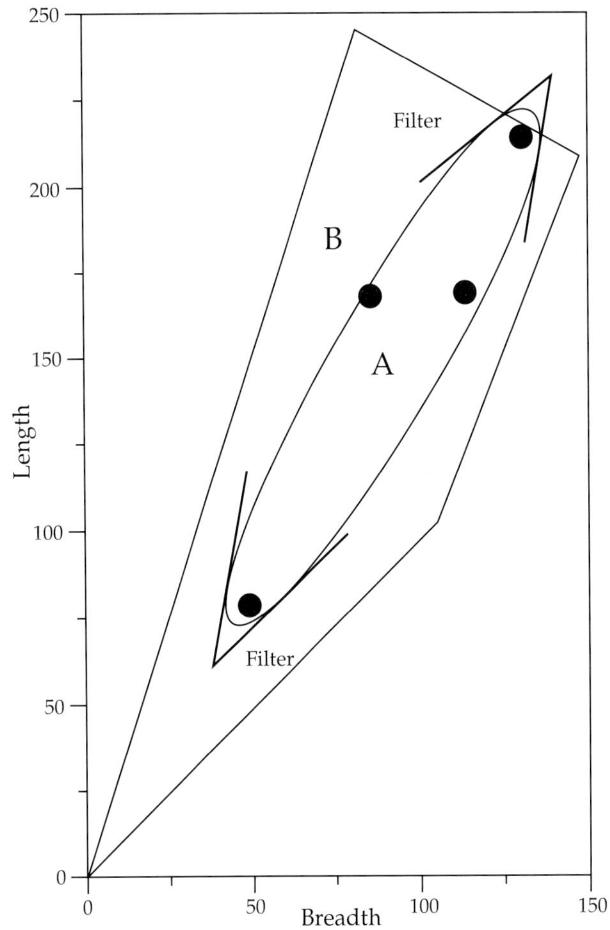

Figure 16.9. *The ovoid distribution and filters (contrast with Fig. 16.8). The variation for breadth/length is not as great for bifaces as for flakes. Biface blanks were subject to a stronger selective process, with filters eliminating the fan of variation especially in long specimens. (A) The ovoid distribution acceptable for bifaces. The data points mark longest and shortest bifaces, and two examples shown in Figure 16.6. (B) The area of flakes apparently excluded from the possibility of being made into bifaces.*

at every size, hence the fan distribution. In comparison the points for a biface distribution show an ovoid (Fig. 16.9). Not only is the breadth/length variation more restricted than in flakes, but it becomes more restricted as the size of the biface increases. There is thus a 'maximum size filter' operating like a kind of fork imposed on the scatter.

(4) Cleavers — subtypes

The number of separate designs utilized by a technology is a matter of interest. In early industries these are relatively few, but it is not easy to measure objectively whether there are also subdesigns in the artefacts, as opposed to a continuous field of variation without multiple modes. Isaac (1977) posed this question in relation to handaxes and cleavers at Olorgesailie. The latter are recognizable through having an axe-like edge, rather than the point of a handaxe, but they may intergrade into handaxes through 'chisel-ended' specimens. Not only is it useful to know how many design modes are present, but if we ignore them and lump together different design forms, we will get a blurred picture of underlying rules. I tried to tackle this problem at Kilombe by seeking to distinguish handaxes and cleavers through using Wishart's Mode analysis (Gowlett 1988). This multivariate analysis attempted to find 'natural' clusters, with suggestive but tantalizing results. Handaxes and cleavers did not appear to represent separate modes, but large and small bifaces did.

Here Figure 16.10 shows Breadth/Length relations from cleavers at Kilombe EH, in relation to all bifaces from the same locality. Cleavers tend to be larger, and tend to be broader. This distinction is not statistically significant for EH alone, but is for larger samples of specimens from the whole site (e.g. 209 handaxes/44 cleavers; the mean length for the latter 14 mm greater, significant at 1 per cent level). Nevertheless, a fitted line has virtually the same slope for cleavers and for handaxes (Fig. 16.10). Possibly the extra width of the cleavers is a feature of the tip only, and a study of the butt and mid-portions would show a greater identity between handaxes and cleavers. The distinction between the two categories in 'classic' form is not really at issue, but for the moment it has not been demonstrated that they represent separate modes. If there were such distinct modes, it would suggest (but not prove) clearly different functions, but perhaps more important it would imply separate instruction sets conceived and manipulated separately by the makers. Arguably, the simple presence/absence of a tranchet edge already documents this point.

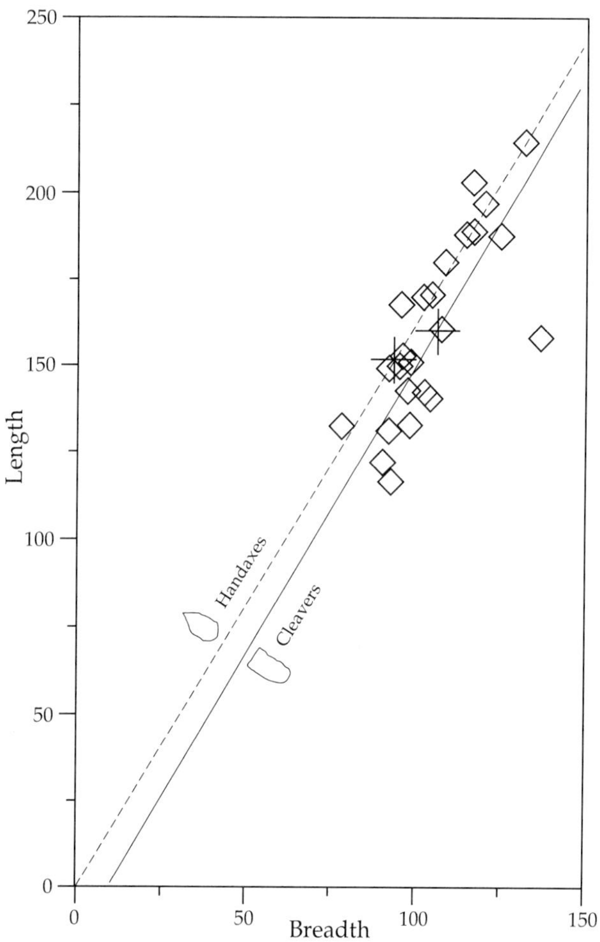

Figure 16.10. *Cleavers and handaxes: Kilombe EH. Cleavers may represent a different design type, and tend to be slightly larger than handaxes, but they appear to have been made with the same considerations of breadth/length relations.*

(5) Trimming characteristics

Some bifaces are better-made than others. They may have been shaped through the removal of more flakes, a feature which can be counted (Isaac 1977). Such flake removals can have a direct effect on plan form, providing a further opportunity to investigate length–breadth relationships. Figure 16.11 shows a plot of lightly trimmed and heavily trimmed specimens from Kilombe area EH. The lightly trimmed specimens retained cortex (original cobble surface) over at least 50 per cent of the dorsal side. The 'well-finished' specimens had at least 16 major flake scars, and only retained small areas of cortex. The two sets (extremes of the range) plainly do overlap, but as a group the heavily trimmed specimens are relatively narrower.

It seems likely that the less-trimmed specimens more closely reflect the proportions of the original blanks. From this we may conclude that in carrying out secondary working the knapper tried to maintain the length of the piece, while reducing breadth. Here at least, a simple *ad hoc* rule may have been isolated.

Table 16.2. *Correlation matrix of variables measured in Kilombe bifaces.*

	T	L	B	BM	BA	BB	PMB1	PMB2	TM	TA	TB
T	1.00										
L	0.57	1.00									
B	0.52	**0.84**	1.00								
BM	0.50	**0.82**	**0.96**	1.00							
BA	0.34	0.57	**0.72**	**0.76**	1.00						
BB	0.46	**0.74**	**0.84**	**0.79**	0.48	1.00					
PMB1	0.36	0.64	0.64	0.68	**0.75**	0.30	1.00				
PMB2	0.36	0.66	0.62	0.65	**0.75**	0.28	**0.87**	1.00			
TM	**0.94**	0.52	0.49	0.49	0.35	0.46	0.35	0.33	1.00		
TA	0.62	0.45	0.44	0.44	0.38	0.33	0.34	0.37	0.67	1.00	
TB	**0.84**	0.31	0.43	0.31	0.09	0.39	0.23	0.14	**0.77**	0.33	1.00

Most Acheulean biface assemblages show a similar set of correlations between measured variables. Thus their significance is not in doubt in a statistical sense. But would the relationships be strong enough to be (a) noticeable to and (b) controllable by the manufacturer? Correlations over 0.7 are marked out as possibly indicating pairs of variables where the relationship received special attention.

(6) Variables and rules

How many rules are there in the instruction sets governing biface production? A principal rationale for multivariate study is that all the measured variables can be taken into account — though analyses such as principal components will tend to eliminate redundant variables. Robin Crompton has made the point that no study has attempted to see how many variables are necessary to specify a biface, and how many of those measured are redundant (Crompton pers. comm.; Crompton & Gowlett 1993). Length and Breadth, for example, are highly correlated, but clearly measure different things. On the other hand, some breadth measurements may be closely predictable from others, and may contribute little information. Table 16.2 shows a correlation matrix for bifaces from Kilombe. Length (L) and Breadth (B) are highly correlated by the maker's intent, whereas Breadth (B) and Breadth at the Middle (BM) are virtually the same measure in this group of bifaces, and so inevitably correlated. Thus some of our measures are isolating real axes of variation, but others may not.

As interesting as the high correlations are the low ones which exist between some variables. If bifaces were identical in shape, and varied only in size, then all the measured variables would have perfect positive correlations (1.00). Bifaces are roughly similar in shape, but the correlations vary from less than 0.10 to about 0.95. We have to square these coefficients to see the proportion of variation in one variable which can be linked with (or at least co-varies with) another. A correlation of 0.70 corresponds roughly to a position where 50 per cent of change in variable 1 is reflected in change in variable 2. This should be observable by and controllable by a human observer, but lower correlations might not be noticeable.

The Kilombe data show a bimodal distribution of correlation coefficients, with peaks at 0.70–0.80 and at 0.30–0.40. There is also a 'tail' of very low correlations. It may well be that this pattern reflects deliberately standardized relationships, incidental relationships, and those of such little importance that the maker had no need to control them at all. Unfortunately I see no way to test the statistical significance of these observations in the near future, since the pattern must partly depend on the variables selected for measurement.

Overall patterns, however, make clear that more rules are needed to specify a set of bifaces than an individual specimen. The allometric adjustments are merely one example. In the Cambridge meeting Robert Foley noted the range of views about rules in early artefact material. Some workers are unwilling to accept the existence of such rules, but they can avoid them only through refusing to analyze the material. If the rules are non-existent or redundant, they will tend to melt away in multivariate analyses such as Principal Components, which reduce dimensionality. In practice, several factors 'survive' such an analysis (Crompton & Gowlett 1993).

I try here to list the rules which underlie a biface assemblage such as that at Kilombe (Table 16.3). One must grant that combinations of simple

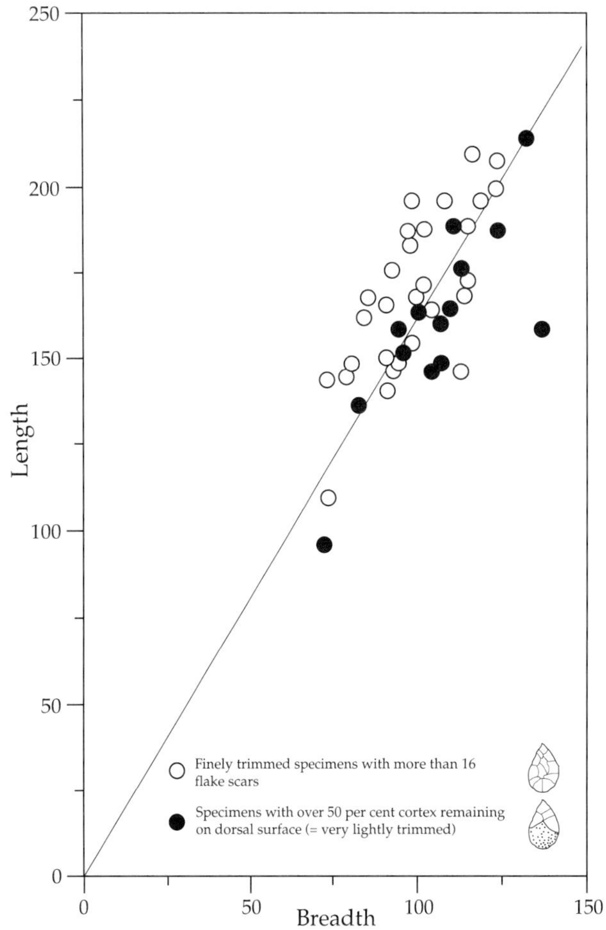

Figure 16.11. *Lightly trimmed and heavily trimmed bifaces from Kilombe EH. A distinct difference in proportion is seen between lightly trimmed blanks (those with more than 50 per cent of cobble cortex remaining on one face) and finely trimmed specimens (with a total of more than 16 flake scars of over 1 cm). The distribution suggests that in trimming a blank the makers were willing to lose about 10–15 per cent of breadth, but were more concerned to maintain length. This adjustment occurs within the constraints of the high correlation of length and breadth demonstrated across the site.*

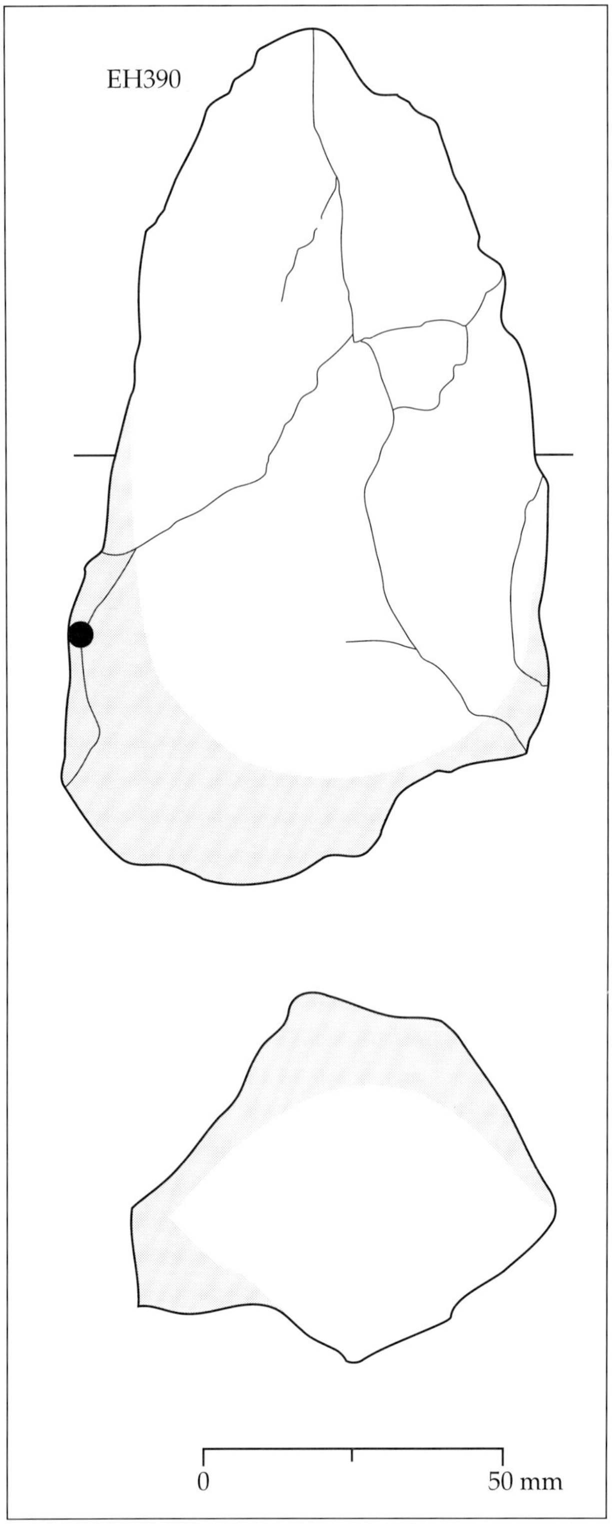

Figure 16.12. *An unfinished biface specimen from Kilombe. It was probably discarded because of a flaw or an error of secondary working. The maker probably intended to trim off the shaded areas shown in plan and mid-section. The dot indicates the highest point of the surface.*

ground rules can add up to an apparently complex picture (*cf.* Sacerdoti 1977), but it would be impossible (I believe) to convey instructions for reproducing a biface assemblage to a being on another planet if one omitted many of these rules.

Overview of the Kilombe evidence

I have tried to use Kilombe as an example of the abilities of *Homo erectus*, who made great numbers of bifaces, constrained no doubt by functional needs and raw materials, but also able to exert controls over these in a fairly systematic way. At one extreme one might imagine that the tool-makers were culturally free to do as they liked, to choose their own style; at the other that they were deterministically constrained by the mechanics of the material and functional needs, and so had no latitude of choice at all. A compromise determined by energetic rewards and penalties seems more likely.

Biface sets from different sites and areas do often have similar limits to their field of variation. Within that zone, there is still much to be said for Isaac's (1972) idea of a random walk within a zone of freeplay. Minor cultural rules (i.e. style) would determine where within the zone of freeplay any particular assemblage fell. Our best hope of further understanding may now lie in attempts to see why the *limits* to the zone of free play — in shape and size —

Table 16.3. *Procedures underlying biface manufacture (necessary elements in an instruction set).*

1. Select (or make) a suitable blank.
- length in range 8–25 cm
- breadth one-half to two-thirds of length
- thickness around one half breadth

2. Select a major plane defined by length and breadth.

3. Select a butt end and a tip end.

4. Decide width at tip end (i.e. handaxe, cleaver or intermediate).

5. Commence bifacial working, striking flakes from both sides of the major plane.

6. Through bifacial flaking create:
- a bifacial edge around the whole of a circumference defining the major plane;
- an outline that is bilaterally symmetric around the long axis in the major plane;
- cross-sections that have bilateral symmetry (i.e. lenticular sections 'skewered' by the long axis);
- a butt which is rounded in plan view;
- a thin tip-end, with point and edges resolved by flaking OR retain a broad tranchet edge.

7. If the specimen is small, make it squat and relatively pointed. If the specimen is large, make it relatively narrow and thin, and emphasize thinness of tip.

Notes: Not all bifaces have all these design features: the class is polythetic in the sense of Clarke (1968). Some bifaces have further design elements not mentioned here; some design elements (e.g. 7) would not be visible in an individual specimen, but could be inferred only from a series.

The instructions could not be followed in strict sequence: some points of outcome could be achieved only by providing for them at an early stage (e.g. the cleaver edge).

lie where they do. The reasons may be primarily mechanical: for example, that specimens thinner than those normally made would tend to break, or that those made thicker would be unmanageable.

Such an explanation leaves open the question of how the maker knew this — by inherited cultural rule, or by personal trial-and-error? Rare specimens at Kilombe give some insight. Figure 16.12 shows an unfinished specimen, where it seems that owing to an error, or flaw in the raw material, the piece could not be finished. On the other hand, returning to Figure 16.9, one can see that specimens of middling length (about 15 cm) were made in some variety of breadth, but that at 23 cm length the variety has been suppressed — if such specimens occurred, they never even reached the site. It seems unlikely that they were ever made, except perhaps by accident, and cultural rules would have had the advantage of saving the wasting of energy on them.

In summary, larger biface specimens are transformations of smaller ones, albeit with allometric adjustments, and this reflects the operation of a rule-system. I argue that the rule-system is culturally maintained, but that it would be regulated by the feedback of accidental contacts with the mechanical and functional limits of the system. One advantage of a cultural system is that through its social regulation it can help to avoid unnecessary 'collisions with the buffers'.

This discussion may be left with the observation that studies of symmetry and handedness also have considerable promise in helping us to disentangle the variables in early artefact production (Toth 1985; 1990; Wynn 1985).

World view of the times of *Homo erectus* and early *Homo sapiens*

On a grand scale, new dates for *Homo erectus* and the Acheulean have redirected emphasis towards a world-wide view (Asfaw *et al.* 1992; Swisher *et al.* 1994). In summary:

(a) *Homo erectus* begins before the Acheulean hand-axe tradition, and is not always associated with it.

(b) The beginning of the Acheulean about 1.5 million years ago is hazy in that bifaces are first found in the Developed Oldowan, and the order in which the design elements of the 'classic' biface first came together is unclear.

(c) *Homo erectus* appeared at least as early in the Far East as in Africa, but apparently never adopted the Acheulean tradition in that region.

(d) *Homo sapiens* appears at about 0.5 Ma and is initially associated with the Acheulean, but perhaps not in all areas where *Homo sapiens* appeared.

(e) At the end of the Acheulean (about 200,000–100,000 years ago) we have more archaeological resolution in the evidence, but not necessarily a better understanding of developments.

There is little dispute that in the earlier Acheulean we are dealing with the artefacts of *Homo erectus*, and it also seems plain that there is continuity of tradition into the times of early *Homo sapiens*. The Acheulean tradition carries on recognizably until between 200,000 and 100,000 years ago. It is easier to see than to understand quite how it ended: larger handaxes were no longer made; flake tools became more standardized; hafting was introduced (Anderson-Gerfaud 1990; Bosinski 1992). The idea of the biface was maintained, but perhaps took on new functional directions: modifications of the one design may have been used in parallel for quite different purposes. Thus in different industries, or even alongside one another, there may have been unhafted hand-held tools, short-hafted hand-held tools, and spear/projectile points, all utilizing the biface pattern. Whether there is truly such a diversification of ideas and practice is a point which requires further research — perhaps starting from simple quantification as employed by Sampson (1974). My diagram presents one possible scenario (Fig. 16.13).

The overall picture is one of continuity at all levels between *Homo erectus* and archaic *Homo sapiens*, but also one of curious regional divergences in archaeological detail. These explorations have been limited to stone artefacts, but I have already mentioned aspects of transport and time-depth, and the use of other materials.

The last hundred thousand years

Early *Homo* shows enough mental and practical abilities that it is now unclear where to place a Rubicon, if we should want one, either between *Homo erectus* and archaic *Homo sapiens*, or between earlier *Homo sapiens* and anatomically modern humans. The recent tendency has been to see a 'hardware improvement' at 100,000, and a 'software explosion' at 40,000 years ago (e.g. papers in Mellars & Stringer 1989; Mellars 1990). The absence of art, however, in the Upper Palaeolithic of the Middle East highlights some of the dangers of negative evidence. Nor is any reason known that 'gracilization' of the skull should affect the intellect. If we seek signs of dramatic

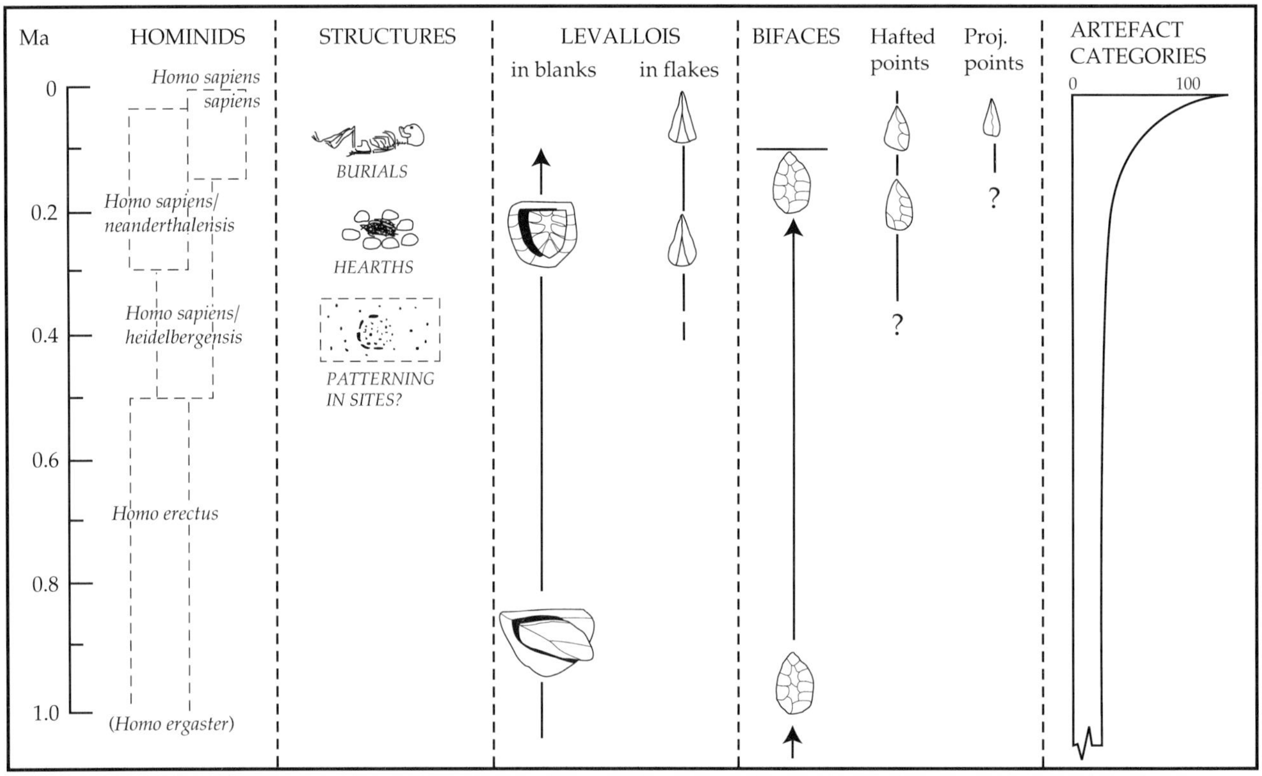

Figure 16.13. *Design traits through time: the accumulation of design features through the Pleistocene.*

technical progress, it would seem that the key developments come earlier, perhaps 300,000 years ago — shown by the introduction of sophisticated preparation routines such as Levallois technique; and by a range of standardized flake tools.

Putting it in deliberately simplistic terms, was the great leap forward early or late; was it declarative or procedural; or both? Or was it an event at all? In terms of procedural skill, the development of Levallois technology seems to be the key. It is incipient in the African Acheulean from about one million years ago, but does not appear in full form, and is not transferred to ordinary flake production, until perhaps 250,000 years ago. In terms of declarative multiplicity (the number of classes of material and ideas contained within them) it would seem that the most abrupt developments come much later, within the last 100,000 years.

Quite when, is difficult to assess in that we cannot easily relate present ideas and categories with those of the makers, and we are also hampered by a lack of dating resolution. Traditional typologies reflect at least partly the ideas of the archaeologist generating them, and only very approximately the number of separate categories known to the makers of a stone industry. Roughly they go as follows:

Oldowan :	10
Developed Oldowan :	12
Acheulean :	15
Middle Palaeolithic :	62
Upper Palaeolithic :	75

The great step from the Acheulean to the Middle Palaeolithic seems misleading. The Bordes typology probably has an excessive number of small divisions, whereas the Kleindienst Acheulean typology is geared towards the Acheulean more than 0.5 Ma ago. In very general terms it might be said that there are over 20 types at 200,000 years ago, and 50 or more by 30,000, but across the Old World there is no clear explosion (a point made by Reynolds 1990; 1991).

In respect of Europe and the Middle East there does seem to be fundamental change at around 40,000. Whether a series of cultural parameters changes together, or at different times, has been considered by many authors (e.g. Rigaud 1989; Mellars 1989; 1993; Otte 1990). Although Reynolds (1991) takes a view of continuity, few authors examining the problem of change at 40,000 years ago stand back to take a longer term view, on the scale of say 400,000 years. On such a longer timescale biological evolution can be admitted: Wynn has stressed memory development, and Holloway the development of factors such as concentration (Wynn 1993; Holloway 1969).

Artefacts incontrovertibly show an increase in multiplicity: not necessarily more complex designs, but more of them. There is thus an increase in the cognitive load of storing so many instruction sets. One way to view this is as an increase in parallel processing; the capacity to load and unload many separate designs; to interweave routines.

Technology does not necessarily pave the way for such developments — some modern societies have used fewer types of stone artefact than are found in the Acheulean — but it probably reflects something that happens at a social level. The sequencing and interweaving of routines is also part of social relationships. Though the idea of separate domains of intelligence has considerable appeal (*cf.* Robson Brown 1993; Mithen this volume) — and indeed is to some extent a necessity of a structured brain — equally the ability to rework and reapply ideas, transferring them from one area to another, must have a great importance.

Conclusions

In this paper I have tried to look at the Stone Age record on a long timescale, centred on the period about one million years ago. There are three principal conclusions:

1. *Although the Stone Age record is fragmented, it presents very similar features from far separated times.* The evidence considered stresses that most of the features which are present at 100,000 years ago can be found at one million years ago. Comparisons are more commonly made over the period 100,000 to 30,000 years ago, but the longer term record emphasized here shows very few evident discontinuities. Complexity of routines, numbers of tool forms, refinement of finish, size of sites — all remain remarkably the same. True, there are new features: burials appear in the record; Levallois technique arrives; smaller points eventually tend to replace handaxes. Rare hearths also appear. But neither the arrival of *Homo sapiens* about 500,000 years ago, nor the first documentation of *Homo sapiens sapiens* at 100,000 years ago is accompanied by discontinuity or even immediate change. If anything, there are pronounced time-offsets of technical change and species change (*cf.* Mellars & Stringer 1989; Reynolds 1991; Gowlett 1990; 1995). In the technical record of Africa/Eurasia, changes in design complexity seem to precede any explosion in numbers of designs.

2. *Elements of constraint and choice do not seem notably different from those confronting more recent toolmakers.* The knappers worked to pattern, but there was normally an area of free-play within which alternatives were viable. Several factors responsible for variation probably combine to mark out and fill such areas of free-play — individual preference, group preference, nature of the immediate task, raw material. Certainly the degree of conscious knowledge implicit in the controls which we see may be different for early times.

One can contrast the situations of an aircraft designer, a Gothic architect and an early stone-knapper. The aircraft designer will have many known constraints, operating so that for example the ratio of fuselage length and wingspan will have strict limits. One of these could be altered only through a compromise involving alterations to other variables. The designer has little free-play, but is fully aware of the theory underlying the permutations. The master-builder constructing a medieval cathedral would similarly be aware of the limits of acceptable variation. He would know that too long an arch, too slender a pier, could spell disaster. But generally he would know this from tradition rather than personal experience, and he would have no theoretical basis for exploring the issues. The early tool-maker would be similarly constrained, and again would have no theoretical basis for understanding this. As suggested above, limits to freeplay might be established by transmitted knowledge (experience of other toolmakers) or by personal experience of advantages and disadvantages of particular approaches. Either way 'knowledge of results' should create a considerable pressure towards eliminating extremes of inefficient or wasteful practice.

On the basis of these observations, I would suggest that some previous analyses, including my own, have been over-simple in separating 'design content' from 'deterministic outcome/paths of least resistance' as exclusive alternatives. There may be a close parallel to the style/function debate of Palaeolithic archaeology. As Sackett notes, in considering active and passive style, these are not separable. A path of least resistance can be best followed if one has learnt the most desirable outcomes from past experience — which amounts to building up an instruction set, which borders on the concept of design content.

3. *The study of* Transformation *offers an important route of analysis, and it presents a useful common theme in the expression of symbolic abilities.* As I have noted, language and number are often regarded as dual

symbolic systems. This tends to leave art somewhere to one side, although art is commonly agreed to be important, and is also commonly visible in the (later) archaeological record. Transformation, which is common to art, mathematics and the constructions of technology, does much to bridge this gap. Transformation is imposed, which means — by dint of cultural transmission — analyzed, and reimposed. I have sought to show that the transformations predate art by hundreds of thousands of years.

On the grounds of these conclusions, I would argue for a continuity in human behaviour through the last million years, which transcends any labels such as *'sapiens'* and *'presapiens'*, wherever we choose to impose them.

In discussing these issues we are often encouraged by habits of language and thought to make dialectical oppositions, such as the following:

language — number
form — transformation
procedural — declarative

But these are verbal distinctions, and concentration on verbal as opposed to visual thinking tends to make us forget that the brain is often many steps ahead of our verbal expressions. The modern human brain possesses about 10^{10} neurones and each neurone 10^4 synapses. Its processing is not limited to linguistic input and output. Terms such as 'presapiens mind' tend to shape our views of earlier human brains, but do little to come to terms with the billions of neurones which operated in them. For the very reason that the brain is a 'clear system', which does not unnecessarily let us know of its operations, archaeology is easily deceived in approaching the problems of its evolution. Careful documentation of actual evidence is almost our only safe step.

The skills necessary for this are more available now than ever before, but in the short-term the prime evidence of early sites has become less accessible. Fewer major sites are now being investigated, especially in Africa, and the primary collections are less well known. There is a broad effort now to replace the static typology of the past with a dynamic approach, but we must not forget that workers such as Isaac (1977) or Cahen (1975) were already concentrating on exploring the dynamics of the Acheulean more than twenty years ago. To introduce the word 'cognitive' does not make up for overlooking their material. Soon, however, we can be sure that improved 3D graphics methods will help to make artefacts much more effectively accessible, even though they may be housed in museums thousands of kilometres apart. But it remains a great challenge to keep our chest of data open, and so to integrate the hard evidence of archaeology with the more general comparative approach based on biology and other disciplines.

Acknowledgements

Fieldwork at Kilombe received support from the Boise Fund and the British Academy. I am grateful to the Ministry of Natural Resources and President's Office of Kenya for research permission, to Dr R.E. Leakey; and to Mrs M. Butler. I am also grateful to my colleague Robin Crompton for our collaboration. The laboratory received support from SERC.

J.A.J. Gowlett
Department of Archaeology
University of Liverpool
Liverpool
L69 3BX

References

Alexander, R.D., 1979. *Darwinism and Human Affairs*. London: Pitman.

Alexander, R.D., 1989. Evolution of the human psyche, in Mellars (ed.), 455–513.

Anderson-Gerfaud, P., 1990. Aspects of behaviour in the Middle Palaeolithic: functional analysis of stone tools from southwest France, in Mellars (ed.), 389–418.

Annett, J., 1969. *Feedback and Human Behaviour*. Harmondsworth: Penguin.

Ardrey, R., 1976. *The Hunting Hypothesis*. London: Collins.

Asfaw, B., Y. Beyene, G. Suwa, R.C. Walter, T.D. White, G. Woldegabriel & T. Yemane, 1992. The earliest Acheulean at Konso-Gardula. *Nature* 360, 732–5.

Bailey, G.N. & P. Callow (eds.), 1986. *Stone Age Prehistory: Studies in Memory of Charles McBurney*. Cambridge: Cambridge University Press.

Bar-Yosef, O. & N. Goren-Inbar, 1992. *The Lithic Assemblages of the Site of Ubeidiya, Jordan Valley*. (Qedem 34.) Jerusalem: Institute of Archaeology, The Hebrew University of Jerusalem.

Belfer-Cohen, A. & N. Goren-Inbar, 1994. Cognition and communication in the Levantine Lower Palaeolithic. *World Archaeology* 26(2), 144–57.

Belitzky, S., N. Goren-Inbar & E. Werker, 1991. A Middle Pleistocene wooden plank with man-made polish. *Journal of Human Evolution* 20, 349–53.

Beynon, A.D. & B.A. Wood, 1987. Patterns and rates of growth in the molar teeth of early hominids. *Nature* 326, 493–6.

Binford, L.R., 1981. *Bones: Ancient Men and Modern Myths*. New York (NY): Academic Press.

Binford, L.R., 1989. Isolating the transition to cultural adaptations: an organizational approach, in Trinkaus (ed.), 18–41.

Blumenschine, R. & M.M. Selvaggio, 1988. Percussion marks on bone surfaces as a new diagnostic of hominid behaviour. *Nature* 333, 763–5.

Blumenschine, R. & M.M. Selvaggio, 1991. On the marks of marrow bone processing by hammerstones and hyenas: their anatomical patterning and archaeological implications, in Clark (ed.), 17–32.

Boëda, E., J.-M. Geneste & L. Meignen, 1990. Identification de chaînes operatoires lithiques du Paléolithique ancien et moyen. *Paleo* 2, 43–80.

Boesch, C., 1993. Aspects of transmission of tool-use in wild chimpanzees, in Gibson & Ingold (eds.), 171–83.

Bosinski, G., 1992. *Eiszeitjäger im Neuwieder Becken*. 3rd edition. AMT Koblenz: Archaologische Denkmalpflege.

Brain, C.K., 1981. *The Hunters or the Hunted: an Introduction to African Cave Taphonomy*. Chicago (IL): University of Chicago Press.

Cahen, D., 1975. *Le site archéologique de la Kamoa (Région du Shaba, République du Zaire)*. Tervuren: Annales du Musée Royal de l'Afrique Centrale.

Chase, P., 1991. Symbols and palaeolithic artefacts: style standardization, and the imposition of arbitrary form. *Journal of Anthropological Archaeology* 10, 193–214.

Chase, P. & H. Dibble, 1987. Middle Paleolithic symbolism: a review of current evidence and interpretations. *Journal of Anthropological Archaeology* 6, 263–96.

Chippindale, C., 1986. Archaeology, design theory, and the reconstruction of prehistoric design systems. *Environment and Planning B: Planning and Design* 1986, 445–85.

Clark, J.D., 1980. The Plio-Pleistocene environmental and cultural sequence at Gadeb, northern Bale, Ethiopia, in *Proceedings of the 7th Panafrican Congress of Prehistory and Quaternary Studies. Nairobi*, eds. R.E. Leakey & B.A. Ogot. Nairobi: TILLMIAP, 189–93.

Clark, J.D. (ed.), 1991. *Approaches to Understanding Early Hominid Life-ways in the African Savanna*. (Römisch-Germanisches Zentralmuseum Forschungsinstitut für Vor- und Frühgeschichte in Verbindung mit der UISSP, 11 Kongress, Mainz, 31 August–5 September 1987, Monographien Band 19.) Bonn: Dr Rudolf Habelt GMBH.

Clark, J.D. & J.W.K. Harris, 1985. Fire and its roles in early hominid lifeways. *African Archaeological Review* 3, 3–27.

Connolly, K., 1979. The development of competence in motor skills, in *Psychology of Motor Behaviour and Sport*, eds. C.H. Nadeau, W.R. Halliwell, K.M. Newell & G.C. Roberts. Champaign (IL): Human Kinetics Publishers, 229–52.

Connolly, K. & M. Dalgleish, 1989. The emergence of a tool-using skill in infancy. *Developmental Psychology* 25(6), 894–912.

Craik, K., 1943. *The Nature of Explanation*. Cambridge: Cambridge University Press.

Crompton, R.H. & J.A.J. Gowlett, 1993. Allometry and multidimensional form in Acheulean bifaces from Kilombe, Kenya. *Journal of Human Evolution* 25, 175–99.

Davidson, I., 1991. The archaeology of language origins: a review. *Antiquity* 65, 39–48.

Davidson, I. & W. Noble, 1993. Tools and language in human evolution, in Gibson & Ingold (eds.), 363–88.

Deacon, T., 1992. Biological aspects of language, in Jones *et al.* (eds.), 128–33.

Dibble, H.L., 1989. The implications of stone tool types for the presence of language during the Lower and Middle Palaeolithic, in Mellars & Stringer (eds.), 415–32.

Duff, A.I., G.A. Clark & T.J. Chadderton, 1992. Symbolism in the early Palaeolithic: a conceptual Odyssey. *Cambridge Archaeological Journal* 2(2), 211–29.

Geertz, C., 1964. The transition to humanity, in *Horizons of Anthropology*, ed. J.M Scher. New York (NY): The Free Press, 713–40.

Gibson, K.R., 1993. Introduction: beyond neoteny and recapitulation: new approaches to the evolution of cognitive development, in Gibson & Ingold (eds.), 273–8.

Gibson, K.R. & T. Ingold (eds.), 1993. *Tools, Language and Cognition in Human Evolution*. Cambridge: Cambridge University Press.

Gowlett, J.A.J., 1982. Procedure and form in a Lower Palaeolithic industry: stoneworking at Kilombe, Kenya. *Studia Praehistorica Belgica* 2, 101–9.

Gowlett, J.A.J., 1984. Mental abilities of early man: a look at some hard evidence, in *Hominid Evolution and Community Ecology*, ed. R.A. Foley. London: Academic Press, 167–92.

Gowlett, J.A.J., 1986. Culture and conceptualisation — the Oldowan–Acheulean gradient, in Bailey & Callow (eds.), 243–60.

Gowlett, J.A.J., 1988. A case of Developed Oldowan in the Acheulean? *World Archaeology* 20 (1), 13–26.

Gowlett, J.A.J., 1990. Technology, skill and the psychosocial sector in the long term of human evolution. *Archaeological Review from Cambridge* 9(1), 82–103.

Gowlett, J.A.J., 1991. Kilombe — Review of an Acheulean site complex, in Clark (ed.), 129–36.

Gowlett, J.A.J., 1993. Le site Acheuléen de Kilombe: stratigraphie, géochronologie, habitat et industrie lithique. *L'Anthropologie* 97(1), 69–84.6 (2).

Gowlett, J.A.J., 1995. Psychological worlds within and without: human–environment relations in early parts of the Palaeolithic, in Ullrich (ed.), 29–42.

Gowlett, J.A.J. & R.H. Crompton, 1994. Kariandusi: Acheulean morphology and the question of allometry. *African Archaeological Review* 12, 1–40.

Gowlett, J.A.J., J.W.K. Harris, B.A. Wood & D. Walton, 1981. Early archaeological sites, hominid remains and traces of fire from Chesowanja, Kenya. *Nature* 294, 125–9.

Graves, P., 1994. Flakes and ladders: what the archaeological record cannot tell us about the origins of language. *World Archaeology* 26(2), 158–71.

Harris, J.W.K., 1983. Cultural beginnings: Plio-Pleistocene archaeological occurrences from the Afar, Ethiopia. *African Archaeological Review* 1, 3–31.

Hassan, F., 1988. Prolegomena to a grammatical theory of lithic artifacts. *World Archaeology* 19(3), 281–96.

Hay, R.L., 1976. *Geology of the Olduvai Gorge*. Berkeley (CA): University of California Press.

Hayden, B., 1993. The cultural capacities of Neandertals: a review and re-evaluation. *Journal of Human Evolution* 24, 113–46.

Hill, A. & A.K. Behrensmeyer, 1985. Natural disarticulation and bison butchery. *American Antiquity* 50(1), 141–5.

Hill, A., S. Ward, A. Deino, G. Curtis & R. Drake, 1992. Earliest *Homo*. *Nature* 355, 719–22.

Hoc, J.-M., 1988. *Cognitive Psychology of Planning*. London: Academic Press.

Holloway, R.L., 1969. Culture: a human domain. *Current Anthropology* 10(4), 395–412.

Humphrey, N.K., 1976. The social function of intellect, in *Growing Points in Ethology*, eds. P.P.G. Bateson & R.A. Hinde. Cambridge: Cambridge University Press, 303–21.

Ingold, T., 1993. Technology, language, intelligence: a reconsideration of basic concepts, in Gibson & Ingold (eds.), 449–72.

Isaac, G.Ll., 1969. Studies of early culture in East Africa. *World Archaeology* 1, 1–28.

Isaac, G.Ll., 1971. Whither archaeology? *Antiquity* 45, 123–9.

Isaac, G.Ll., 1972. Early phases of human behaviour: models in Lower Palaeolithic archaeology, in *Models in Archaeology*, ed. D.L. Clarke. London: Methuen, 167–99.

Isaac, G.Ll., 1976. The activities of early African hominids: a review of archaeological evidence from the time span two and a half to one million years ago, in *Human Origins: Louis Leakey and the East African Evidence*, eds. G.Ll. Isaac & E. McCown. Menlo Park (CA): Benjamin, 483–514.

Isaac, G.Ll., 1977. *Olorgesailie: Archaeological Studies of a Middle Pleistocene Lake Basin*. Chicago (IL): University of Chicago Press.

Isaac, G.Ll., 1981. Stone Age visiting cards: approaches to the study of early land-use patterns, in *Patterns in the Past*, eds. I. Hodder, G. Isaac & N. Hammond. Cambridge: Cambridge University Press, 131–55.

Isaac, G.Ll., 1984. The archaeology of human origins: studies of the lower Pleistocene in East Africa, 1971–1981. *Advances in World Archaeology* 3, 1–87.

Isaac, G.Ll., 1986. Foundation stones: early artifacts as indicators of activities and abilities, in Bailey & Callow (eds.), 221–41.

James, S.R., 1989. Hominid use of fire in the Lower and Middle Pleistocene: a review of the evidence. *Current Anthropology* 30(1), 1–26.

Johanson, D.C., F.T. Masao, G.C. Eck, T.D. White, R.C. Walter, W.H. Kimbel, B. Asfaw, P. Manega, P. Ndessokia & G. Suwa, 1987. New partial skeleton of *Homo habilis* from Olduvai Gorge, Tanzania. *Nature* 327, 205–9.

Jones, S., R. Martin & D. Pilbeam (eds.), 1992. *The Cambridge Encyclopedia of Human Evolution*. Cambridge: Cambridge University Press.

le Gros Clark, W.E., 1967. Human food habits as determining the basic patterns of economic and social life, in *Proceedings of the 7th International Congress of Nutrition*, vol. IV, ed. J. Kuhnau. Braunschweig: Viewig & Sohm, 18–24.

Leakey, M.D., 1971. *Olduvai Gorge*, vol. III: *Excavations in Beds I and II, 1960–1963*. Cambridge: Cambridge University Press.

McBrearty, S., 1989. Cutlery and carnivory (Lead Review). *Journal of Human Evolution* 18, 277–82.

McGrew, W.R., 1992. Comment on, Was there no place like home? A new perspective on early hominid archaeological sites from the mapping of chimpanzee nests, by J. Sept. *Current Anthropology* 33(2), 197–8.

McGrew, W.R., 1993. *Chimpanzee Material Culture: Implications for Human Evolution*. Cambridge: Cambridge University Press.

Mania, U., 1995. The utilisation of large mammal bones in Bilzingsleben — a special variant of Middle Pleistocene Man's relationship to his environment, in Ullrich (ed.), 239–46.

Mellars, P., 1989. Major issues in the emergence of modern humans. *Current Anthropology* 30, 349–85.

Mellars, P. (ed.), 1990. *The Emergence of Modern Humans: an Archaeological Perspective*. Edinburgh: Edinburgh University Press.

Mellars, P., 1993. Archaeology and the population-dispersal hypothesis of modern human origins in Europe, in *The Origin of Modern Humans and the Impact of Chronometric Dating*, eds. M.J. Aitken, C.B. Stringer & P.A. Mellars. Princeton (NJ): Princeton University Press, 196–216.

Mellars, P. & C. Stringer (eds.), 1989. *The Human Revolution: Behavioural and Biological Perspectives on the Origins of Modern Humans*. Edinburgh: Edinburgh University Press.

Miller, G.A., E. Galanter & K.H. Pribram, 1960. *Plans and the Structure of Behaviour*. New York (NY): Holt, Rinehart & Winston.

Oakley, K.P., 1981. Emergence of higher thought 3.0–0.2 Ma BP. *Philosophical Transactions of the Royal Society of London* Series B292, 205–11.

Otte, M., 1990. From the Middle to the Upper Palaeolithic: the nature of the transition, in Mellars (ed.), 438–56.

Passingham, R.E., 1981. Broca's area and the origin of human vocal skill. *Philosophical Transactions of the Royal Society of London* Series B292, 167–75.

Pelegrin, J., C. Karlin & P. Bodu, 1988. Chaînes operatoires: un outil pour le préhistorien. Technologie préhistorique. (Notes et Monographies Techniques 25.) Paris: Editions du CNRS, 55–62.

Potts, R., 1992. The hominid way of life, in Jones *et al.* (eds.), 325–34.

Reynolds, T.E.G., 1990. The Middle–Upper Palaeolithic transition in southwestern France: interpreting the lithic evidence, in Mellars (ed.), 262–75.

Reynolds, T.E.G., 1991. Revolution or resolution? The archaeology of modern human origins. *World Archaeology* 23(2), 155–66.

Rigaud, J.-Ph., 1989. From the Middle to the Upper

Paleolithic: transition or convergence, in *The Emergence of Modern Humans*, ed. E. Trinkaus. Cambridge: Cambridge University Press, 142–53.

Robson Brown, K., 1993. An alternative approach to cognition in the Lower Palaeolithic: the modular view. *Cambridge Archaeological Journal* 3(2), 231–45.

Roche, H., 1980. *Premiers outils taillés d'Afrique*. Paris: Société d'Ethnographie.

Roe, D.A., 1964. The British Lower and Middle Palaeolithic: some problems, methods of study, and preliminary results. *Proceedings of the Prehistoric Society* 30, 245–67.

Roe, D.A., 1968: British Lower and Middle Palaeolithic handaxes groups. *Proceedings of the Prehistoric Society* 34, 1–82.

Roe, D.A., 1981. *The Lower and Middle Palaeolithic Periods in Britain*. London: Routledge & Kegan Paul.

Ronen, A., 1991. The Yiron-Gravel lithic assemblage: artifacts older than 2.4 My in Israel. *Archäologisches Korrespondenzblatt* 21, 159–64.

Sacerdoti, E.D., 1977. *A Structure for Plans and Behaviour*. New York (NY): Elsevier.

Sackett, J.R., 1982. Approaches to style in lithic archaeology. *Journal of Anthropological Archaeology* 1, 59–122.

Sackett, J.R., 1986. Isochrestism and style: a clarification. *Journal of Anthropological Archaeology* 5, 266–77.

Sampson, C.G., 1974. *The Stone Age Archaeology of South Africa*. New York (NY): Academic Press.

Savage-Rumbaugh, E.S., 1986. *Ape Language: From Conditioned Response to Symbol*. Oxford: Oxford University Press.

Schick, K.D., 1987. Modelling the formation of stone artifact concentrations. *Journal of Human Evolution* 16, 789–807.

Schick, K.D., 1991. On making behavioral inferences for early sites, in Clark (ed.), 79–107.

Schick, K.D. & N. Toth, 1993. *Making Silent Stones Speak: Human Evolution and the Dawn of Technology*. New York (NY): Simon & Schuster.

Schmidt, R.R., 1936. *The Dawn of the Human Mind*. London: Sidgwick & Jackson.

Sept, J., 1992. Was there no place like home? A new perspective on early hominid archaeological sites from the mapping of chimpanzee nests. *Current Anthropology* 33(2), 187–207.

Sinclair, A. & N. Schlanger (eds.), 1990. Technology in the humanities. *Archaeological Review from Cambridge* 9(1), 1–157.

Steele, J., 1994. Dispersal as a mechanism of cultural diffusion: a revised equilibrium basin model of the persistence of the Acheulian. *World Archaeology* 26(2), 126–43.

Swisher, G.C., G.H. Curtis, T. Jacob, A.G. Getty, A. Suprijo & Widiasmoro, 1994. Age of the earliest known hominids in Java, Indonesia. *Science* 263, 1118–21.

Tobias, P.V., 1981. The emergence of man in Africa and beyond. *Philosophical Transactions of the Royal Society of London* Series B 292, 43–56.

Tobias, P.V., 1991. The emergence of spoken language in hominid evolution, in Clark (ed.), 67–78.

Toth, N., 1985. The Oldowan reassessed: a close look at early stone artefacts. *Journal of Archaeological Science* 12, 101–21.

Toth, N., 1990. The prehistoric roots of a human concept of symmetry. *Symmetry: Culture and Science* 1(3), 257–81.

Toth, N. & K.D. Schick, 1993. Early stone industries and inferences regarding language and cognition, in Gibson & Ingold (eds.), 346–62.

Toth, N., K.D. Schick, E.S. Savage-Rumbaugh, R.A. Sevcik & D.M. Rumbaugh, 1993. Pan the tool-maker: investigations into the stone tool-making and tool-using capabilities of a Bonobo (*Pan paniscus*). *Journal of Archaeological Science* 20(1), 81–91.

Trinkaus, E. (ed.), 1989. *The Emergence of Modern Humans*. Cambridge: Cambridge University Press.

Ullrich, H. (ed.), 1995. *Man and Environment in the Palaeolithic*. (Etudes et Recherches de l'Université de Liège 62.) Liège: Université de Liège.

Washburn, S.L. & C.S. Lancaster, 1968. The evolution of hunting, in *Man the Hunter*, eds. R.B.Lee & I. DeVore. Chicago (IL): Aldine Press, 293–303.

White, T.D., G. Suwa & B. Asfaw, 1994. *Australopithecus ramidus*, a new species of early hominid from Aramis, Ethiopia. *Nature* 371, 306–12.

Wiener, N., 1954. *The Human Use of Human Beings: Cybernetics and Society*. Reprint. Garden City (NY): Doubleday Anchor.

Wilson, E.O., 1983. Sociobiology and the Darwinian approach to mind and culture, in *Evolution from Molecules to Men*, ed. D.S. Bendall. Cambridge: Cambridge University Press, 545–53.

Wilson, L., 1988. Petrography of the Lower Palaeolithic tool assemblage of the Caune de l'Arago (France). *World Archaeology* 19(3), 376–87.

Wobst, H.M., 1974. Boundary conditions for Palaeolithic social systems: a simulation approach. *American Antiquity* 39, 147–78.

Wood, B. 1992. Origin and evolution of the genus *Homo*. *Nature* 355, 783–90.

Wrangham, R., 1992. Comment on, Was there no place like home? A new perspective on early hominid archaeological sites from the mapping of chimpanzee nests, by J. Sept. *Current Anthropology* 33(2), 201–2.

Wynn, T., 1979. The intelligence of later Acheulian hominids. *Man* 14, 371–91.

Wynn, T., 1981. The intelligence of Oldowan hominids. *Journal of Human Evolution* 10, 529–41.

Wynn, T., 1985. Piaget, stone tools and the evolution of human intelligence. *World Archaeology* 17, 32–43.

Wynn, T., 1993. Layers of thinking in tool behaviour, in Gibson & Ingold (eds.), 389–406.

Wynn, T. & W.C. McGrew, 1989. An ape's-eye view of the Oldowan. *Man* 24, 383–98.

Wynn, T. & F. Tierson, 1990. Regional comparison of the shapes of later Acheulean handaxes. *American Anthropologist* 92, 73–84.

17
Domain-Specific Intelligence and the Neanderthal Mind

Steven Mithen

During the last decade archaeologists have paid considerable attention to the evolution of language. The weight of evidence currently suggests that this has a long evolutionary history and the Neanderthals are likely to have possessed complex linguistic capacities. Consequently, if we believe that the marked contrasts in their behaviour patterns to those of modern humans derive from a difference in mentality, archaeologists need to look at non-linguistic aspects of cognition. By drawing on recent work in evolutionary pyschology I suggest that the Neanderthal mind may have been constituted by 'multiple intelligences' with low degrees of accessibility. In other words, although we find evidence that their stone tool making, social and natural history knowledge and thought were as advanced as that of modern humans, they were unable to integrate their knowledge and thought located within each of these domains as is characteristic of modern humans. The model for the Neanderthal mind summarized here has been developed at greater length, and placed within a long-term evolutionary context in Mithen (1996).

Archaeologists need to make explicit reference to past minds when interpreting the material residues of past behaviour. It is perhaps unnecessary to stress the challenge this poses to the aspiring cognitive archaeologist. Few entities differ more greatly than the physical reality of the stone tools which constitute the bulk of their data, and the abstract quality of past 'minds'; the first can be measured, weighed and drawn, the other defies definition by philosophers and psychologists, let alone by archaeologists.

Academics in disciplines other than archaeology also have vested interests in the development of models for the early human mind as these impinge upon our knowledge of the modern mind. For instance, our understanding of the nature of language, creativity, and aggression will be furthered if we understand their evolutionary histories (e.g. Pinker & Bloom 1990). Indeed, the development of models for the early human mind can only be a multi-disciplinary endeavour. Specialists in any one field will inevitably need to draw on work in related disciplines. Recent publications (e.g. Gibson & Ingold 1993), and the conference leading to this volume, show an increasing awareness of the need for psychologists, primatologists, anthropologists, linguists and archaeologists to attempt an integration of their material for the identification of key concepts and approaches. But the task cannot be easy, for even the modern mind remains so poorly understood and terminological distinctions plague the literature.

A considerable proportion of the recent literature about the early human mind has focused on language (e.g. Davidson & Noble 1989; Davidson 1991; Whallon 1989; Mellars 1991). This is neither surprising nor inappropriate for our linguistic capacity is critical to the behavioural gulf between ourselves and other species. The discussion about language has become polarized around two positions. One of these sees language appearing suddenly and late in human evolution, around the time of the transition from the Middle to the Upper Palaeolithic (e.g. Davidson 1991). This stresses the differences between human language and the vocalizations of other animals. It contrasts with the second position which recognizes continuity and is more ready to see a gradual evolution of the linguistic capacity during the Lower and Middle Pleistocene (e.g. Parker & Gibson 1979; Gibson 1983; 1985; Aiello & Dunbar 1993).

A problem that may arise with the current almost overwhelming concern with language is that other aspects of cognition will be neglected. In the archaeological literature there has been very limited discussion about features such as intelligence, creativity, emotions and consciousness, even though these are critical for our understanding of past behaviour (Mithen 1991a,b) and are currently much discussed in the cognitive sciences (e.g. Dennett 1992; Humphrey 1992). To understand the Neanderthal mind we need not only to model early linguistic capacities but also these other aspects of the mind. Indeed, if the evolution of modern linguistic capacities did occur relatively early in human evolution, then we may be forced to look towards other cognitive changes to explain the transition to modern behaviour marking the start of the Upper Palaeolithic. I have recently argued that this is indeed the case and this transition is essentially concerned with a change in the character of intelligence from one that can be described as 'domain-specific' to one that can be described as 'generalized' (Mithen 1994a). My intention in this paper is to further that argument by outlining a model for the early human mind in terms of a 'domain-specific' intelligence. For this I will focus on one type of early human — Neanderthals.

Intelligence: domain-specific and generalized

The term 'intelligence' is as difficult to define as that of 'language'. As Byrne (1994) has recently described, attempting to reach consensus about the meaning of the term 'intelligence' risks a loss of its meaning altogether. He quotes Wechsler, writing in 1944, who suggests that intelligence is 'the aggregate or global capacity of the individual to act purposefully, think rationally, and to deal effectively with his environment'. This simply approaches the notion of biological adaptedness, while our more general usage of the term implies something rather different. Humphrey (1976) drew this distinction in his seminal paper on the 'social function of intellect'. His concern was not with the ability for flexible behaviour or that which attains rationally defined goals, but with what we recognize as a specifically human form of intelligence, involving creativity for no immediately apparent functional ends; that which can result in the mind of an Einstein or a Michaelangelo.

This drew Humphrey to focus on what has become known as 'social intelligence' — a suite of cognitive processes which evolved, and which perhaps continue to function, with regard to the domain of social behaviour. Notions of planning, alliance formation, deception and empathy are critical to social intelligence. Humphrey (1984, 4–5) suggests it involves 'being sensitive to other people's moods and passions, appreciative of their waywardness and stubborness, capable of reading signs in their faces and equally the lack of signs, capable of guessing what each person's past experience holds hidden in the present for the future'. The essence of the idea of a social intelligence is that the selective pressure for the evolution of capacities for intellectual thought arose from the complex patterns of social interaction required to live in large groups. As a consequence, our intellectual abilities may be more adapted to solving problems about the social rather than the non-social world. The notion of a 'social intelligence' has been central to the development of the 'Machiavellian' model for intelligence (Byrne & Whiten 1988), which in turn has led to a greater concern with the notion of mind reading and deception (e.g. Whiten 1991).

The notion of a distinctly social intelligence can, however, be questioned. Among humans and non-human primates, a considerable proportion of social interaction is directly related to the non-social world. Non-human primates form large groups as a means of predator defence or to improve resource acquisition (Dunbar 1988) and Neanderthals are likely to have done likewise. The boundary between a social and non-social world is vague (Cords 1992). Nevertheless, if only due to the amount of literature that has been inspired by the notion of social intelligence, the idea that it is productive to consider specific domains of intellectual behaviour, rather than intelligence as an all-embracing concept appears worthwhile.

Gardner (1983) has made one of the few attempts to identify domains of intelligence other than that relating to social behaviour. In his proposal that the mind is composed of 'multiple intelligences' he suggested that one can identify a linguistic, musical, mathematical, spatial, and bodily-kinesthetic intelligence in humans (the last relating to the control of body movements). He suggested that although each of these interact with each other, at the core of each is a computational capacity or information-processing mechanism unique to that particular intelligence. Other than social/personal intelligence, that of language is most widely accepted as relating to domain-specific cognitive processes (e.g. Fodor 1983).

Cheney & Seyfarth (1990) have made an explicit comparison between social and non-social intelligence for a single species, vervet monkeys. While the monkeys appeared very competent at drawing inferences about the behaviour of con-specfics, their ability at drawing inferences about the non-social world appeared less developed. For instance, they appeared limited in their ability to draw inferences from tracks about the presence of potential predators. Cheney & Seyfarth also argued that monkeys have a far greater capacity to solve problems concerning social interaction, than those concerning the non-social world, even when the structure of those problems appear to be very similar. For instance although a male may have no difficulty in ranking the social stature of other males, he may be unable to rank the relative amounts of water in a series of containers. Cosmides (1989) has argued that a very similar phenomena can be found among modern humans. When problems which share the same logical structure are framed in social terms rather than with reference to abstract entities, people appear to solve them with much greater ease.

More generally Cosmides & Tooby (1987; Tooby & Cosmides 1989; Barkow *et al.* 1992) have argued that the modern mind is composed of a multitude of domain-specific psychological mechanisms which they term, 'Darwinian algorithms'. These evolved, they suggest, in response to specific selection pressures while people lived as hunter-gatherers during the Pleistocene. One might conceive of several, perhaps many, separate Darwinian algorithms composing a particular domain of intelligence. The essence of their argument is that it is wrong to conceive of the mind as a general purpose learning mechanism. They argue that this is incompatible with our understanding of biological evolution and that natural selection could only have led to a highly modular mentality.

One of the problems with this approach is that it appears very easy to propose a specific module for almost any domain of behaviour (e.g. see Tooby & Cosmides 1992, 113; Cosmides & Tooby 1994, 35; Pinker 1994, 420). It remains unclear as to how large we should consider modules to be (in terms of the scope of the domain to which they apply), how many exist and how they interact.

Three approaches have been used to identify the existence of specific domains/modules: experimental psychology, the study of child development and cross-cultural ethnography. Cosmides (1989) used the experimental approach (based on the Wason selection test) to identify a 'social exchange' module and further work based on this approach would be worthwhile. By the study of child development areas of 'intuitive knowledge' can be identified, i.e. knowledge about the world which could not have been acquired from experience and must derive from a content-rich cognitive module. The classic example of this is the intuitive knowledge about language that children appear to possess (Pinker 1994). Recent studies have also suggested that children possess intutive knowledge about psychology (Whiten 1991; Wellman 1991; Gopnik & Wellman 1994; Harris 1994), physics (e.g. Spelke *et al.* 1992) and biology (Atran 1990; 1994; Keil 1994). Cross-cultural ethnographic studies can provide further evidence for specific modules or cognitive domains by demonstrating the existence of behavioural and cognitive universals (Brown 1991). Again the classic example is the 'universal grammar' as identified by Chomsky. Atran's (1990) work in this area is of particular importance as he demonstrates people from all cultures adopt the same essential categories for living things; i.e. a pan-human cognitive domain for natural history.

With regard to the manner in which modules/domains interact there are a wide range of views. Cosmides & Tooby adopt a strictly modular view and effectively deny any interaction. An increase in behavioural complexity, they argue, derives from an increase in the number of modules (Tooby & Cosmides 1992, 113). In contrast, Gardner (1983) stressed the importance of interaction between the multiple intelligences of his model. It is through such interaction that human creativity and wisdom arise. Rozin (1976; Rozin & Schull 1988) also stressed the importance of 'accessibility' between domains. He used the term 'adaptive specialization' for a specific module and suggested that the course of human evolution has involved an increase in the accessibility between what had originally evolved as domain-specific cognitive processes. Similarly, Gibson (1993) has suggested that cognitive evolution has involved a process of hierarchization — the construction of more complex cognitive processes from the integration of those which become placed at a lower level in a cognitive hierarchy. Yet another approach has been that of Sperber (1994). He suggests that there exists a 'second order' metarepresentational module. This contains representations from all other, 'first order' modules and consequently enables people to integrate their knowledge acquired in different domains by domain-specific cognitive processes. The evolution of such a metarepresentational module, he suggests, would have caused a cultural explosion. I have previously suggested that increased accessibility between cognitive domains is indeed the explanation for the cultural explosion of the Middle/Upper Palaeolithic transition (Mithen 1994a).

In that paper I argued that the idea of a shift

from a domain-specific intelligence to a generalized intelligence, though one still based on a modular architecture, appears to provide a very powerful framework for the interpretation of the Middle–Upper Palaeolithic transition. It is surely one of the quintessential features of the modern mind that we combine and share knowledge and ways of thinking from different domains, even though our cognitive architecture may be modularized. Anthropomorphic thinking, the attribution of mental states of humans to non-human animals, is universal among hunter-gatherers (and perhaps people in general) and is a combination of social and non-social thought. More generally, analogical thinking which pervades modern thought reflects the distinct lack of a domain-specific intellect. In this regard the model for the human psyche that Cosmides & Tooby present appears more suitable for archaic rather than modern humans. In the following I expand on this interpretation by attempting to build a model for the Neanderthal mind as a series of intellectual domains with limited cognitive accessibility. This model is further expanded in Mithen (1996).

Multiple domains of intelligence for early humans

For a study of Neanderthals it is productive to explore those domains of intelligence which would be essential to a hunter-gatherer life style. In light of the data available in the archaeological record and our general understanding about early hominid communities, three forms of intelligence (other than that concerning language) appear to be most worthwhile to explore for the Early Palaeolithic: social, technical and natural history intelligence. My argument will be that for each of these we can infer fairly advanced intellectual capacities; but we can also recognize a lack of accessibility or integration between them.

Social intelligence

Social intelligence is without doubt one of the most important domains of intelligence, but equally one of the most difficult to study using archaeological data. The notion of a social intelligence has been one of the most important developments in primate studies during the last decade and thrown substantial new light on primate behaviour. It may be of less significance, however, for those studying the latter stages of human evolution as the major evolutionary phase of social intelligence is likely to have been completed by the time of *Homo erectus*.

The recent studies of apes and monkeys have shown their levels of social intelligence to be highly developed. Indeed, Byrne & Whiten (1992) have argued that by 5–8 Ma, the last common ancestor of apes and humans had sophisticated abilities in the attribution of intention to others and the imagination of alternative possible worlds. The enlargement of the brain during human evolution may provide a direct monitor of the evolution of social intelligence. This notion derives from the positive correlation between group size and brain size among non-human primates, and the assumption that living in a large group requires a high degree of social intelligence due to the large number of relationships that require monitoring (Dunbar 1992; Aiello & Dunbar 1993). In light of this, the similar brain size between Neanderthals and modern humans may indicate similar group sizes (with an approximate mean of 150) and essentially the same degree of social intelligence.

Circumstantial evidence for large group size for Neanderthals supports the idea of a well developed social intelligence. As I will describe below, Neanderthals were able to make complicated stone tools, such as Acheulean handaxes and Levallois cores. The knowledge to make such tools is most likely to have been acquired by social learning, possibly involving instruction (perhaps with the use of gestural or spoken language). This suggests intense social interaction and the existence of large groups. Similarly, on the basis of comparative studies of group size among modern primates (Dunbar 1988) the efficient exploitation of resources and effective defence against predators in the glaciated northern environments inhabited by Neanderthals is likely to have required large group sizes (Mithen 1994b). Of course, this need not always have been the case. Pleistocene environments switched repeatedly between relatively warm and relatively cold states, and when temperatures were sufficiently high to allow woodland to become substantial, the ecological pressures to form large groups are likely to have been alleviated.

If we take into account the high levels of social intelligence that have been described for monkeys and apes, and the various lines of evidence suggesting large group sizes for Neanderthals, it seems difficult to arrive at a conclusion other than Neanderthals possessed a degree of social intelligence not significantly different from that of modern humans.

Technical intelligence

By technical intelligence I am referring to the sensorimotor skills to make stone tools. These would appear to involve aspects of the spatial intelligence and the bodily-kinesthetic intelligences as defined by Gardner. The principle justification for defining a 'technical' intelligence, with specific regard to stone tools, is that these are likely to have played a critical role in the early stages of human evolution. The success of *Homo habilis* during the Plio-Pleistocene in African savanna environments is likely to have derived from the ability to produce artefacts with sharp, robust edges. As such, we should expect 'Darwinian Algorithms' to have evolved with specific reference to the sensorimotor skills required to flake stone. Similarly, the dominance of stone tools in the archaeological records of all early humans suggests they remained essential to their life styles.

For assessing the degree of technical intelligence attained by Neanderthals we must focus on the most complex of the artefacts they produced; handaxes and Levallois cores. That handaxes imply a high degree of intelligence has previously been argued at length. Wynn (1979) suggested that if a Piagetian framework is used to describe the evolution of intelligence, handaxes indicates that an operational level of intelligence had been attained, i.e. one that is essentially the same as modern humans (but see Parker 1985). A similar argument was made by Gowlett (1984), who stressed the complexity of the manufacturing process. In his later work, Wynn (1989; 1993) has been more cautious, adopting a domain-specific approach, and suggesting that handaxes imply a high degree of spatial thought the relationship of which to other aspects of the mind remain unclear.

The most important features of handaxes with regard to technical intelligence are the imposed form and the symmetry, which is often found in three dimensions. The difficulty in achieving this has been stressed by Pelegrin (1993), who has many years experience of replicating handaxes. As he notes (1993, 310–11), the goal of the knapper making a handaxe is not to simply obtain a sharp cutting edge but to extricate an integral shape from a nodule, independent from the starting shape of that nodule. This involves the rough shaping of a nodule into a bifacial preform and then its completion using bifacial thinning and sharpening. Both of these presuppose a calculated and critically determined organization of the numerous flakes to be detached. Planning ahead is essential if symmetry is to be achieved, and maintained as the piece is developed. The knapper needs to assess both what is desirable and what is possible, using a sequence of mental images, and achieve such ends by blows of a specific force and direction at specific points on the artefact. Each nodule worked by a knapper will have its own unique characteristics and consequently to produce standardized forms, the knapper needs to derive a particular set of actions to work that nodule from a generalized knowledge about making stone tools, rather than following a fixed set of rules in a rote fashion. A similar degree of technical intelligence is required for making Levallois cores. When excavated, these do not look as impressive as handaxes, but when substantial refits are made, the complexity of the knapping process becomes apparent (e.g. Roebroeks 1988).

While the presence of handaxes and Levallois cores indicate that Neanderthals had the potential to make such tools, such abilities were not always expressed. Many industries and assemblages from Middle and early late Pleistocene Europe lack such artefacts, such as the Clactonian and denticulate Mousterian assemblages. There are likely to be several reasons for this. One of the most significant may have been the social context of learning technical skill (Mithen 1994b). If Neanderthals were living in relatively small groups which lacked intense social interaction, as is likely when their environments were relatively thickly wooded (Mithen 1994b), then the opportunities for social learning about technology (which is likely to have been a by-product of social learning about other issues) will have been limited. In this regard Neanderthals would have acquired technical knowledge principally by individual trial-and-error learning, resulting in a low degree of technical skill. Functional factors may also have played a role, but in light of the generalized use of technology during the Early Palaeolithic (as discussed below) these are unlikely to have been of major significance.

In summary, when the social context was adequate, Neanderthals were able to develop very high levels of technical intelligence. It appears just as complex to manufacture a handaxe or a Levallois core, as to make the blade cores that dominates the stone tool technology of modern humans during the Upper Palaeolithic. In this regard, as with social intelligence, Neanderthals, appear to have possessed equivalent levels of technical intelligence to modern humans.

Natural history intelligence

If the stone tools manufactured by Neanderthals are impressive, then so too is the fact that Neanderthals survived in glaciated northern landscapes for many millennia. The demands of living in such landscapes cannot be underestimated. As Gamble (1986) has shown, for the majority of the Middle and early Late Pleistocene, these environments would have been in neither a fully glacial nor interglacial condition. Consequently the success of Neanderthals would have depended upon their adaptation to predominantly open and highly seasonal environments.

The faunal remains from caves and open sites indicate very diverse animal communities (Gamble 1986; Stringer & Gamble 1993). Among the herbivores were mammoth and woolly rhino, bovids, deer and horse, reindeer, ibex and chamois. The relative frequencies of these species would have varied with environmental variables across time and space. In general, the animal communities appear to have been substantially more diverse than any in the modern world. Similarly the carnivore element included species that today are found in diverse environments, such as cave bear, hyena, lion and wolf. Although poor preservation prevents a direct assessment of the role of plants in Neanderthal subsistence, it is generally assumed that this would have been minimal and Neanderthals survived by exploiting game.

With this diversity of game, it might initially appear that Neanderthals were living in a Garden of Eden; but far from it. Acquiring the necessities of life in such an environment would have been immensely challenging. The animal and plant resources may have been diverse, but they are not likely to have been abundant. Each animal is likely to have been linked into a complex foodweb resulting in frequent, and unpredictable fluctuations in their numbers. The availability of plants and game would also have shown dramatic variations with the seasons, with a marked deterioration during the winter months. Moreover, if Guthrie's (1984) argument that prior to the end of the Pleistocene, different types of vegetation (e.g. boreal, temperate species) showed a markedly more patchy rather than zoned distribution is correct, then the predictability of game species may have been substantially lower than that in high latitude environments today. In these ever-changing, fluctuating and unpredictable environments, Neanderthals would have been competing with carnivores for game and possibly relied on scavenging from carcasses during the leanest, winter months (Gamble 1987). Competition with carnivores would not only have been for food, but also shelter. In many areas caves were in short supply and would have been as attractive for carnivores and hibernating bears, as for people.

The difficulties that Neanderthals faced living in such environments — and their achievement at successfully doing so — is exacerbated by their technology, or rather their lack of technology. As I have discussed, Neanderthals appear to have mastered very complex sequences of stone tool production. Making handaxes and Levallois cores involves understanding fracture dynamics, 'reading' a nodule of stone, planning sequences of knapping actions and various mental manoeuvres, such as 'picturing' a finished artefact forming. Yet in spite of this technical proficiency, the range of tool types appears remarkably narrow, and to have made a limited contribution to coping with the glaciated landscapes.

Certainly a range of stone tools were made to act as spear points, knife blades, and scraping tools. These must have played an essential part in killing and processing animal resources. But Neanderthals lacked the highly specialized tools, dedicated to specific tasks, which have characterized modern humans when exploiting open environments. The most complex tool that Neanderthals made is unlikely to have had more than two or three parts. Modern hunter-gatherers who exploit northern latitudes rely on tools with multiple components and various complex facilities unknown for the Middle Palaeolithic (Oswalt 1976; Churchill 1993). Indeed the reliance of modern humans on a complex and diverse technology to exploit glaciated landscapes makes the technologically simple Neanderthal achievement particularly impressive. One further factor adds to this: Neanderthals were considerably less agile than modern humans. The skeletal anatomy of Neanderthals suggests that while they were physically strong, they were certainly not fleet of foot (Aiello & Dean 1993).

So if their technological and anatomical resources were so limited, and if the glaciated landscapes were so challenging, how did Neanderthals manage to survive and flourish in northern Europe during the Pleistocene? The answer must lie within their minds — a sophisticated understanding of their environments and the animals within it, which can be termed a 'natural history intelligence'.

If we once again consider modern hunter-gatherers, we can see that their ability to survive in harsh environments is highly dependant upon a detailed knowledge base of the spatial and temporal distribution of animals and plants, and of their behaviour (Mithen 1990). Their technological sophistication and physical agility, both lacking among Neanderthals, are important, but an ability to 'read' landscapes, monitor the movement of animals, and make predictions about their behaviour is essential. Modern children appear to be born with an intuitive biological knowledge and the universal pattern of zoological classification employed throughout the world suggest a pan-human natural history intelligence (Atran 1990). So too must it have been for Neanderthals. Indeed, without the use of complex tools and a physical agility, Neanderthals must have relied even more heavily on a 'natural history intelligence' than do modern humans, to have 'thought' their way through the hazards of living by hunting and gathering in a glaciated landscape.

What are the essential elements of a natural history intelligence? Probably the most important is an ability to simply sit quietly and watch the natural world. Any would-be natural historian knows that patient, quiet observation, is the key to success. To this must be added an inquisitiveness about animal and plant behaviour; an interest in that behaviour almost for its own sake. From these a natural historian can accumulate a knowledge base which is continually expanded and modified. And this knowledge base is not just about individual species, but about their interactions and relationships.

This becomes very evident when we consider modern hunter-gatherers. Mithen (1990, 55–66) summarized the diverse sources of information in the natural world used by modern foragers to plan subsistence behaviour. These involve the use of animal tracks, from which inferences about the makers age, sex, state of health and movement can be drawn (e.g. Sullivan 1942, 68; Nelson 1973, 200; Lee 1979, 212–13); faeces which may provide information about current feeding patterns and movement (e.g. Blurton-Jones & Konner 1976, 342; Lee 1979, 212; Hill & Hawkes 1983, 152); marks left on the ground (such as from sleeping) and on vegetation, such as from feeding (e.g. Sullivan 1942, 71; Jennes 1977, 55; Goodale 1971, 161); smells (such as of urine) and sounds (e.g. Hill & Hawkes 1983, 147; Nelson 1973, 91). Some of the most impressive uses of information in the natural world are so-called calendar plants by groups such as the Groote Eylandt islanders (Levitt 1981). Changes in these plants, such as the flowering or the ripening of seeds, is used as an indication that resources in a different part of the landscape are ready for exploitation. Similarly, many modern hunter-gatherers demonstrate an understanding of ecological relationships between species and use this knowledge when making foraging decisions (e.g. Flood 1980, 66; Marks 1976, 116; Nelson 1983, 211).

While there is no direct evidence that Neanderthals used such sources of information the circumstantial evidence must be conclusive. If Neanderthals were unable to read the natural world in this manner, how could they have been able to survive in such challenging northern landscapes in light of their limited technology and agility? Indeed, a good case can be made for Neanderthals having had a rather more developed degree of natural history intelligence and knowledge than that possessed by modern humans. Perhaps this is the reason why the interest in the natural world of at least some modern hunter-gatherers appears to far beyond functional necessity — this may be an evolutionary hangover from a time before complex technology and physical agility relieved the dependency on information about the natural world.

Accessibility between cognitive domains for early humans

My argument has been that with regard to social, technical and natural history intelligence, Neanderthals appear to have been very similar to modern humans. I now want to turn to the manner in which these intelligences were related, or in Rozin's terms the degree of cognitive accessibility. To do this I will consider the relationships between technology and the natural and social environments.

Technology and the natural environment

In this section I will contrast the relationship between technical and natural history intelligence for the Early and the Upper Palaeolithic. I will start by drawing a gross contrast: for Neanderthals technological variability provides a passive reflection of environmental variables, mediated by mobility patterns, while for behaviourally modern humans technology actively structures human–environment interaction. These reflect, I suggest, relatively low and relatively high degrees of cognitive accessibility between domains of intelligence for Neanderthals and behaviourally modern humans respectively. Two examples of inter-assemblage variability in the Middle Palaeolithic serve to illustrate its relationship with ecological variables.

Stiner & Kuhn (1992) have explored the connection between variability in lithic and faunal assemblages from four Mousterian cave sites in west-central Italy: Grotta Breuil, Grotta Guattari, Grotta dei Moscerini and Grotta di Sant'Agostino. The faunal assemblages prior to *c.* 55,000 BP suggested the procurement of animals by scavenging which is interpreted as being one element of a resource-gathering strategy including the low level exploitation of shell-fish, and aquatic tortoises. The tools and flakes associated with these assemblages are relatively large and were predominantly produced with a centripetal core reduction technique. The artefacts exhibit evidence of intense, prolonged use and relatively high frequencies of transport. In contrast, faunal assemblages after 55,000 BP reflect the hunting of ungulates, with an age structure biased towards prime aged adults and the transport of body parts with high meat yields. The associated lithic assemblages are dominated by platform core reduction techniques that yielded relatively large numbers of small flakes per core and the artefacts were neither worked intensively nor transported over large distances.

While there is clear patterning between the variation in lithic and faunal assemblages there is no direct or obvious functional link between these. In assemblages from either pre- or post-55,000 BP the majority of tools are sidescrapers, and 'points' remain at about the same low frequency. Stiner & Kuhn (1992, 326) argue that

> the patterns of covariation express the influence of differing patterns of mobility and land use, revolving around how hominids searched for resources in space. Shifts in the frequency of residential movement and duration of particular occupations would have resulted in changing constraints over where and when lithic raw materials could be gathered and where tools could be made and maintained. The advantages of alternative tactics of tool manufacture and renewal would vary accordingly.

Scavenging and gathering implies a relatively wide-ranging search pattern. The production of large flakes by centipetal core reduction techniques would be advantageous as such blanks would be suitable when gathering/scavenging in raw material poor areas. Similarly, persistent resharpening is a way of coping with the uncertainty about when and where raw materials for manufacturing new tools will be available. After 55,000 BP hunting is likely to have involved targetting concentrated food patches, reducing the ranging area of the hominids, increasing the duration of occupations within the caves and reducing uncertainties about raw material availability. As a consequence, greater numbers of light tools were made using a platform core reduction technique producing blanks which could be used unmodified or after light retouching for processing animal tissue.

A similar set of relationships between mobility patterns and core/artefact reduction can be seen in the classic region of southwest France. Ever since Bordes defined the five Mousterian variants in this area there has been considerable discussion as to their interpretation, epitomized by the classic Bordes–Binford, culture-function debate. Many factors, including chronological change through time (Mellars 1969; 1986) and 'cultural' factors are likely to have contributed to the frequencies of different artefact types in different assemblages, but a substantial proportion of the variability appears to be accountable to variability in raw material availability and in the intensity of reduction (Rolland & Dibble 1990). The second of these factors is related to variation in mobility patterns and the duration of occupation at cave sites, which in turn reflects environmental factors.

Rolland & Dibble (1990) show that the less intensively reduced Mousterian variants, the Typical and Denticulate assemblages, appear to have been produced under milder climatic conditions while the heavily reduced Charentian assemblages coincide with severe climatic phases. They argue that the long winters of these cold phases would have forced groups to minimize journeys to quarries resulting in a need to more frequently resharpen tools. In contrast residential mobility would have been more frequent during milder conditions allowing for the replenishment of raw materials and resulting in less intensively reduced assemblages.

Variation in reduction techniques relating to the costs of acquiring raw materials during the Middle Palaeolithic have also been recognized in other areas of southwest France. For instance in the Aquitatine area Geneste (1985) has recognized that the 'utilization' index (UI) of raw materials progressively increases with the distance from the sources at which it was procured: in a 5 km radius, the UI is only 5 per cent; for sources between 5–20 km from the sites the UI was 10–20 per cent while for exotic materials from 50–80 km distance the UI was 75–100 per cent. Moreover the form in which the raw material entered the sites systematically varied with only prepared blocks for the 5–20 km distance sources and pieces at the final stage of the reduction sequence for these from exotic sources.

In both west-central Italy and southwest France we see that cost-risk-benefit considerations appear to have been used by Neanderthals when making decisions about the production of stone tools resulting in the covariation between mobility patterns and the reduction methods of cores and artefacts. But such cost-risk-benefit assessments need have been no more complex than those used by a variety of animals in their foraging behaviour whether these involve mobility patterns, foraging strategies or indeed the use of tools. As Neanderthals were products of biological evolution, in which natural selection is likely to have played a prominent role, it would indeed be very surprising if these patterns were *not* apparent in their technological behaviour. For as the procurement of raw material and the reduction of cores and artefacts involves time and energy, they ultimately contribute to reproductive fitness and we should expect natural selection to have resulted in decision-making algorithms which result in a high level of technological efficiency. Certainly this appears to be the pattern that is emerging from the studies which show covariation between technology, mobility and environments as described in the studies above.

Cost-risk-benefit considerations also influenced the technological behaviour of modern humans as can be seen in the variations in tools with mobility patterns and environments among modern hunter-gatherers (e.g. Torrence 1983; 1989; Bleed 1986; Churchill 1993) and those of the Upper Palaeolithic and Mesolithic (e.g. Straus 1990). But there is a fundamental difference between the technology of early and modern humans for that of the latter clearly plays an active role in structuring mobility and foraging behaviour, rather than providing merely a passive reflection of these behaviours.

The most direct examples of this is the use of facilities such as fish traps, drive lines and portable fences (Straus 1993). These influence the behaviour of game, effectively structuring their mobility patterns to allow effective slaughter. Such facilities are amply described in the ethnographic literature. The best preserved archaeological examples are fish traps from the Mesolithic (e.g. Larsson 1983). For the Upper Palaeolithic direct archaeological evidence is often sparse, but the circumstantial evidence for the use of facilities in terms of the evidence of mass kills is conclusive.

The construction of facilities requires a very high level of integrated thought about natural history and technology. This is also required for the manner in which projectile point technology developed during the Upper Palaeolithic. Whereas Middle Palaeolithic points have a very limited degree of variability, appearing not to vary with the type of ungulate being exploited (Kuhn 1993), those of the Upper Palaeolithic are very diverse with distinct patterning across space and time. There is evidence for the continuous development and experimentation in use of different projectile technologies (e.g. Straus 1990; Pike-Tay & Bricker 1993), and for associations between specific type of points and resources (e.g. Clark 1989; Peterkin 1993). As with the use of facilities, this can only derive from a capacity to integrate understanding of animal behaviour and the possibilities of technology in a fashion that appears to have been absent from Neanderthal minds.

Technology and the social environment

The Neanderthal mind appears to have had a limited degree of accessibility between the technological and the natural history domains of intelligence. Similarly, while it is most likely that their social intelligence was at an equivalent level of complexity to modern humans this too appears relatively isolated from other types of thought. Social relationships do not appear to have been mediated by the use of material culture. During the Middle Palaeolithic social and symbolic information was not invested in tools (Chase & Dibble 1987; Chase 1991; Stringer & Gamble 1993). Items of personal adornment used to convey information and manipulate the behaviour of others are also absent, first appearing with the start of the Upper Palaeolithic (White 1989). There can be no doubt that Neanderthals were engaging in complex social strategies, manipulative and Machiavellian in character. But unlike modern humans, these were not played out with the use of material culture.

Related to this is the social use of space. As Stringer & Gamble (1993, 155–8) have recently discussed, features are extremely rare on Middle Palaeolithic sites and this cannot be accounted for by poor preservation. While Neanderthals certainly used fire, hearths are extremely rare. In addition, there appears to be limited, if any, spatial patterning in the distribution of artefacts and faunal remains on sites (Farizy & David 1992). The significance of this negative evidence is that among modern humans space is a 'tool' used to conduct social relationships. The positions in which people sit around hearths, the distance between family groups, spatial associations between different activities all reflect and creates social relationships (e.g. Whitelaw 1991). It is unlikely that the absence of this social use of space reflects different types of social interaction between Neanderthals and behaviourally modern humans; it is most likely to be a further reflection of a domain-specific intelligence in which thought about social interaction was only marginally integrated with thought about the use of space.

Conclusion: the Neanderthal mind

Having considered different domains of intelligence for Neanderthals, and the limited degree of cognitive accessibility between them, what model for the early human mind has been created? On the one hand, the Neanderthal mind appears very similar to the modern mind. The levels of social, technological and natural history intelligence are likely to be equivalent to that of modern humans. Indeed, a good case can be made for a rather more developed form of natural history intelligence. On the other hand, these domains of intelligence appear to be relatively isolated from each other in the Neanderthal mind. Cognitive accessibility appears weak and this is in very marked contrast to the minds of modern humans which might be referred to as possessing a 'domain-sharing' rather than 'domain-specific' intelligence (Fig. 17.1). Modern humans have a propensity to think about the natural world in social terms, or to think about other social beings as 'objects' (as in racist ideologies), or to use technology to structure relationships with the social and natural worlds. All of these reflect a high degree of cognitive accessibility, a fluidity of thought between the domains of social, technological and natural history intelligence. Moreover, it is likely to be from this cognitive fluidity that new behaviours such as art and ritual arose at the start of the Upper Palaeolithic (Mithen 1994a).

We cannot, of course, know what it was like to have had the mind of a Neanderthal. Just as the philosopher Thomas Nagel wanted to know what it is like for a bat to be a bat, we would like to know what it was like for a Neanderthal to be a Neanderthal. As Nagel wrote:

> . . . if I try to imagine this, I am restricted to the resources of my own mind, and those resources are inadequate to the task. I cannot perform it either by imagining additions to my present experience, or by imagining segments gradually subtracted from it, or by imagining some combinations of additions, subtractions and modifications. (1979, 169)

Being phylogentically closer to Neanderthals than to bats, we may believe we can make greater progress in knowing what it was like to have the mind of a Neanderthal. I suspect that in many ways it would have been a far richer experience than having the mind of a modern human. Among modern humans social thought appears to dominate so much: landscapes, animal behaviour, tools, and spatial relationships

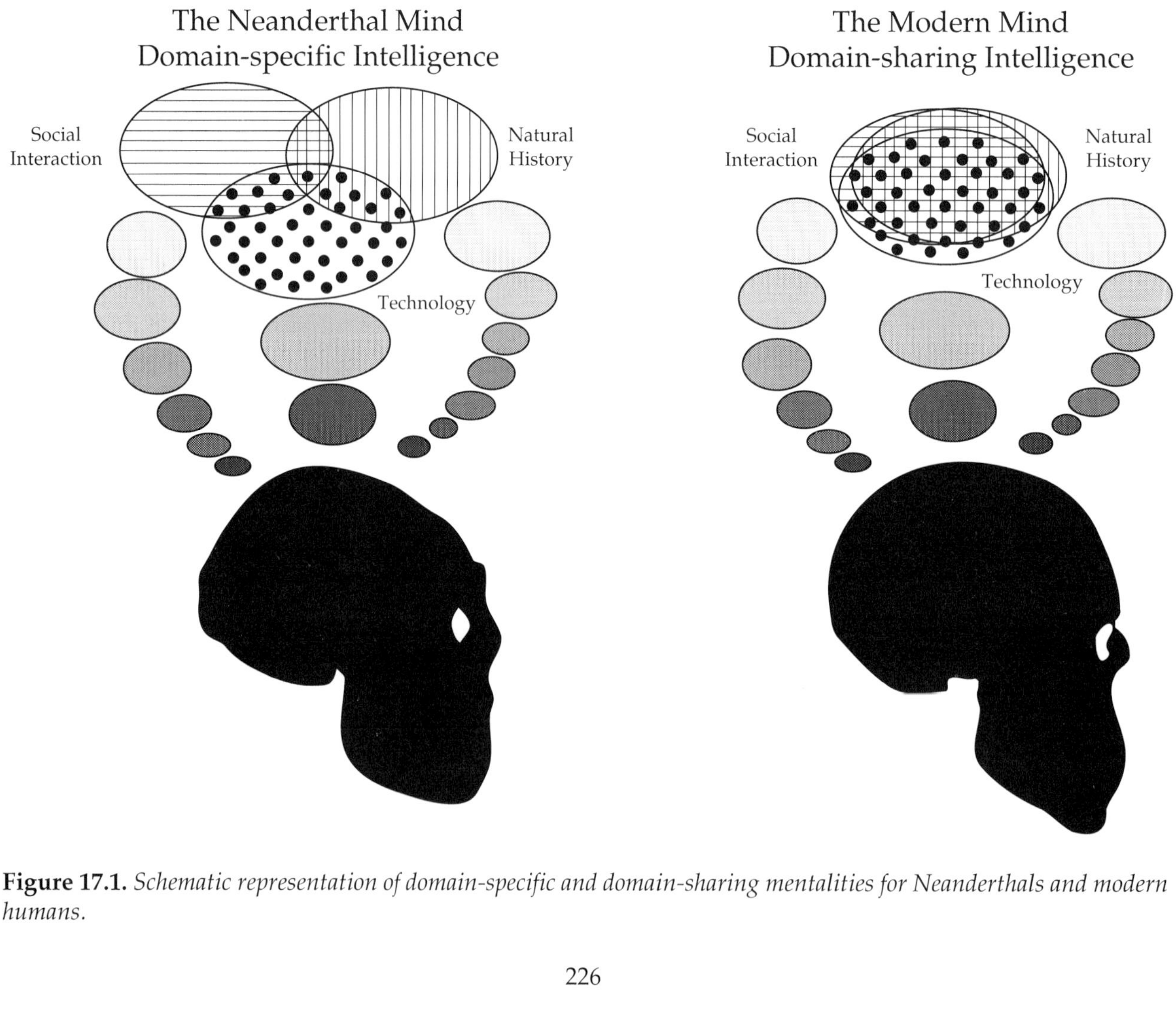

Figure 17.1. *Schematic representation of domain-specific and domain-sharing mentalities for Neanderthals and modern humans.*

are continually interpreted as metaphors for human social relationships or actively used to mediate those relationships. Neanderthals may well have been able to appreciate the natural world or a handaxe for its own sake, and consequently had a much more diverse experience than that which we can grasp through our own muddled thought of today.

Steven Mithen
Department of Archaeology
University of Reading
Whiteknights
P.O. Box 218
Reading
RG6 2AA

References

Aiello, L. & M.C. Dean, 1990. *An Introduction to Human Evolutionary Anatomy*. London: Academic Press.

Aiello, L. & R.I.M. Dunbar, 1993. Neocortex size, group size and the evolution of language. *Current Anthropology* 34, 184–93.

Atran, S., 1990. *Cognitive Foundations of Natural History: Towards an Anthropology of Science*. Cambridge: Cambridge University Press.

Atran, S., 1994. Core domains versus scientific: evidence from systematics and Itza-Maya folkbiology, in Hirschfeld & Gelman (eds.), 316–40.

Barkow, J., L. Cosmides & J. Tooby (eds.), 1992. *The Adapted Mind: Evolutionary Psychology and the Generation of Culture*. New York (NY): Oxford University Press.

Bleed, P., 1986. The optimal design of hunting weapons. *American Antiquity* 51, 737–47.

Blurton-Jones, H. & M.J. Konner, 1976. !Kung knowledge of animal behaviour, in *Kalahari Hunter-Gatherers*, eds. R. Lee & I. DeVore. Cambridge (MA): Harvard University Press.

Brown, D.E., 1991. *Human Universals*. New York (NY): McGraw-Hill.

Byrne, R.W., 1994. The evolution of intelligence, in *Behaviour and Evolution*, eds. P.J.B. Slater & T.R. Halliday. Cambridge: Cambridge University Press, 223–65.

Byrne, R.W. & A. Whiten (eds.), 1988. *Machiavellian Intelligence: Social Expertise and the Evolution of Intellect in Monkeys, Apes and Humans*. Oxford: Clarendon Press.

Byrne, R.W. & A. Whiten, 1992. Cognitive evolution in primates: evidence from tactical deception. *Man* 27, 609–27.

Chase, P., 1991. Symbols and Palaeolithic artefacts: style, standardization and the imposition of arbitrary form. *Journal of Anthropological Archaeology* 10, 193–214.

Chase, P. & H. Dibble, 1987. Middle Palaeolithic symbolism: a review of current evidence and interpretations. *Journal of Anthropological Archaeology* 6, 263–93.

Cheney, D.L. & R.M. Seyfarth, 1990. *How Monkeys See the World*. Chicago (IL): University of Chicago Press.

Churchill, S., 1993. Weapon technology, prey size selection and hunting methods in modern hunter-gatherers: implications for hunting in the Palaeolithic and Mesolithic, in Peterkin *et al.* (eds.), 11–24.

Clark, G.A. ,1989. Romancing the stones: biases, style and lithics at La Riera, in *Alternative Approaches to Lithic Analysis*, eds. D.O. Henry & G.H. Odell. (Archaeological Papers 1.) Washington (DC): American Anthropological Association, 27–50.

Cords, M., 1992. Social versus ecological intelligence (comment on D.L. Cheney & R.M. Seyfarth, 1992. Prévis of *How Monkeys See the World*). *Behavioural and Brain Sciences* 15, 151.

Cosmides, L., 1989. The logic of social exchange: has natural selection shaped how humans reason? Studies with the Wason selection task. *Cognition* 31, 187–276.

Cosmides, L. & J. Tooby, 1987. From evolution to behaviour: evolutionary psychology as the missing link, in *The Latest on the Best: Essays on Evolution and Optimality*, ed. J. Dupre. Cambridge (MA): The M.I.T. Press.

Cosmides, L. & J. Tooby, 1994. Origins of domain specificity: the evolution of functional organization, in Hirschfeld & Gelman (eds.), 85–116.

Davison, I., 1991. The archaeology of language origins: a review. *Antiquity* 65, 39–48.

Davidson, I. & W. Noble, 1989. The archaeology of perception: traces of depiction and language. *Current Anthropology* 30, 125–55.

Dennett, D., 1992. *Consciousness Explained*. London: Allen Lane.

Dunbar, R.I.M., 1988. *Primate Social Systems*. London: Croom Helm.

Dunbar, R.I.M., 1992. Neocortex size as a constraint on group size in primates. *Journal of Human Evolution* 20, 469–93.

Farizy, C. & F. David, 1992. Subsistence and behavioral patterns of some Middle Palaeolithic local groups, in *The Middle Palaeolithic, Adaptation, Behaviour and Variability*, eds. H. Dibble & P. Mellars. Philadelphia (PA): The University Museum, University of Pennsylvania, 87–96.

Flood, J., 1980. *The Moth Hunters: Aboriginal Prehistory of the Australian Alps*. Canberra: Australian Institute for Aboriginal Studies.

Fodor, J., 1983. *The Modularity of Mind*. Cambridge (MA): M.I.T. Press.

Gamble, C., 1986. *The Palaeolithic Settlement of Europe*. Cambridge: Cambridge University Press.

Gamble, C., 1987. Man the shoveler: alternative models for Middle Pleistocene colonization and occupation in northern latitudes, in *The Pleistocene Old World*, ed. O. Soffer. New York (NY): Plenum Press, 81–98.

Gardner, H., 1983. *Frames of Mind: the Theory of Multiple Intelligences*. New York (NY): Basic Books.

Geneste, J.M., 1985. Analyse lithique d'Industries Moustériennes du Perigord: une approche technologique du comportement des groupes humaines au Paléolithique moyen. Thèse Univérsite Bordeaux I.

Gibson, K., 1983. Comparative neurobehavioural ontogeny: the constructionist perspective in the evolution of language, object manipulation and the brain, in *Glossogenetics: the Origin and Evolution of Language*. ed. E. de Grolier. New York (NY): Harwood Academic Publishers, 41–66.

Gibson, K., 1985. Brain size and the evolution of language, in *The Genesis of Language: a Different Judgement of Evidence*. ed. M. Landsberg. Berlin: Mouton de Gruyter, 149–72.

Gibson, K., 1993. Tool use, language and social behaviour in relationship to information processing capacities, in Gibson & Ingold (eds.), 251–70.

Gibson, K. & T. Ingold (eds.), 1993. *Tools, Language and Cognition in Human Evolution*. Cambridge: Cambridge University Press.

Goodale, J.C., 1971. *Tiwi Wives: a Study of the Women of Melville Island, North Australia*. Seattle (WA): University of Washington Press.

Gopnik, A. & H.M. Wellman, 1994. The theory theory, in Hirschfeld & Gelman (eds.), 257–93.

Gowlett, J., 1984. The mental abilities of early man: a look at some hard evidence, in *Human Evolution and Community Ecology*, ed. R. Foley. London: Academic Press, 167–92.

Guthrie, D., 1984. Mosaics, allelochemicals and nutrients: an ecological theory of late Pleistocene extinctions, in *Quaternary Extinctions: a Prehistoric Revolution*, eds. P.S. Martin & R.G. Klein. Tucson (AZ): University of Arizon Press, 259–98.

Harris, P., 1994. Thinking by children and scientists; false analogies, in Hirschfeld & Gelman (eds.), 294–315.

Hill, K. & K. Hawkes, 1983. Neotropical hunting among the Ache of Eastern Paraguay, in *Adaptive Responses of Native American Indians*, eds. R. Hames & W. Vickers. New York (NY): Academic Press, 139–88.

Hirschfeld, L.A. & S.A. Gelman (eds.), 1994. *Mapping the Mind: Domain Specificity in Cognition and Culture*. Cambridge: Cambridge University Press.

Humphrey, N., 1976. The social function of intellect, in *Growing Points in Ethology*, eds. P.P.G. Bateson & R.A. Hinde. Cambridge: Cambridge University Press, 303–17.

Humphrey, N., 1984. *Consciousness Regained*. Oxford: Oxford University Press.

Humphrey, N., 1992. *A History of the Mind*. London: Chatto & Windus.

Jennes, D., 1977. *The Indians of Canada*. 7th edition. Ottawa: University of Toronto Press.

Keil, F.C., 1994. The birth and nurturance of concepts by domains: the origins of concepts of living things, in Hirschfeld & Gelman (eds.), 234–54.

Larsson, L., 1983. *Ageröd V. An Atlantic Bog Site in Central Scania*. (Acta Archaeologica Lundensia 12.)

Lee, R.B., 1979. *The !Kung San: Men, Women and Work in a Foraging Society*. Cambridge: Cambridge University Press.

Levitt, D., 1981. *Plants and People: Aboriginal Uses of Plants on Groote Eylandt*. Canberra: Australian Institute of Aboriginal Studies.

Marks, S.A., 1976. *Large Mammals and a Brave People: Subsistence Hunters in Zambia*. Seattle (WA): University of Washington Press.

Mellars, P., 1969. The chronology of the Mousterian industries in southwest France. *Proceedings of the Prehistoric Society* 35, 134–71.

Mellars, P., 1986. A new chronology for the French Mousterian Period. *Nature* 322, 410–11.

Mellars, P., 1991. Cognitive changes and the emergence of modern humans. *Cambridge Archaeological Journal* 1(1), 63–76.

Mellars, P. & C. Stringer (eds.), 1989. *The Human Revolution: Behavioural and Biological Perspectives on the Origins of Modern Humans*. Edinburgh: Edinburgh University Press.

Mithen, S.J., 1990. *Thoughtful Foragers: a Study of Prehistoric Decision Making*. Cambridge: Cambridge University Press.

Mithen, S.J., 1991a. A cybernetic wasteland? Rationality, emotion and Mesolithic foraging. *Proceedings of the Prehistoric Society* 57(I), 9–14.

Mithen, S.J., 1991b. Ecological interpretations of Palaeolithic art. *Proceedings of the Prehistoric Society* 57(II), 103–14.

Mithen, S.J., 1993. Individuals, groups and the Palaeolithic record: a reply to Clark. *Proceedings of the Prehistoric Society* 59, 393–8.

Mithen, S.J., 1994a. From domain specific to generalised intelligence: a cognitive interpretation of the Middle/Upper Palaeolithic transition, in *The Ancient Mind*, eds. C. Renfrew & E. Zubrow. Cambridge: Cambridge University Press, 29–39.

Mithen, S.J., 1994b. Technology and society during the Middle Pleistocene. *Cambridge Archaeological Journal* 4(1), 3–32.

Mithen, S.J., 1996. *The Prehistory of the Mind: a Search for the Origins of Art, Religion and Science*. London: Thames & Hudson.

Mithen, S.J., n.d. Evolutionary psychology and cognitive archaeology: an archaeological critique of Cosmides and Tooby's model for the human psyche. (Submitted to *Cognition*.)

Nagel, T., 1979. *Mortal Questions*. Cambridge: Cambridge University Press.

Nelson, R.K., 1973. *Hunters of the Northern Forest: Designs for Survival among the Alaskan Kutchin*. Chicago (IL): University of Chicago Press.

Nelson, R.K., 1983. *Make Prayers to the Raven: a Koyukon View of the Northern Forest*. Chicago (IL): Chicago University Press.

Oswalt, W., 1976. *An Anthropological Analysis of Food-Getting Technology*. New York (NY): Wiley.

Parker, S.T., 1985. A social-technological model for the evolution of language. *Current Anthropology* 26, 617–39.

Parker, S.T. & K.R. Gibson, 1979. A model for the evolution of language and intelligence in early hominids. *Behavioural and Brain Sciences* 2, 367–407.

Pelegrin, J., 1993. A framework for analysing prehistoric

stone tool manufacture and a tentative application to some early stone industries, in *The Use of Tools by Human and Non-human Primates*, eds. A. Berthelet & J. Chavaillon. Oxford: Clarendon Press, 302–14.
Peterkin, G.L., 1993. Lithic and organic hunting technology in the French Upper Palaeolithic, in Peterkin *et al.* (eds.), 49–68.
Peterkin, G.L., H.M. Bricker & P. Mellars (eds.), 1993. *Hunting and Animal Exploitation in the Later Palaeolithic and Mesolithic of Eurasia.* (Archaeological Papers 4.) Washington (DC): American Anthropological Association.
Pike-Tay, A. & H.M. Bricker, 1993. Hunting in the Gravettian: an examination of evidence from southwestern Europe, in Peterkin *et al.* (eds.), 127–46.
Pinker, S., 1994. *The Language Instinct*. New York (NY): William Morrow & Company.
Pinker, S. & P. Bloom, 1990. Natural language and natural selection. *Behavioural and Brain Sciences* 13, 585–642.
Roebroeks, W., 1988. From flint scatters to early hominid behaviour: a study of Middle Palaeolithic riverside settlements at Mastrict-Belvedere. *Analecta Praehistorica Leidensai* 1988, 1–196.
Rolland, N. & H. Dibble, 1990. A new synthesis of Middle Palaeolithic variability. *American Antiquity* 55, 480–99.
Rozin, P., 1976. The evolution of intelligence and access to the cognitive unconscious, in *Progress in Psychobiology and Physiological Psychology*, eds. J.N. Sprague & A.N. Epstein. New York (NY): Academic Press, 245–77.
Rozin, P. & J. Schull, 1988. The adaptive-evolutionary point of view in experimental psychology, in *Stevens' Handbook of Experimental Psychology*, vol. 1: *Perception and Motivation*, eds. R.C. Atkinson, R.J. Herrnstein, G. Lindzey & R.D. Luce. New York (NY): John Wiley & Sons, 503–46.
Spelke, E.S., K. Breinlinger, J. Macomber & K. Jacobsen, 1992. Origins of knowledge. *Psychological Review* 99, 605–32.
Sperber, D., 1994. The modularity of thought and the epidemiology of representations, in Hirschfeld & Gelman (eds.), 39–67.
Stiner, M. & S. Kuhn, 1992. Subsistence, technology and adaptive variation in Middle Palaeolithic Italy. *American Anthropologist* 94, 12–46.
Straus, L., 1990. The original arms race: Iberian perspectives on the Solutran phenomenon, in *Feuilles du Pierre: les Industries à Pointes Foliacées du Paléolithique Supérieur Européen*, ed. J. Kozlowski. (ERAUL 42.) Liège: Université de Liège, 425–47.
Straus, L., 1993. Upper Palaeolithic hunting tactics and weapons in Western Europe, in Peterkin *et al.* (eds.), 83–94.
Stringer, C. & C. Gamble, 1993. *In Search of the Neanderthals*. London: Thames & Hudson.
Sullivan, R.J., 1942. *The Ten'a Food Quest*. Washington (DC): The Catholic University of America Press.
Tooby, J. & L. Cosmides, 1989. Evolutionary psychology and the generation of culture, part I. *Ethology and Sociobiology* 10, 29–49.
Tooby, J. & L. Cosmides, 1992. The psychological foundations of culture, in Barkow *et al.* (eds.), 19–136.
Torrence, R., 1983. Time budgeting and hunter-gatherer technology, in *Hunter-Gatherer Economy in Prehistory*, ed. G.N. Bailey. Cambridge: Cambridge University Press, 11–22.
Torrence, R. (ed.), 1989. *Time Energy and Stone Tools*. Cambridge: Cambridge University Press.
Wellman, H.M., 1991. From desires to beliefs: acquisition of a theory of mind, in Whiten (ed.), 19–38.
Whallon, R., 1989. Elements of cultural change in the Later Palaeolithic, in Mellars & Stringer (eds.), 433–54.
White, R., 1989. Production complexity and standardization in early Aurignacian bead and pendant manufacture: evolutionary implications, in Mellars & Stringer (eds.), 366–90.
Whitelaw, T., 1991. Some dimensions of variability in the social organisation of community space among foragers, in *Ethnoarchaeological Approaches to Mobile Campsites*, eds. C. Gamble & W. Boismier. Ann Arbor (MI): International Monographs in Prehistory, 139–88.
Whiten, A. (ed.), 1991. *Natural Theories of Mind: Evolution, Development and Simulation of Everyday Mindreading.* Oxford: Basil Blackwell.
Wynn, T., 1979. The intelligence of later Acheulian hominids. *Man* 14, 371–91.
Wynn, T., 1989. *The Evolution of Spatial Competence*. Urbana (IL): University of Illinois Press.
Wynn, T., 1993. Two developments in the mind of Early *Homo*. *Journal of Anthropological Archaeology* 12, 299–322.